Dynamic Modeling and Econometrics in Economics and Finance

Volume 17

Editors
Stefan Mittnik
University of Munich
Munich, Germany

Willi Semmler
Bielefeld University
Bielefeld, Germany
and
New School for Social Research
New York, USA

For further volumes:
www.springer.com/series/5859

Frauke Schleer-van Gellecom
Editor

Advances in Non-linear Economic Modeling

Theory and Applications

Springer

Editor
Frauke Schleer-van Gellecom
Centre for European Economic Research
ZEW
Mannheim, Germany

ISSN 1566-0419
ISBN 978-3-642-42038-2 ISBN 978-3-642-42039-9 (eBook)
DOI 10.1007/978-3-642-42039-9
Springer Heidelberg New York Dordrecht London

Library of Congress Control Number: 2013956767

Printed on acid-free paper

Springer is part of Springer Science+Business Media (www.springer.com)

Preface

In recent years non-linearities have gained increasing importance in economic and econometric research, particularly after the outbreak of the financial crisis in 2007–2008 and the ensuing economic downturn in many developed countries. The financial crisis has led many researchers to rethink macroeconomics and macroeconomic modeling. First, it is widely acknowledged that financial sector dynamics have not been fully understood in theoretical and applied work. Hence, there is a need to improve our understanding of the destabilizing effects arising from the financial sector and spilling over to economic activity. Second, the existing literature investigating the link between the macroeconomy and the financial sector shows that non-linearities play a central role and are becoming increasingly relevant. Theoretical and empirical studies typically discover highly non-linear amplifying and destabilizing effects on the real economy which come from the financial sector. As a consequence, theoretical models and econometric approaches, which are able to capture this kind of asymmetric dynamics, multiple regimes, and non-linearities, are coming to the fore.

This was the motivation to compile this book with a focus on non-linearities in a broad sense. A lot of research is being conducted to analyze the link between the financial sector and the real economy. I am convinced that it is also important to point out that non-linear relations might be a major factor—albeit neglected in research—in a wide range of economic problems. Disregarding non-linearities could severely distort outcomes.

The book intends to emphasize others areas of research where non-linear features also play a crucial role. Next to contributions that deal with the interaction between financial markets and the real economy, the volume includes studies focusing on non-linearities in other fields of research. The contributions develop novel or use advanced recent methods for working with non-linearities: the construction of new unit-root tests for a non-linear smooth transition type model, the application of Wavelets as a powerful empirical method, a duration analysis approach, a bilinear process, as well as a multi-regime VAR.

This book contains theoretical, computational and empirical contributions that incorporate non-linearities to address various economic problems. It serves as an inspiration to take potential non-linearities into account in model specification. Researchers should be careful not to employ linear model-types spuriously to problems which include non-linear features. Using the correct model type is indispensable in order to interpret economic processes properly and derive valid policy conclusions. Ignoring this aspect may result in biased economic policy recommendations.

This book is associated with the SEEK workshop "Non-linear Economic Modeling: Theory and Applications" held at ZEW in Mannheim in December 2012. The publication received funding from the SEEK Research Programme ("Strengthening Efficiency and Competitiveness in the European Knowledge Economies") of the Centre for European Economic Research (ZEW).

Part I is devoted to non-linearities within financial sector applications. The contributions in this chapter explore instabilities of the banking sector resulting in harmful macrodynamics, quantitatively assess the nexus between monetary policy and financial frictions, and construct a financial stress index. Part II shows the broad variety of applications of non-linearities in different research fields. This book highlights potential research fields in economics where non-linearities may play a crucial role. Hence, it also features analyses which deal with the Beveridge curve, exchange rate behavior and regimes, and forecasts of economic activity or inflation.

It has been a great pleasure for me to edit this book. First of all, I would like to thank the contributors for their highly interesting and excellent research. Without their effort this book would not exist. I also would like to thank Professor Stefan Mittnik and especially Professor Willi Semmler who encouraged me and gave me the opportunity to publish this book. I am grateful to Dr. Martina Bihn and Ruth Milewski from Springer for their patience, friendly co-operation and organizational support. I am also indebted to Stephan Reichert, Eric Retzlaff and Patrick Pilarek from ZEW who supported the completion of this volume in various ways.

Mannheim, Germany
September 2013

Frauke Schleer-van Gellecom

Contents

Contributors

Luca Benati Department of Economics, University of Bern, Bern, Switzerland

Wojciech W. Charemza University of Leicester, Leicester, UK; AFiBV, Warsaw, Poland

Marco Gallegati Department of Economics and Social Sciences, Faculty of Economics "G. Fua", Polytechnic University of Marche, Ancona, Italy

Mohammad Karimi Department of Economics, Ottawa University, Ottawa, ON, Canada

Yuriy Kharin Belarusian State University, Minsk, Belarus

Thomas A. Lubik Research Department, Federal Reserve Bank of Richmond, Richmond, VA, USA

Vladislav Maevskiy EPAM-Systems, Minsk, Belarus

Enrique Martínez-García Federal Reserve Bank of Dallas, Dallas, TX, USA

Stefan Mittnik Department of Statistics and Center for Quantitative Risk Analysis, University of Munich, Munich, Germany

Rickard Sandberg Department of Economic Statistics, Department of Economics, Stockholm School of Economics, Stockholm, Sweden

Willi Semmler Department of Economics, New School for Social Research, New York, NY, USA

Marcel-Cristian Voia Department of Economics, Carleton University, Ottawa, ON, Canada

Part I
Non-linearities Related to the Financial Sector

Estimating a Banking-Macro Model Using a Multi-regime VAR

Stefan Mittnik and Willi Semmler

Abstract This paper indroduces a Banking-Macro Model and estimates the linkages using a Multi-Regime Vector Auto Regression (MRVAR). The model of the banking-macro link is a simplified version of the Brunnermeier and Sannikov (Am. Econ. Rev., 2014) model. The banking sector is represented as a wealth fund that accumulates capital assets, can heavily borrow and pays bonuses. We presume that the banking sector faces not only loan losses but is also exposed to a deterioration of its balances sheets due to adverse movements in asset prices. In contrast to previous studies that use the financial accelerator—which is locally amplifying but globally stable and mean reverting—our model shows local instability and globally multiple regimes. Whereas the financial accelerator leads, in terms of econometrics, to a one-regime VAR, we demonstrate the usefulness of the MRVAR approach. We estimate our model for the U.S. with a MRVAR using data on a constructed financial stress index and industrial production. We also conduct impulse-response analyses which allowing us to explore regime dependent shocks. We show that the shock profiles depend on the regime the economy is in and the size of the shocks. As to the recently discussed unconventional monetary policy of quantitative easing, we find that the relative effects of monetary shocks depend on the size of the shocks.

1 Introduction

As many of the historical financial crises have shown, the crises may have originated from adverse shocks to firms, households, foreign exchange, stock market, or sovereign debt. Yet, as Reinhart and Rogoff (2009) and Gorton (2009, 2010) have

S. Mittnik
Department of Statistics and Center for Quantitative Risk Analysis, University of Munich, Munich, Germany

W. Semmler (✉)
Department of Economics, New School for Social Research, 79 Fifth Avenue, New York, NY 10003, USA

F. Schleer-van Gellecom (ed.), *Advances in Non-linear Economic Modeling*, Dynamic Modeling and Econometrics in Economics and Finance 17,
DOI 10.1007/978-3-642-42039-9_1, © Springer-Verlag Berlin Heidelberg 2014

demonstrated, the banking sector could rarely escape the crises. In fact, most crises ended up as a meltdown of the banking sector, and the banking sector has often exacerbated and amplified the crisis whatever origin it had. As Gorton (2010) shows, in the past, loan losses and bank runs where the conventional mechanisms by which the crises where triggered, but more recently, banking crises seem to be strongly related to adverse shocks in asset prices and financial stress.

Here we study how this channel may have some exacerbating or even destabilizing effects on the macroeconomy. An issue is: do we have proper models to explain this? Do we have models that help to understand this central aspect of sudden financial meltdowns? There are earlier, non-conventional studies by Kindleberger and Aliber (2005) and Minsky (1975, 1982) that view the role of credit as a significant amplifying force. In Kindleberger it is the instability of credit, and in Minsky it is the way financing becomes de-linked from collaterals that contributes to a downward spiral once large real or financial shocks occur. This is surely an important tradition that captures many of the aspects of past boom-bust scenarios.

Recently, there has been significant work by Greenwald and Stiglitz (1993), Bernanke et al. (1999) that shows that the financial sector can have amplifying effects. In the DSGE tradition there is only a locally magnifying effect, through collaterals.[1] Collaterals rise at high levels of economic activity, making credit available and cheap, and the reverse happens at low levels of economic activity. The debt-to-asset-value ratio is predicted to fall in a boom and rise in recessions.[2] Yet, in most models of this type, there is no tracking of the debt dynamics, so that the fragility of the debt dynamics does not come into play. The models are solved through local linearizations about a unique and stable steady state and departures from the steady state are eventually mean reverting. Although the economy is accelerating, it will revert back to the steady state. Empirically, this is often shown by fitting a conventional (one-regime) VAR; see Gilchrist et al. (2009), Gilchrist and Zagrajsek (2012), Christensen and Dib (2008), and Del Negro et al. (2010).

Many studies of the great depression developed the perception that locally destabilizing effects, arising from the banking sector, are missing in modern macroeconomic modeling. So far, the financial accelerator theory has mainly been applied to firms and households. Yet, as the recent meltdown of the years 2007–2009 has demonstrated, the financial accelerator also applies to the balance sheets of banks and seems to be destabilizing rather then mean reverting. Work by Adrian and Shin (2009), Adrian et al. (2010) and Brunnermeier and Pedersen (2009) suggests that

[1]Theoretical literature has studied the amplifying effect of shocks near the steady state. See for example, see Carlstrom et al. (2009), and Curdia and Woodford (2009a, 2009b).

[2]This is, for example, empirically stated in Gilchrist et al. (2009). Yet, as Geanakoplos (2010) mentions, the empirical measure is distorted through the way the debt-asset ratio is measured, namely as total assets over equity. Equity value rises during the boom and falls in a recession.

financial intermediaries[3] are not able to fulfill their functions as intermediaries. Often they have to liquidate their capital, when asset prices get depressed and margin requirements in the money market rise. This forces financial intermediaries to take a hair cut and to liquidate further, with another subsequent fall of asset prices reinforcing the downward trajectory. The depressed asset prices, generated by firesales of assets by some intermediaries, have external effects on the industry, and bank runs can exacerbate this effect, see Gorton (2010). There is a large body of literature showing that there might be a downward spiral through interconnectedness, interlinkages and contagion. Such studies have started with Greenwald and Stiglitz (1993) and continued with Adrian and Shin (2009), Adrian et al. (2010), Gorton (2010), Geanakoplos (2010), Geanakoplos and Farmer (2009), and Brunnermeier and Sannikov (2014).

Following this line of research, we examine how this process works through the balance sheets of banks. Banks, in the first instance, usually have extensive loan losses. This may be arising from default in the firm and household sectors, the foreign sector or resulting from sovereign debt. The shocks to asset prices will so affect the banks balance sheets first through loan losses, but also—and substantially—through the asset and liability side of the balance sheets.[4] This, in turn, affects the banks' availability of credit in the interbank credit market and the price of credit, i.e., the actual interest the banks have to pay.

Usually, with deteriorating balance sheets of the financial intermediaries, due to loan losses and falling asset prices, the risk premium they are asked to pay in the interbank loan market rises rapidly. Frequently, banks have to liquidate more assets to stay liquidity and keep payment obligations. With the value and of their capital basis shrinking, banks have to sell assets, and this might trigger a firesale of assets by some intermediaries, making their capital basis even weaker. This has effects on other intermediaries (as well as firms and households). The repo rate, the TED spread, and credit spreads, indicating in general that financial conditions, will rise. Thus, one would expect low financial stress and narrow credit spreads in a period of high economic activity, and high financial stress and widened credit spreads in a period of low economic activity. Hence, the dissipating liquidity, the credit constraints and credit spreads might be locally destabilizing rather than locally mean reverting.

The remainder of the paper is organized as follows. Section 2 builds up a model that reflects those features. Section 3 discusses some extensions. Section 4 solves numerically some model variants using dynamic programming. Section 5 employs a financial stress variable—that captures the financial conditions of banks—and industrial production to empirically estimate the model using a Multi-Regime VAR (MRVAR). Section 6 concludes the paper.

---

[3]We also include here what has been called by Gorton (2010) the shadow banking system, such as investment firms, brokers and money market dealers. Those have been growing rapidly in the U.S. during the last 15 to 20 years.

[4]As Gorton (2010) shows, this has often been magnified through bank runs.

Table 1 The balance sheet of banks

Assets	Liabilities
$p_t k_t$	d_t
	$n_t = p_t k_t - d_t$
Total assets	Total liabilities

2 The Basic Model

Next, let us present the above developed ideas in a more formal model. The best way to explain the model is to refer to the balance sheets of the financial intermediaries (see Table 1).

On the left hand side there are assets, valued at current asset prices. On the right hand side there is debt d_t and net worth $n_t = p_t k_t - d_t$. Next, let us introduce the dynamics of the variables. The asset price, the capital stock and the debt may evolve as follows

$$dp_t = \mu_t p_t dt + \sigma_t p_t dZ_t \tag{1}$$

$$dk_t = \big(\varphi(i_t/k_t) - \delta\big)k_t dt + \sigma_t k_t dZ_t \tag{2}$$

$$dd_t = \big(rd_t - (ak_t - i_t)\big)dt \tag{3}$$

The growth rate of asset prices follows a geometric Brownian motion. In fact, since it is formulated here as social planning or monopoly problem, prices will be implicit in the solution, given by the preferences and the Eqs. (2)–(3), see Brunnermeier and Sannikov (2014). But actual price movements can affect the dynamics when we consider the role of asset price shocks to financial intermediaries. The asset price shocks will reduce the collateral value of the financial intermediaries, and the fast depreciation of asset prices—possibly triggered by a firesale of assets—will have extensive externality effects on other intermediaries, leading to a general loss of net worth. This may magnify the downward movement of the spiral. Though, at first sight, the asset prices do not have a special role in the above equation, they will come into play below.

The assets of the financial intermediaries will increase with investment, i_t/k_t, the function $\varphi(i_t/k_t)$ includes some adjustment cost which is concave in the argument, and δ is an operating cost. The actual gross capital of the bank increases at the rate i_t/k_t. The debt evolves at a rate that is essentially determined by the excess spending of investment over capital income, which is defined here as ak_t. Investment in the second equation will generate a greater stock of assets for financial intermediaries, but the high rate of purchase of assets will increase their debt, once the investment spending exceeds their income. We have taken here the interest rate, r, to be paid on debt, as a constant, it may later be made endogenous depending on net worth. Note that only the first and second equations are stochastic.

So far we have neglected the bonuses of the executives which can be viewed to serve the consumption stream of the executives.[5] We can define the executives bonuses as an optimal consumption stream, to be derived optimally through some intertemporal decision making process. We can also have the investment being computed as optimal, with $g_t = i_t / k_t$. Then we have the dynamic decision problem:

$$V(k,d) = \max_{c_t, g_t} \int_0^{\infty} e^{-rt} U(c_t) dt \tag{4}$$

$$\text{s.t.} \quad dp_t = \mu_t p_t dt + \sigma_t p_t dZ_t \tag{5}$$

$$dk_t = \big(\varphi(i_t / k_t) - \delta\big) k_t dt + \sigma_t k_t dZ_t \tag{6}$$

$$dd_t = \big(rd_t - (ak_t - i_t - c_t)\big) dt \tag{7}$$

The latter model includes now bonus payments of the executives, c_t, which is used for a consumption stream.[6] Note that we have here $g_t = i_t / k_t$. Note also that in Eq. (7) if the excess of spending for new assets and bonus payments exceeds the income generated, then the debt of the financial intermediary will rise. As mentioned before, for the problem of a social planner, which is equivalent to a monopoly problem[7] of the financial intermediary, the prices are endogenous and do not play a role at first.

We want to remark that the above is a standard model of wealth management as it is now commonly used for wealth management of financial intermediaries, see He and Krishnamurthy (2008). If we replace the constant income for a unit of wealth, a in ak_t, by a weighted average of risky and risk free returns of a wealth fund k_t, then the remaining parts of the equations above are reasonably familiar from the wealth management literature, see also Semmler et al. (2009). Yet the explicit equation for the evolution of debt of the financial intermediary, as represented in Eq. (7), is missing. This reflects the innovative part of the model by Brunnermeier and Sannikov (2014) and other recent literature:[8] here then, the financial intermediaries are encouraged by more risk taking through transfer of risk to outside investors, consequently the financial intermediaries will build up their debt and thus their default risk.

[5] In Semmler and Bernard (2012) bonus payments of the six largest US investment banks are computed. Bonus payment, as a percent of revenues, went up from roughly 10 percent in 2000 to 35 percent in 2007, see Figure 17 in Appendix B.

[6] In recent attempts of financial market reforms in Europe the cash payment of bonus payments is planned to be restricted to 20 percent of total bonus payments, the remaining part is only allowed to be paid out in subsequent years via common stocks. In our model we leave aside those complications.

[7] See Brunnermeier and Sannikov (2014).

[8] See for example Hall (2010) who also includes an equation for the evolution of debt.

Now let us derive a dynamic equation for the debt-asset ratio.[9] Let us take as the debt-asset ratio: d_t/k_t: We can rewrite this, for convenience, as $\omega = -(d_t/k_t)$.[10] Taking log and time derivative of this, we can write the asset accumulation and debt dynamics with the previous objective function of the financial intermediaries as:[11]

$$V(\omega_t) = \max_{\tilde{c}_t, g_t} \int_0^\infty e^{-rt} U(\tilde{c}_t) dt \tag{8}$$

$$d\omega_t = \big((g_t - r - \delta + \sigma^2)\omega_t + a - \tau(g_t)\big)dt - \tilde{c}_t + \sigma_t \omega_t dZ_t \tag{9}$$

Hereby $\tilde{c}_t$ is the new control variable.[12] Term $\tilde{c}_t$ is the consumption wealth ratio, $\frac{c}{k}$. The expression $\tau(g_t)$ represents a convex adjustment cost which is affecting the size of borrowing to achieve a growth rate g_t. This is modeled by following the capital adjustment cost literature. Yet, of course only the growth of wealth g_t appears in the equation for the evolution of assets k_t. The other expressions in the latter equation are straight forward derivations from the negative of the growth rate of the debt-asset ratio as stated above.

3 Extensions of the Basic Model

In the next section we want to treat several extensions that amplify some of the mechanisms studied above in the basic model. We might think about four types of extensions.

First, in the context of the above model, the effect of asset price changes can easily be discussed. Note that fundamental asset price movements are implicitly contained in the above model. So far we have presented a model where asset prices are endogenous. Brunnermeier and Sannikov (2014) provide basic proofs of the probability of instability with endogenous asset prices. But what one could be interested in, as indicated above, are potential significant deviations from the fundamental asset price movements. Those can result from market price movements (driven by trading strategies of agents with different expectations, sentiments and opinion dynamics) which we could model as

$$dp_t = (\mu_t + z_t) p_t dt + \sigma_t p_t dZ_t \tag{10}$$

Hereby z_t could represent such market price movements.[13]

---

[9]Note that we use stocks of assets and debt, in contrast to Geanakoplos (2010) who uses flows as leverage measure, hereby then leveraging is highly positively correlated with booms.

[10]See Brunnermeier and Sannikov (2014).

[11]For a similar approach, see Brunnermeier and Sannikov (2014) and Hall (2010).

[12]A derivation of a dynamic equation in the stochastic case, using Itoh's lemma, is given in Brunnermeier and Sannikov (2014). The term σ^2 comes in through Itoh's lemma.

[13]Such market price movements, such as z_t, resulting from sentiments are, for example, studied in Lux (2009). A market sentiment is also at play in the theory of Geanakoplos (2010), where

When we consider the role of asset price shocks for financial intermediaries, we might posit that the asset price movements are likely to affect the balance sheet dynamics of the financial intermediaries.[14] Considering net worth, such as $p_t k_t - d_t$ on the balance sheets of banks, an adverse asset price shock reduces the collateral value of the financial intermediaries, its equity, and thus since the bank has to offer less collateral, it will face a greater haircut and higher repo rate or greater default premium. It will thus demand less capital, and with demand for capital falling asset prices will fall further. This has contagion effects: classes of similar types of assets will fall in price too, and a fast depreciation of asset prices—possible triggered by a firesale of assets—will have extensive externality effects for other intermediaries in the form of increased financial stress and credit spreads.[15] It might also affect the asset holdings and activities of firms and households that subsequently will have to sell assets to meet liquidity or payment requirements. The distinct contagion and externality effects are that a general loss of net worth could occur that may magnify the downward spiral. Though, at first sight, the asset price did not have a special role in the model above, it is easy to see how it might magnify the downward spiral.[16] Furthermore, it is very likely that positive and negative asset price shocks may have asymmetric effects.[17]

The second type of extension pertains to the bonus payments. We could assume if the net worth, as a ratio of net worth to total assets, falls below a certain safe threshold, then the bonus payments are reduced. Equivalently we could postulate that if the debt to asset ratio rises above some threshold, lets say $\omega = -(d_t/k_t) \leq \omega^*$, then the bonus payments are cut or reduced to zero. It could hold that bonus payments are used to give the managers an incentive to reduce leverage, so when the leverage is lower, a higher bonus payments could be allowed.[18] This might be considered as a type of penalty on risk taking and high indebtedness—the latter resulting from leveraged asset purchases. The dynamics of the debt-wealth ratio, once those policies are introduced, might be of interest. This modification will also be studied in our numerical section.

---

leveraging drives asset prices. The role of heterogeneous expectations and trading strategies for market price movements are explored in Chiarella et al. (2009: Chaps. 6–9). Yet, there are more general effects that can make the market price of the asset deviate from its fundamental price, as present value of future cash flow, for example liquidity problems, fire sales of assets and market dysfunctions, see Geneva Report (2009).

[14]See also Stein (2012), who shows that with an asset price boom and capital gains debt can be serviced and net worth and leverage rises. On the other hand when asset prices fall, capital gains become negative, the source for servicing debt dissipates and net worth falls.

[15]Those positive feedback effects are extensively studied in Geanakoplos (2010) and Gorton (2010).

[16]For further details and a number of other effects that falling asset prices may have, see Brunnermeier and Sannikov (2014).

[17]This is also discussed in Basel III.

[18]This is for example planned by Basel III, where it refers to "linkages of the total variable compensation pool to the need ... to maintain a sound capital base".

One could consider a third extension that takes into account the availability of funds for the financial intermediaries. There might be a fraction of households that accumulate risky assets,[19] which will provide funds for the financial intermediaries. A fraction of funds could also come from capital inflows, see Caballero and Krishnamurthy (2009). To be more formal, with ψ the fraction of assets being held by financial intermediaries,[20] the evolution of their capital assets and debt would be formulated as follows:

$$dk_t = \big(\psi(i_t/k_t) - \delta - (1-\psi)\sigma\big)k_t dt + \sigma_t k_t dZ_t \tag{11}$$

$$dd_t = \big(rd_t - (ak_t - \psi i_t - c_t)\big)dt \tag{12}$$

In this context, the inflow of funds from the Central Bank could be considered, which for example took place in the US in the years 2008 and thereafter when the Fed employed an unconventional monetary policy, called quantitative easing, buying bad—and rapidly declining—assets from the financial intermediaries. The latter would have more of a mitigating effect on the unstable forces generated by the banking system. An estimation of this effect in a multi-regime setting will be presented in Section 5. On the other hand, the precautionary motives of households (and firms), the "run into high quality assets",[21] would lead to a reduction of financial funds for the financial intermediaries.

A fourth type of extension could relate to the interest rate paid by the financial intermediaries. So far, in our basic model, the interest rate paid is a constant, r, but one could assume, as in Brunnermeier and Sannikov (2014), that there is a cost of state verification which will depend on net worth of the financial intermediary. In fact, it is likely that adverse asset price shocks, as discussed in our basic model, will affect borrowing cost by banks through the LIBOR, TED spread, or margin requirements, as discussed in the first type of extension above. Therefore, when there is a shock to asset prices and a magnification of a downward spiral, the credit spread and thus borrowing cost for financial intermediaries will rise. We thus will have:

$$d\omega_t = \big((g_t - r(\omega) - \delta + \sigma^2)\omega_t dt + a - \tau(g_t)\big)dt - \tilde{c}_t + \sigma_t\omega_t dZ_t \tag{13}$$

Brunnermeier and Sannikov (2014) have also included the effect of a rising volatility σ_t on the spread.[22] The above variant, with $r(\omega)$ and $r_\omega < 0$[23] can also be

[19]See He and Krishnamurthy (2008) and Brunnermeier and Sannikov (2014).

[20]For further details of this effect on the stability of the banking system, see Brunnermeier and Sannikov (2014, Section 3).

[21]Gorton (2010) calls this the run into "information insensitive assets", since one does not need acquire much information when one wants to hold them like treasury bonds.

[22]The importance of the effect of a rising volatility has also been indicated by the financial stress index developed by the Fed of Kansas City. It will be relevant in a distance to default model where it is shown that the distance to default shrinks with rising volatility.

[23]Note that we have the derivative $r_\omega < 0$, since we have the negative of the debt asset ratio as argument.

numerically solved and it might be very important to study its effect on the overall stability of the banking system. On the other hand, one might argue that the financial intermediaries have in fact transferred risk to outside investors through securitization, i.e. through pooling and tranching of mortgage debt or other kind of liabilities, through MBSs or CDOs. Successfully undertaking the transfer of risk encourages them to take on more risk, but passes the verification cost on to someone else. The verification cost usually defines the amount that financial intermediaries have to pay, but if it is passed on, they can generally borrow at a lower risk premium, and their evolution of debt is determined by an almost constant interest rate as defined in our basic model.[24] A model with state or time depending credit spread can also be solved by our numerical procedure.

Overall, as we can see from the above considerations, some of the extensions may have further destabilizing effects, and some may have more stabilizing effects.

4 Solution Method and Numerical Results

As Brunnermeier and Sannikov (2014) correctly state, the dynamics for a model such as represented by Eqs. (8)–(9) should not be studied by common linearization techniques. The first or even second order Taylor approximations to solve for the local dynamics of a model such as (4)–(7) or (8)–(9) will not properly capture the global instabilities of the model in particular in some regions of the state space. We have used the dynamic programming method by Gruene and Semmler (2004) to study the dynamics of the stochastic version of the basic model (8)–(9) and some extensions. Here, the debt to asset ratio is the state variable, and the control variables are the growth rate of assets and consumption, which can be interpreted as bonus payments.

The dynamic programming method can explore the local and global dynamics by using a coarse grid for a larger region of the state space, and then employing grid refinement for smaller region. Here we use dynamic programming, which can provide us with the truly global dynamics in a larger region of the state space without losing much accuracy (see Becker et al. 2007). In contrast, local linearization, as has been argued there, and also in Brunnermeier and Sannikov (2014), does not give sufficient information on the global dynamics. A more detailed description of the method is given in Appendix A. We want to study two major cases, a model with large bonuses and a model with small bonuses, and explore the stability properties of each variant.

Note that in both cases prices are implicitly given by the solution of the dynamic decision problem, in our case by the derivative of the value function.[25] Here, their

[24] See Brunnermeier and Sannikov (2014: Section 4) for details of such considerations.

[25] Note that the derivative of the value function is equivalent to the co-state variable using the Hamiltonian, or the Lagrangian multiplier, using the Lagrangian approach. The latter two are usually used in asset pricing theories as the shadow price for capital.

effects are not separately considered, an issue we will make comments on further below. We also do not consider specifically the inflow of funds from households, the public (for example, through TARP) or from abroad (as, for example in the case of Citigroup in 2008, after the insolvency of Lehman Brothers). At first we also neglect state or time depending credit spreads for the financial intermediaries, since we first abstract from large non-fundamental asset price movements, and also the cost of state verification resulting in cyclically varying credit spreads. The latter will be added later.

4.1 Solution with Large Bonuses

In the first variant of our model we allow for negative and positive growth rates of the assets purchased by the financial intermediaries. We constrain the growth of the assets to $-0.1 < g_t < 0.1$, and the consumption capital ratio by $0.01 < \tilde{c}_t < 0.8$. The latter is always positive but is allowed to be rather large.[26] The growth rates of assets and consumption can be chosen optimally, and the latter is permitted to be rather large.

Brunnermeier and Sannikov (2014) conjecture that when the bonus payouts are chosen endogenously "the system is relatively stable near its "steady state" ... but becomes unstable below the steady state..." (Brunnermeier and Sannikov 2014: 17). Moreover they state: "Papers such as BGG (Bernanke, Gertler and Gilchrist) and KM (Kiyotaki and Moore) do not capture the distinction between relative stable dynamics near the steady state, and much stronger amplification loops below the steady state..." (Brunnermeier and Sannikov 2014: 18).

The reason for the different result is "With endogenous payout, the steady state naturally falls in the relative unconstrained region where amplification is low, and amplification below the steady state is high" (Brunnermeier and Sannikov 2014: 18). Brunnermeier and Sannikov make this statement with respect to the ratio of net worth to assets. Since we take the negative of the debt to asset ratio, the statements can be immediately translated into the properties of our model using the debt to asset ratio.

As to the parametrization of our model we take: $a = 0.5$, $\alpha = 0.3$, $\sigma = 0.008$, $\gamma = 0.03$, and $r = 0.03$.

The Figure 1 shows on the horizontal axis the state variable ω and on the vertical axis the stochastic path for state variable ω. Since we have stochastic shocks, with pre-defined standard deviation $\sigma = 0.008$, the path of ω varies in the state space, and thus there is no unidirectional vector field, i.e. the path of ω_t is not a straight

[26]Note that the low bonus payments reflect roughly the time period until 2002 and then large bonus payments set in, see Figure 17, Appendix B. In the solution method using DP we have allowed for a larger maximum bonus payment in order to make the change of the subsequent dynamics visible. Note also that we could allow for dividend payments, in fact as our model is constructed the bonus payments can encompass dividend payments.

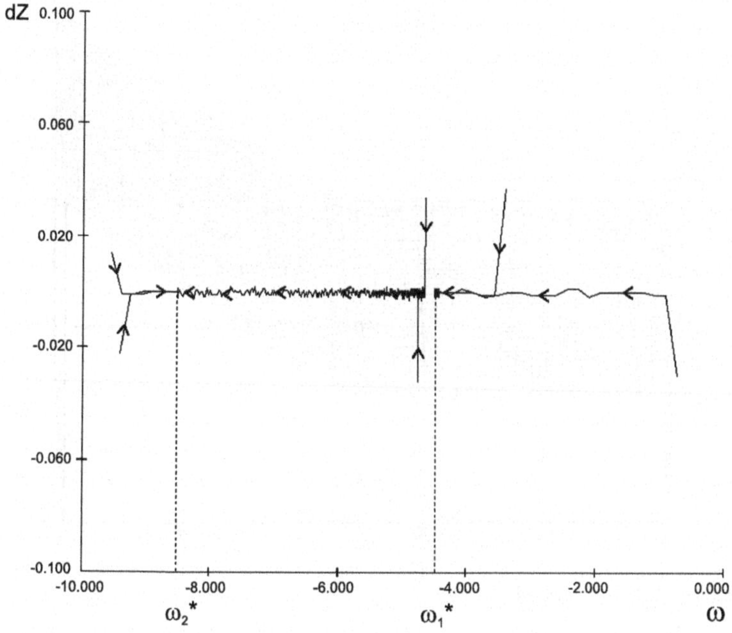

Fig. 1 Trajectories for large bonuses

line. In our numerical procedure the shocks are drawn from a distribution having a pre-defined standard deviations $\sigma = 0.008$. As visible from the numerical solution path in Figure 1, for different initial conditions there is a dynamic path from a low level of debt to asset ratio (ω close to zero) to roughly $\omega = -4.5$, see the first steady state ω_1^*. This indicates that large bonuses will make the debt to asset ratio rise, moving the ω_t toward the first steady state of $\omega_1^* = -4.5$.[27]

To the left of this first steady state, the debt to asset ratio rises, moving roughly to $\omega = -8.5$. This is a second, high debt to asset value steady state $\omega_2^* = -8.5$, which is stable and thus attracting, but it might be considered much too high. For the possibly high bonuses the first steady state ω_1^* of about -4.5, is attracting only from the right and repelling starting from the left of -4.5. That means that with large bonuses in the interval $0.01 < \tilde{c}_t < 0.8$, as optimal choices, the debt to asset ratio will rise even if the debt to asset ratio is low.

As Figure 2 shows, the value function, computed through our numerical solution procedure, increases once the bonuses start rising, see the upward slope of the value function to the right of about $\omega = -4.5$. The rise of the value function is reasonable,

[27]Note that we do not pursue the issue here at what leverage ratio bankruptcy would occur. This depends on the distance to default, which is defined by the KMV model by the distance of the asset value of the bank to the debt, divided by the standard deviation (volatility) of the asset value. We are not pursuing this question here, since we do not explicitly computing the asset value of the financial firm. This is issue is pursued in Gruene and Semmler (2005).

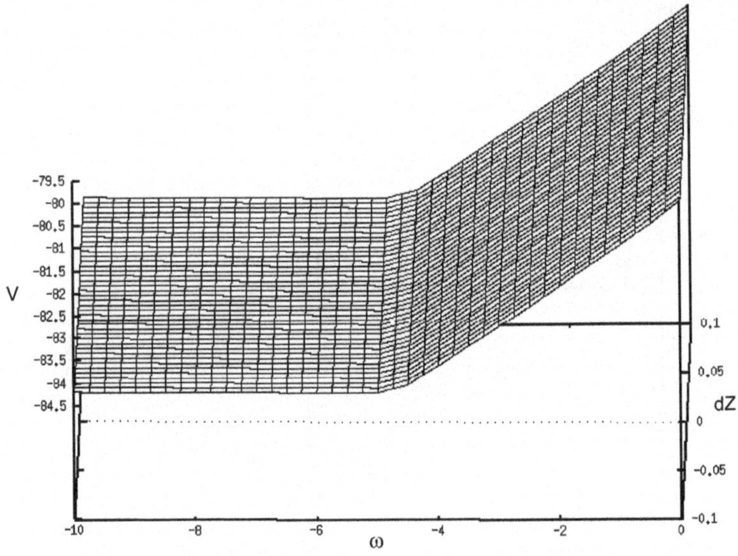

Fig. 2 Value function for large bonuses

since it is the welfare from the rising bonuses that make the value function increase. At the same time, in the region to the right of $\omega = -4.5$, higher (optimal) bonus payment are allowed. Yet this eventually gives rise to a higher debt to asset ratio.[28]

4.2 Solution with Small Bonuses

Brunnermeier and Sannikov (2014: 32) state further that allowing the debt to asset ratio rise too much, driven by the incentives of the intermediaries to take on too much risk for the sake of short term profits, paying out high bonuses, and neglecting externalities may lead to damages and downturns. In their view the triggering of the downturn in the financial, product and labor markets results from not taking into account the full extent of the externalities, and they argue that a competitive financial sector is likely to trigger such events even more frequently.

They thus state that limiting bonuses should be welfare improving. More explicitly they say: "We would like to argue that a regulator can improve social welfare by a policy that limits bonus payments within the financial sector. Specifically, suppose that experts are not allowed to pay themselves as long financial intermediaries are not sufficiently capitalized" (Brunnermeier and Sannikov 2014: 32). This type of

[28]Note that the shape of the value function is roughly the same as shown in Brunnermeier and Sannikov (2014) in their Figure 7, though we have negative values on the vertical axis, since we are taking log $\tilde{c}_t$, not $\tilde{c}_t$, in the preferences.

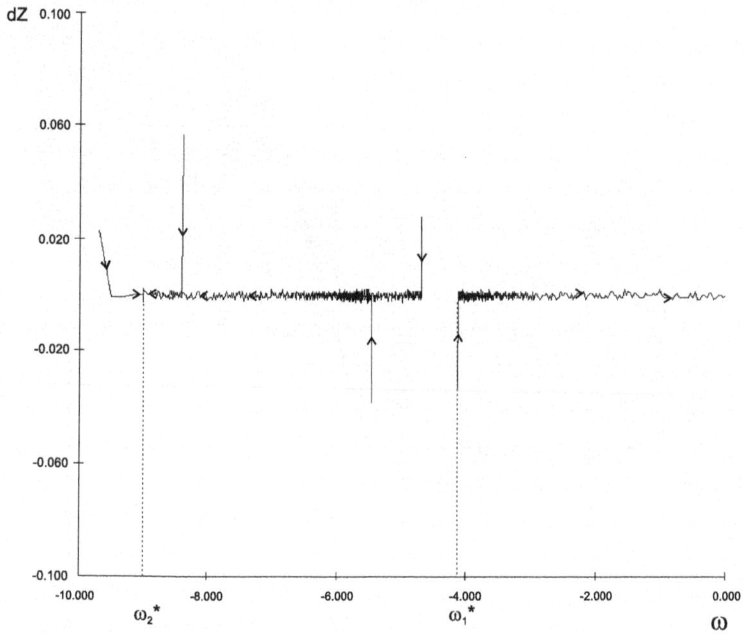

Fig. 3 Trajectories for small bonuses

regulatory effort would keep sufficient capital within the financial system and make it more stable.[29] Actually, this conjecture can also be shown to hold using our DP solution algorithm.

In order to explore this variant we, as before, allow for negative and positive growth rates of the assets purchased by the financial intermediaries to be in the range $-0.1 < g_t < 0.1$,[30] but we constrain the consumption to capital ratio by $0.01 < \tilde{c}_t < 0.1$. Again, the latter is always positive but it is constrained not to be too large. Under the condition that the growth rate of assets and the consumption rate can be chosen optimally, consumption will be constrained to be low. Figure 3 shows that now the first steady state, $\omega_1^* = -4.5$ becomes a complete repeller: with lower debt and low bonus payouts the debt to asset ratio will go to zero. No dangers of large externalities, financial stress and meltdowns will appear. Again to the left of $\omega_1^* = -4.5$[31] the debt to asset ratio will rise, possibly going up to roughly -9, see $\omega_2^* = -9$. Thus, as before, the second high level debt to asset ratio is still there, but a lower initial debt to asset ratio, with low payouts, will produce stability.

[29] A similar view is present in the Geneva Report (2009: Section 6.2) and Basel III.

[30] In the subsequent use of our DP algorithm we have used a maximum bonus payment of 0.1, reflecting roughly the time period before 2003, see Figure 17, Appendix B.

[31] Actually the rise of the debt to asset ratio will start immediately to the left of $\omega_1^* = -4.5$; the trajectories with initial conditions immediately to the left of $\omega_1^* = -4.5$, not shown here, would also go to $\omega_2^* = -9$.

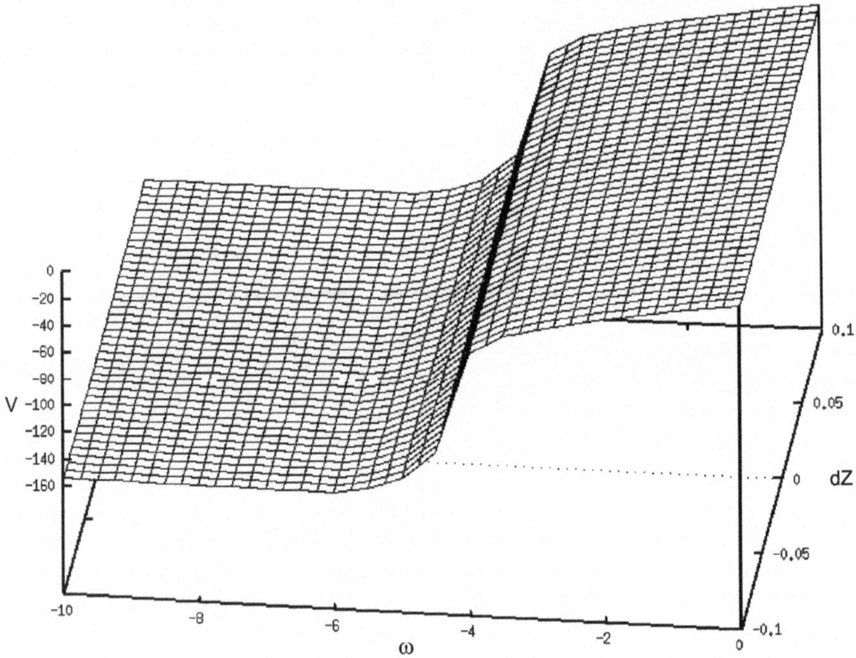

Fig. 4 Value function for small bonuses

Figure 4 shows the corresponding value function, revealing again that total welfare (for the financial intermediaries) is rising with lower debt to asset ratio, but it is, in terms of level, also higher as compared to the variant with a weak constraint on the payouts.

4.3 Solution with a Varying Risk Premium

Since the publication of the financial accelerator principle by Bernanke et al. (1999) the economists have been greatly concerned with the fact that borrowing cost moves counter-cyclically, and the ease of lending standards cyclically. Accordingly, in Section 3, in Eq. (13) we have proposed that we might have not a constant interest rate for the debt dynamics but a state dependent interest rate $r(\omega_t)$ with $r_\omega < 0$. Thus the interest payment could be made state dependent, such as[32]

$$r(\omega_t) = \frac{\alpha_1}{(\alpha_2 + (1+\omega_t))^{\mu}} r d_t \tag{14}$$

The risk premium, and thus the credit spread, is hereby made state dependent thus $r(\omega_t)$ rises with the leveraging. The parameters α_1, α_2 are positive constants.

[32]For further details, see Gruene et al. (2004, 2007).

If they are appropriately chosen, the risk premium goes to zero and a constant (risk free) interest rate will re-emerge with no leverage. The constant interest rate, as assumed in the previous version of the model, see Eq. (9), is a limit case of the above scenario.[33]

Using information economics and the theory of costly state verification, we can say that Eq. (14) reflects the standard case of the financial accelerator, according to which the risk premium rises with leverage, since a greater cost of state verification is needed with higher leverage. If there are no possible losses and no verification cost, the constant (risk free) interest rate will be charged. A case close to this may emerge, according to the argument by Brunnermeier and Sannikov (2014: Section 4), if the financial intermediaries can transfer risk through the securitization of loans and selling them as CDOs to a secondary risk market. This will not only reduce their risk exposure, but also give them less incentives for monitoring loans and increase leveraging and thus increase systemic risk: if idiosyncratic shocks are fully hedged out through securitization, the financial intermediaries then "face the cost of borrowing of only r ... Lower cost of borrowing leads to higher leverage and quicker payouts. As a result the system becomes less stable" (Brunnermeier and Sannikov 2014: 39).[34]

The cases of a constant interest rate, with larger and smaller bonus payments have already been numerically solved in Sections 4.1 and 4.2. Now it may be of interest to solve the case of a state dependent interest rate $r(\omega_t)$. Yet, the results of this case of a debt dynamics of Eq. (14) with $r(\omega_t)$ are easy to anticipate. The dynamics to the left of ω_1^* in Figure 3 will be more unstable, debt and the leverage ratio will increase faster to the left of ω_1^* and the leverage will decrease faster to the right of ω_1^*. Since the results are rather obvious we do not explore this case numerically here.[35]

On the other hand, one might argue that risk premia embodied in credit spreads cannot solely be measured by leverage ratios. There are other factors affecting risk premia, such as financial stress being built up through externalities and contagion effects generated by financial intermediaries, as well as resulting from macro economic risk. Adrian et al. (2010) have defined such a risk premium as a macro economic risk premium. They summarize the macro risk in one indicator using principle component analysis. So, we might argue that, in fact, we should have a risk premium varying with leverage, as well as other risk factors such as externality and contagion effects and asset price volatility. We might expect then a varying risk premium that is impacted by several factors and exhibits some periodic movements.

We can estimate such periodic movements in risk premia by the estimation of harmonic oscillations in the data using Fast Fourier Transform. This has been done in Hsiao and Semmler (2009) to estimate time varying asset returns. One of the

---

[33]See Gruene et al. (2004, 2007). For a similar formulation, but in terms of net worth, see Christensen and Dib (2008).

[34]They further argue that though in principle securitization may be good, since it allows for sharing of idiosyncratic risk, it also leads to the creation of severe leverage and the amplification of systemic risk.

[35]This case is further explored in Mittnik and Semmler (2013).

most important measure for macro risk is the BAA/AAA spread or the BAA/T-Bill spread. Many studies have worked with the former measure.[36] We employ here time series data from 1983.1 to 2009.4 to estimate periodic components in such a macro risk premium.[37]

The periodic component of our risk measure, which is estimated from empirical data, can be employed in our dynamic programming algorithm as presented in Appendix A. Doing so we have the following extended system to be solved on risk premia. We can write

$$V(\omega_t) = \max_{\tilde{c}_t, g_t} \int_0^\infty e^{-rt} U(\tilde{c}_t) dt \tag{15}$$

$$d\omega_t = ((g_t - r(x_t) - \delta + \sigma^2)\omega_t + a - \tau(g_t))dt - \tilde{c}_t + \sigma_t \omega_t dZ_t \tag{16}$$

$$dx_t = 1dt \tag{17}$$

The latter dynamic equation above creates a time index x_t through which the actual periodic components in the credit cost, including a risk premium,[38] can be read into the DP algorithm. If we use the notation $r(x_t)$ in Eq. (16), this indicates that there are risk factors at work that make the credit cost $r(x_t)$ time varying.[39] Formally our stochastic dynamic decision problem will have two decision variables and three state variables, the leverage ratio ω_t, the time index $x_t = t$ and the stochastic term dZ_t.

As shown in detail in Appendix B, the estimated time varying credit cost represented by the BAA bond yield, which includes a premium, takes on the form:

$$\begin{aligned} x_t &= 0.0862 - 0.0022(t - t_0) \\ &\quad + \sum_{i=1}^{n} \left(a_i \sin\left(\frac{2\pi}{\tau_i}(t - t_0) \right) + b_i \cos\left(\frac{2\pi}{\tau_i}(t - t_0) \right) \right) \end{aligned} \tag{18}$$

Note that the first two terms in the above equation represent the time trend of credit cost, the next terms the periodic variations. Appendix B also discusses how many periodic components are needed to properly replicate the actual time series of the credit cost that includes a risk component. Since we are only interested in

---

[36] See for example, Gilchrist and Zagrajsek (2012).

[37] We want to note that one might also take the periodic components of the BAA/AAA spread as measure of the time variation of financial stress. Actually our measure of financial stress is highly correlated with other measures, see Section 5.

[38] Note that in Eq. (16) we have, with $r(x_t)$, the actual credit cost modeled that includes a risk premium.

[39] Of course, the interest rate set by the Central Bank also affects $r(x_t)$.

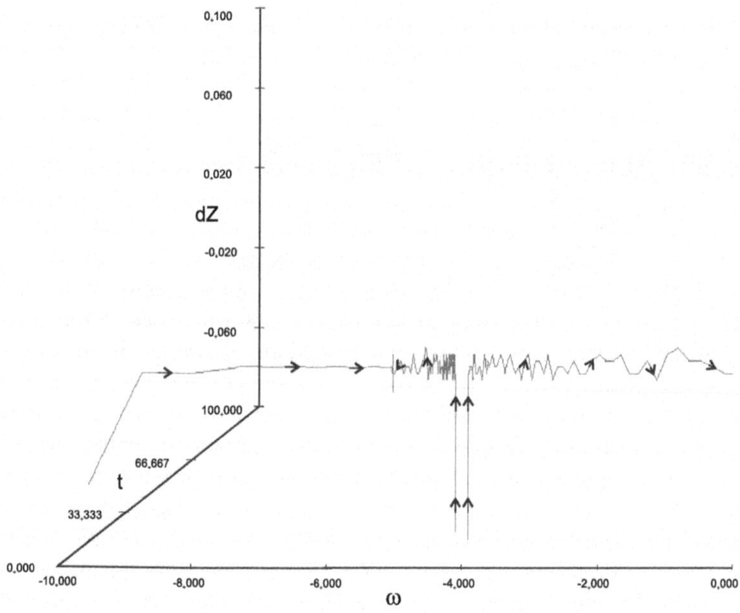

Fig. 5 Trajectories with time varying risk premium

low frequency components, it turns out that in our case we need three oscillatory components.[40]

Using (18) and solving the problem (15)–(17) through DP generates the trajectories as shown in Figure 5.

Using our set up of low bonus payments of Section 4.2, there are again two steady states, a stable one around ω_2^*, close to -5, and a repelling steady state, close to -4. Yet, compared to the fluctuations seen in Figure 3, the periodic fluctuations of the credit cost (including a risk premium) increase the volatility of the state variable ω_t. The shock dZ_t, as before, moves the trajectories, along the vertical axis, whereas the periodic fluctuations of the credit spread moves the trajectories along the time axis t. Note that here again we have the shock drawn from the range $-0.1 < dZ_t < 0.1$, and we have used $-10 < \omega_t < 0$. For our third dimension, our time index, the range is defined as $0 < t < 100$.[41]

As one can observe from Figure 5, to the right of the middle steady state, -4, the inclusion of the time varying risk premium in the credit cost, the use of $r(\omega)$, not only amplifies the volatility of ω_t, but moves the leverage ratio to zero faster as

[40] From the estimation procedure, as discussed in Appendix B, we get the following parameters for Eq. (18): $a_{11} = 0.006$, $a_{12} = 0.0062$, $a_{13} = 0.0063$, $b_{11} = 0.0049$, $b_{12} = 0.006$, $b_{13} = 0.0016$, $c_{11} = -0.002$, $c_{12} = 0.0862$, $\tau_{11} = 305$, $\tau_{12} = 152.5$, $\tau_{13} = 101.5$.

[41] We do not pursue here to compute and graph the value function, since the 3-dim problem (15)–(18) makes it more cumbersome to present the value function.

compared to a constant credit cost. On the other hand it will also built up financial stress faster, once the middle steady state, −4, is surpassed.[42]

5 Financial Intermediaries and Financial Stress Measures

In the previous section we have postulated that the financial intermediaries are not only exposed to the tides of the financial market, but also amplify them. As we saw, a variety of constellations are feasible, depending on potential restrictions on bonus payments, and whether there is a constant or state and time varying credit spread. Yet the different variants show that a shock to the banks' balance sheets can entail a considerable instability.

Thus, shocks to asset prices, and therefore to capital assets, $p_t k_t$, and net worth, $n_t = p_t k_t - d_t$, (increasing debt) will be amplifying, particularly in the case of large bonus payments, represented by Figure 1, where initial leverage ratios to the right of ω_1^* will be increasing, and even a very low or zero leverage will create higher leverage and financial stress. The trajectories will be attracted to a high steady state leverage ratio for high bonus payments. Yet, the leverage may fall in the case of the low bonus payment scenario, see Figure 3. So we may observe some superior stability properties of the leverage ratio for small bonus payments. The strength of the local instabilities will be the empirical issue to be explored next.

The problem is, however, what measures can one utilize to empirically evaluate the predictions of the model and undertake empirical estimates. What actual measurements should one take to evaluate how financial stress of banks is interlinked with the financial market, or more specifically, to asset prices? In the context of our model in Sections 2–4, one could take leverage ratios stemming from the balance sheets of the financial intermediaries as measuring this linkage: high leverage implying high financial stress and low leverage the reverse.

However, there is an issue whether the ratio of net worth to capital assets, or the reverse measure, the leveraging ω can be good measures of financial stress. First, both are greatly affected by the market valuation of assets as well as liabilities, which is not easy to undertake. In particular, asset valuation is heavily impacted by the confidence and the estimate of income streams the asset generate, as well as presumed discount rates, and the liabilities such as bonds or short and long term loans are strongly affected by their corresponding risk premia.[43] Moreover, credit constraints, for example, as measured by the Fed index of changes in credit standards to determine the ease and tightness of obtaining credit as well as credit spreads and short term liquidity, are also important financial stress factors for financial intermediaries. All this will affect credit demand and supply of financial intermediaries. We thus need more extensive measures than only leverage to evaluate financial stress.

[42] A varying risk premium might thus prevent the agents to go to a too high leverage, but on the other hand if the middle steady state is surpassed it creates a vicious cycle of credit spreads and higher indebtedness, as often has been observed for companies as well as sovereign debt.

[43] This is implicit in Merton's risk structure of interest rates, see Merton (1974).

The Federal Reserve Bank of Kansas City and the Fed St. Louis have thus developed a general financial stress index, called KCFSI and STLFSI respectively. The KCFSI and the STLFSI,[44] take into account the various factors generating financial stress. They can be taken as substitutes for the net worth or leverage ratios as measuring financial stress of financial intermediaries. Other financial stress indices have been developed before, for example the Bank of Canada index[45] for Canada, and the IMF (2008) index. Both of them include a number of variables that are included in the KCFSI and STLFSI but are less broad. Both the KCFSI and STLFSI assert that the times of stress: (1) increase the uncertainty of the fundamental value of the assets, often resulting in higher volatility of the asset prices, (2) increase uncertainty about the behavior of the other investors, (3) increase the asymmetry in information, (4) increase the flight to quality, (5) decrease the willingness to hold risky assets, and (6) decrease the willingness to hold illiquid assets.[46]

Following this characterization of the period of financial stress, the above mentioned FSIs take the following variables: The TED spread (spread between the 3 month LIBOR/T-bill), the 2 year swap spread, the AAA/10-year Treasury spread, the BAA/AAA spread, the high yield bond/BAA spread, Consumer ABS/5 year Treasury spread, the correlation between returns on stocks and Treasury bonds (a measure for the flight to quality), the VIX (implied volatility of bank stocks) and the cross dispersion of banks stocks. As one can see here, spreads, volatility and dispersion measures are taken as variables for a financial stress index. The principle component analysis is used to obtain the FSI.[47] We want to note that most of the above variables are highly correlated and the leading variables are the spread variables.[48]

Combining all the variables with appropriate weight in a stress index produces a clearly counter-cyclical behavior. This is illustrated in Figure 6.

As the comparison of the smoothed growth rate of the production index and the stress index in Figure 6 shows there is less financial stress in good times, but more in bad times. Although we are not using balance sheet variables directly, nevertheless we can safely presume that financial intermediaries are clearly doing better in economic booms then in recessions.[49] Given the linkages between the financial stress index and economic activity, we would also expect a strong linkage between net

[44]The KC index is a monthly index, the STL index a weekly index, to capture more short run movements. Another recent work on financial stress indexes can be found in Hatzius et al. (2010).

[45]See Illing and Liu (2006).

[46]The latter tendencies have described by Gorton (2010) as a flight from information sensitive to information insensitive assets.

[47]This is done as follows. Linear OLS coefficients are normalized through their standard deviations and their relative weights computed to explain the index. A similar procedure is used by Adrian et al. (2010) to compute a macro economic risk premium.

[48]In the sense that they have the highest weight in the index, for details see Hakkio and Keeton (2009: Tables 2–3).

[49]This coincides also with the empirical study by Gorton (2010) that there is more insolvency of financial institutions in bad times, see also Figure 15 in Appendix B.

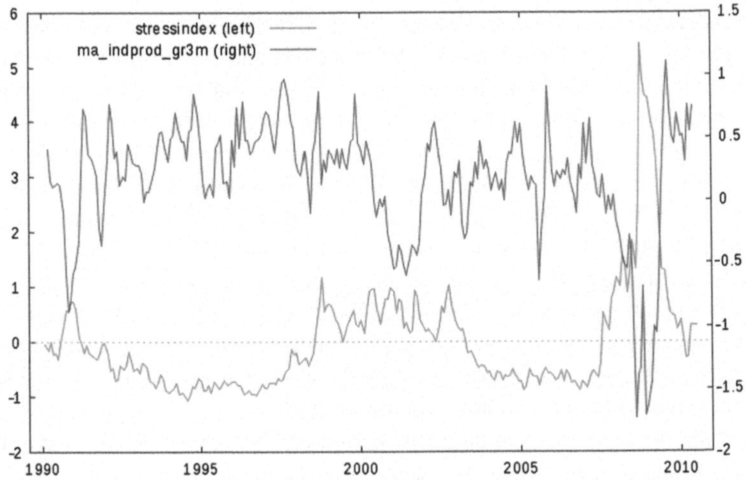

Fig. 6 Financial stress index (KCFSI, *lower graph*) plotted against growth rates of industrial production (3 month moving average, *upper graph*) (Color figure online)

worth, or leveraging, of financial intermediaries and economic activity, since the financial stress is affecting the balance sheets of financial intermediaries.[50] We want to note that the financial stress index is also highly linked to some broader index of economic activity.[51]

A "one-regime VAR" has been used many times to study the financial accelerator.[52] Yet those "one-regime VAR" studies presume only local instability, symmetry effects of shocks and mean reversion after the shocks. What we will pursue here is an MRVAR. Our MRVAR[53] takes the aforementioned stress index KCFSI as empirical measure of financial stress, and the growth rate of the monthly production index as a threshold variable to define regimes.

6 Empirical Analysis Using a MRVAR

To empirically investigate how strong the local instabilities are—and whether one finds sufficient empirical evidence for instabilities at all—requires an empirical approach that can accommodate varying dynamic patterns across alternative states of

[50]The fact that the leverage ratio is rising in recessions and falling in booms, is documented in Gilchrist et al. (2009).

[51]See Hakkio and Keeton (2009).

[52]Estimating the financial accelerator for the macroeconomy with a "one regime VAR", see Christensen and Dib (2008) and for the application of the financial accelerator to study financial intermediaries in a "one regime VAR", see Hakkio and Keeton (2009) and Adrian and Shin (2009), Adrian et al. (2010).

[53]For using a MRVAR, see Mittnik and Semmler (2012) and Ernst et al. (2010).

the economy. For this reason, we adopt a multi-regime modeling strategy, which allows us to explore regime-dependencies of responses to shocks to the system. A related question is whether there might be local stability for small disturbances, but not for larger shocks. Thus, we will also explore whether the size of shocks matters.[54]

6.1 Methodology

To assess the dependence of the responses to shocks to the stress index, we employ a multi-regime VAR (MRVAR) approach. A major limitation of conventional linear VAR models is that shock responses are independent of the economy's state at the time a shock occurs. Also, VAR response profiles are invariant with respect to the sign and size of a shock. That is, responses to positive and negative shocks are sign-symmetric; and the response to shocks of different sizes are simply scaled versions of the response to a shock of size one. To capture state dependencies and asymmetries of shock responses, a nonlinear model or a linear model with state dependencies needs to be specified. The mildest form of generalizing a linear, constant–parameter VAR is to adopt a piecewise linear VAR, such as Markov switching autoregressions (Hamilton 1989) or threshold autoregressions (Tong 1978, 1983). A characteristic of Markov switching autoregressions is that the states are unobservable and, hence, do not necessarily have a clear interpretation. Also, a given observation cannot directly be associated with any particular regime. Only conditional probabilistic assignments are possible via statistical inference based on past information.

For our purposes, namely, state-dependent response analysis, states are associated with specific stages of the business cycle as measured, for example, in terms of output growth. Multi-regime vector autoregression (MRVAR) models in the form of threshold autoregression models of Tong (1978, 1983) or, in a vector setting, of multivariate threshold autoregressions (Tsay 1998) are obvious candidates. In contrast to Markov switching autoregressions or standard multivariate threshold autoregressions, our approach assumes that we can, based on some observable variable, define upfront a meaningful set of regimes, which are not a result of some estimation procedure, but rather motivated by the objective of the empirical analysis. This is preferable in our setting, where we are interested in evaluating the potential effectiveness of policy measures for a particular state of the economy.

The MRVAR specification adopted here is given by

$$y_t = c_i + \sum_{j=1}^{p_i} A_{ij} y_{t-j} + \varepsilon_{it}$$

$$\text{if } \tau_{i-1} < r_{t-d} \leq \tau_i, \ \varepsilon_{it} \sim \text{NID}(0, \Sigma_i), \ i = 1, \ldots, M \tag{19}$$

[54] Since we work with historical data since the 1990s it is probably realistic to assume a historical period of large bonus payments, for data and computation on this, see Semmler and Lucas (2009) and Figure 17.

where r_{t-d} is the value of the threshold variable observed at time $t-d$; and regimes are defined by the (prespecified) threshold levels $-\infty = \tau_0 < \tau_1 < \cdots < \tau_M = \infty$. In the following analysis we estimate a two-regime VAR, with the output-growth rate as the threshold variable, and the average growth rate delineating the threshold for the sample.

In addition to the more straightforward regime interpretation, MRVAR models are also more appealing than Markov switching autoregressions as far as estimation is concerned. Rather than EM-estimation, MRVARs with predefined threshold levels resemble conventional VARs and can be estimated regime by regime, using standard common least-squares-provided the regime-specific sample sizes permit this, or using Bayesian techniques.

Response analysis for linear VAR models is straightforward. Point estimates and asymptotic distributions of shock response can be derived analytically from the estimated VAR parameters (cf. Mittnik and Zadrozny 1993). In nonlinear settings, this is, in general, not possible, and one has to resort to Monte Carlo simulations. Following Koop et al. (1996), the so-called generalized impulse responses, which depend on the overall state, z_t, type of shock, v_t, and the response horizon, h, are defined by

$$\mathrm{GIR}_h(z_t, v_t) = \mathrm{E}(y_{t+h} \mid z_t, u_t + v_t) - \mathrm{E}(y_{t+h} \mid z_t, u_t) \tag{20}$$

where the overall state, z_t, reflects the relevant information set. For a Markov-switching VAR process, z_t comprises information about the past realizations of y_t and the states; for an MRVAR process with known threshold levels, only information about past realizations $y_{t-1}, \ldots, y_{t-p_{\max}}$, with $p_{\max} = \max(p_1, \ldots, p_M)$, is required.

To understand the differences in the dynamic characteristics between the different regimes, regime-specific response analysis as in Ehrmann et al. (2003) is helpful. Regime-specific responses of MRVAR models assume that the process remains within a specific regime during the next h periods. This is particularly reasonable when regimes tend to persist or when we are interested in short-term analysis, and helps to understand regime-specific dynamics.

6.2 Estimation

For our bivariate analysis, we use monthly data on U.S. industrial production and the KCFSI stress index covering the period February 1990 to June 2010.[55]

We estimate a standard VAR and an MRVAR model for the IP growth rate and changes in the stress index, by defining $y_t = (100\Delta \log \mathrm{IP}_t, \Delta \mathrm{KCFSI}_t,)'$. We use

[55] Seasonally-adjusted industrial-production (IP) data (Series ID INDPRO) come from Board of Governors of the Federal Reserve System; the KCFSI stress-index data were obtained from the Federal Reserve Bank of Kansas City.

the AIC for model selection. For MRVAR model (19), the AIC is given by

$$\mathrm{AIC}(M, p_1, \ldots, p_M) = \sum_{j=1}^{M} \left[T_j \ln |\hat{\Sigma}_j| + 2n\left(np_j + \frac{n+3}{2}\right)\right] \tag{21}$$

where M is the number of regimes; p_j is the autoregressive order of Regime j; T_j reflects the number of observations associated with Regime j; $\hat{\Sigma}_j$ is the estimated residual covariance matrix for Regime j; and n denotes the number of variables in vector y_t. Formulation (21) differs from that in Chan et al. (2004) in that we account for possible heterogeneity in the constant terms, c_j, and residual covariance, Σ_j, across regimes.[56]

Based on the AIC, a VAR of order $p = 4$ is suggested. Specifying a two-regime MRVAR with the threshold, τ, set to the sample mean of the IP-growth rate given by 0.165, we assign observations associated with below-mean (above-mean) growth rates to Regime 1 (Regime 2). Then, the AIC suggests an autoregressive order of four for Regime 1 and order of three for Regime two. Although the MVAR has quite a few more free parameters than the fitted VAR (35 vs. 21 parameters), the AIC favors the two-regime MRVAR with AIC ($M = 2$, $p_1 = 4$, $p_2 = 3$) $= -1084.9$ (and regime-specific sample sizes $T_1 = 112$ and $T_2 = 126$) over a standard VAR with AIC ($M = 1$, $p = 4$) $= -842.1$.

6.3 *Response Analysis*

To assess the effects of linear versus nonlinear model specification, we first look at the estimates of the cumulative unit-shock responses for the VAR model and the regime-specific responses for the MRVAR model. To derive structural responses, we assume that a shock to IP simultaneously affects the stress index, whereas IP reacts with a one-period delay to a stress shock. As Figures 15 and 16 in Appendix B show, when compared to Figure 6, a positive stress shock reveals a high co-movement with a rise in the overall leverage ratio and the rise of banking failures. We can thus view the positive stress shock as measuring the deterioration of the balance sheets of banks and a rise of insolvency risk of banks.

The cumulative responses to unit shocks are shown in Figure 7.

In our analysis we will focus on the responses of IP due to shocks in the financial stress index, since we want to evaluate the impact of the change of financial stress on the banking system and the macroeconomy. In our case, the latter is measured by output growth.

The results for the conventional VAR model (Figure 7) suggests that a positive one-standard-deviation stress-index shock has an increasingly negative cumulative

[56] When employing (21) to discriminate between an MRVAR and a standard VAR specification (i.e., a one-regime MRVAR), we need to include the n parameters in the intercept vector, c, and the $n(n+1)/2$ parameters in the residual covariance matrix for an equivalent parameter count.

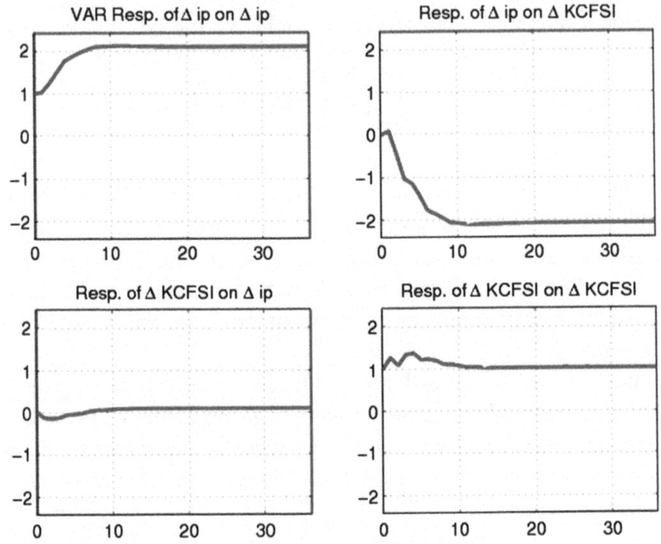

Fig. 7 Cumulative responses from a one-regime linear VAR model

IP growth effect which settles at −2 % after about 10 months. Additionally, the output responds strongly to output, but the stress index responds weakly to the output change and the stress shock itself. These are all responses that one would expect from macroeconomic one-regime VARs.

The symmetry and size independence of responses of the shock in the one-regime linear VAR is shown in Figure 8. As one can observe, the positive and negative stress shocks have symmetric effects and the effects are linearly dependent on the size of the shocks. There is no difference whether there is a positive or negative shock that hits the economy.

Next we want to explore regime-specific responses that may help to understand the dynamic properties of the estimated regimes. Assuming that one stays within one regime after a shock will not be very realistic. Results of such an exercise are not convincing. Such an exercise will be of limited use when trying to assess the overall impact of a shock for two reasons. First, the process is not expected to stay within a given regime for an extended period of time; it will rather switch between regimes. Second, by looking at the within-regime dynamics, we would solely focus on the regime-specific autoregressive parameters and ignore the level effects induced by the differences in the regime intercepts. They will induce additional variation in the dynamics as the process switches between regimes.

In order to investigate the economy's overall reaction to shocks, we simulate generalized cumulative response functions to unit-impulse shocks. We do this for specific states at which the shock is assumed to occur. The two states we select are given by the sample averages we observe for the two regimes and are, thus,

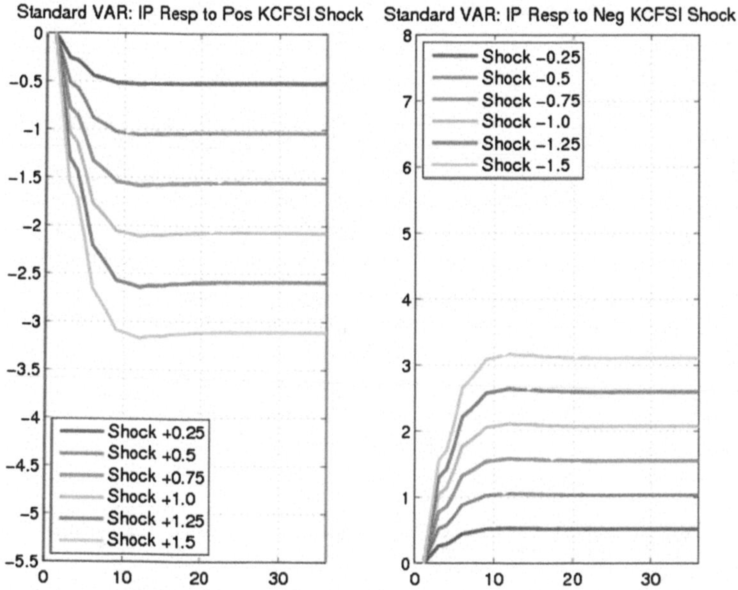

Fig. 8 Cumulative responses from a one-regime linear VAR with negative and positive shocks of different size (Color figure online)

representative for low- and high-growth states of the economy. In the mean in low-growth regime is $\bar{y}_{low} = (-0.3463, -0.0294)'$, and that for the high-growth one is $\bar{y}_{high} = (0.6137, 0.0296)'$. For each case we simulate two shocks to the stress-index: a positive and a negative unit-shock. The mean cumulative responses to output and one-standard deviation confidence bands are shown in Figure 9.[57]

The responses strongly suggest that the impact of a stress-shock on output varies with the state of the economy. A *positive* unit-shock in the average high-growth state (top left plot in Figure 9) causes IP to drop by about -4 % within about 36 months. The same shock applied in the average low-growth state (bottom left plot), results in a less severe output contraction, namely, -2 % after three years. Thus, in a boom period an increase in financial stress curbs growth more strongly than in a recession. This presumably comes from the fact that an increase of financial stress, and thus a deterioration of the balance sheets of banks and a rising insolvency risk in the high growth regime, can trigger sharp downturns. This might come from the tendency that even our risk measure (and any risk measure) declines during the boom, even though risk might build up in the background. This often leads to the paradox that while the risk is rising in terms of higher leveraging in high growth regimes, the financial risk measure (for example due to falling credit spreads) shows a decline, see Figure 6.

[57]The generalized cumulative responses were simulated based on 100 replications, which were repeated 200 times to approximate the standard errors of the responses.

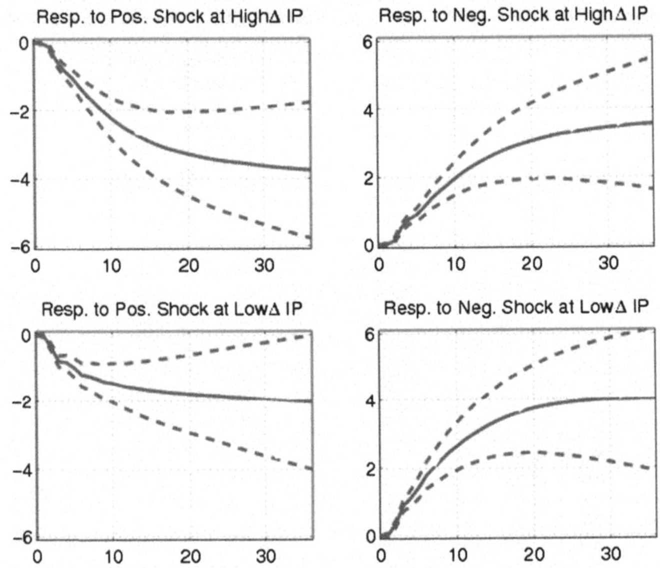

Fig. 9 Cumulative MRVAR responses to positive (*left panel*) and negative stress-index shocks (*right panel*) in high- (*upper panel*) and low-growth states (*lower panel*)

On the other hand, a reversion of the sign of the stress-shock, a negative stress-shock, also indicates some asymmetries in the IP response. A *negative* unit stress-shock in the high-growth state (top right plot) produces a 3.5 % IP increase in the long-run. In contrast, in the depressed state (bottom right plot) the negative shock boosts IP by 4 %. Thus, a negative stress shock in the low growth regime seems to be more effective in boosting the economy. We will come back to this issue when we consider larger shocks.

Moreover, the state-dependent response analysis indicates that in a negative growth period the IP responds very differently with the sign of the stress-shock: A positive unit-shock reduces IP by about −2 %, whereas a stress reduction by a unit induces an IP increase of 4 %. This cannot be observed for the linear one regime VAR as shown in Figure 8. There is virtually no such asymmetry: The size (not the sign) of the IP response is virtually identical for positive and negative shocks to financial stress.

Next, we investigate to what extent the *size* of the shock to financial stress matters. Instead of a *unit* shock to the stress index we simulate the cumulative IP responses to stress shocks with different sizes. Specifically, we impose positive and negative shocks of sizes 0.25, 0.5, 0.75, 1.0, 1.25, and 1.5, always measured in terms of standard deviations of the stress index.

The consequences of positive shocks after 36 months differ quite dramatically with the magnitude of the shocks. Figure 10 compares the response profiles. Whereas the responses to a small (+0.25 std. dev.) shock are similar in both regimes,

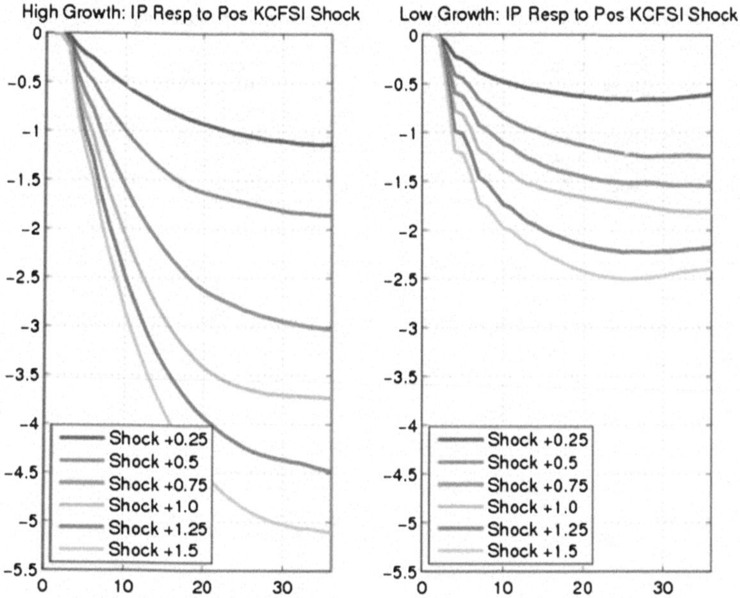

Fig. 10 Cumulative MRVAR responses to positive stress-index shocks in average high- (*left*) and low-growth states (*right*) (Color figure online)

namely, −0.8 % in the average high- and −0.7 % in the average low-growth regime, this does not hold for larger shocks. For larger shocks, 0.5 standard deviations and more, IP drops roughly about twice as much in the high-compared to the low-growth state. One can see here not only that large positive shocks have quite a different final impact than small shocks, but that the effects of the large shocks are quite different in a high growth regime as compared to low growth regime. As already argued above, given the fragility of the finance sector likely to be built up during the boom, a sudden large financial stress shock in the high growth regime will likely trigger a significant deterioration of balance sheets through externality effects in the interconnectedness of the financial firms and a rise of a credit risk spread (and possibly a cascade of insolvencies), generating strong chain effects of shocks in the positive regime.

As Figure 11 shows, we find an analogous but somewhat less extreme divergence for negative shock scenarios. For small negative shocks (−0.25 and −0.5) IP responds more or less identically. Larger stress reductions, however, have a much stronger positive effect on IP growth when the economy is in a recessionary rather than a boom period, a phenomenon observed earlier for the unit shock. In case of larger shocks (−1.25 and −1.5), the impact in low-growth is about 50 % larger than in the high-growth.

A comparison of the left plots in Figures 10 and 11 reveals that—as in the unit-shock experiment—there is also not much asymmetry in the IP responses when

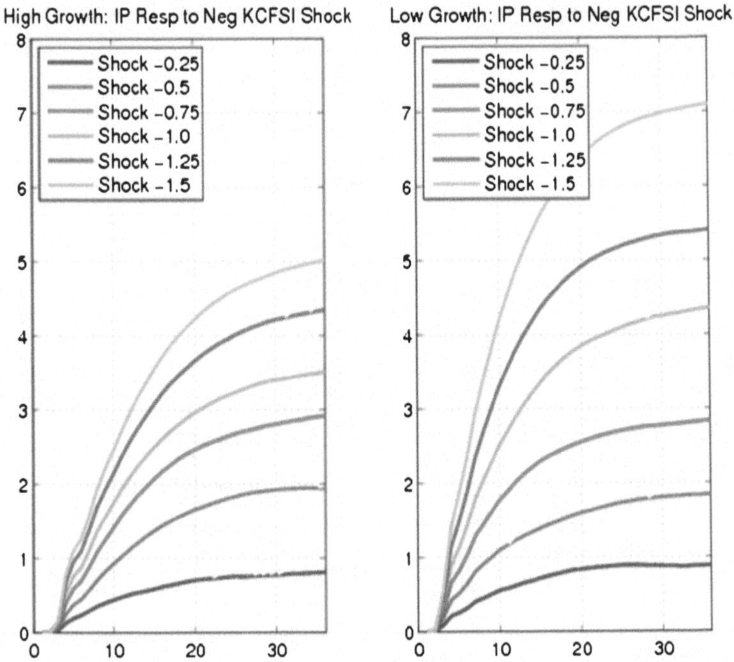

Fig. 11 Cumulative MRVAR responses to negative stress-index shocks in average high- (*left*) and low-growth states (*right*) (Color figure online)

the shock size varies. In a low-growth state, however, there is a strong asymmetry. This holds especially, for large shocks. A 1.5 standard-deviation reduction in the stress index raises IP by about three times as much as a stress-increase of the same size would lower IP. This seems to us a very relevant observation concerning the asymmetric impact of monetary policy on the economy. Monetary policy shocks—in particular what has recently been called unconventional monetary policy[58]—is likely to have large effects if the shocks are large.

We can stress that our empirical results suggest that the timing of policy actions affecting financial stress can be crucial to the success of such measures. The findings are compatible with recent studies which argue that unconventional monetary policy is needed in a depressed economy that is accompanied by a sharp rise in credit spreads, which, more so than asset-price volatility, constitute the dominant component of the stress index.[59] The results suggest that not only a decrease in the interest rate but also in credit spreads are required to induce significant expansionary effects.

[58]For example of quantitative easing, as pursued by the Fed since 2008.

[59]See for example Curdia and Woodford (2009a, 2009b) and Gertler and Karadi (2011).

7 Conclusions

Though most of the historical economic crises ended up as a meltdown of the banking sector, the banking sector has usually exacerbated and amplified the crisis whatever origin it had. To investigate those feedback effects, we have studied the linkage of asset prices and the financial intermediaries' balance sheets. We have modeled the financial intermediaries as they are affected by adverse asset price shocks, but we also considered the reverse effect from the instability of the banking system to the macroeconomy, which is here represented by the growth rate of output. In particular we studied the issue of local instability of the banking sector that is exposed to asset price shocks, credit spread shocks and financial stress.[60] We modeled financial intermediaries as wealth fund that accumulates capital assets, can heavily borrow and pays bonuses. When the banking sector is exposed to a deterioration of its balances sheets, it turns out that the size of the bonus payments plays an important role for the dynamics of the leverage ratio, the financial stress and the local instability.

In contrast to previous studies that use the financial accelerator—which is locally amplifying but globally stable and mean reverting—our model admits local instability and globally multiple regimes. Whereas the financial accelerator leads, in terms of econometrics, to a one-regime VAR, the multi-regime dynamics studied here requires to use a multi-regime VAR (MRVAR). Using a financial stress index as evaluating the financial intermediaries' stress and output growth, measuring the state of the macroeconomy, our method of a MRVAR estimate permits us to undertake an impulse-response study which lets us explore regime dependent shocks.

We show that the shocks have asymmetric effects depending on the regime the economy is in, but we also show that the effects of the shocks are dependent on the size of the shocks.[61] Large positive financial stress shocks in booms seem to have a stronger contractionary effect than in a recessions, but large negative stress shocks in recessions appear to have a stronger expansionary effect in recessions than in booms. The latter result seems to us very important for the evaluation of an unconventional monetary policy, since frequently not only the timing, but also the strength of policy actions matter.

Acknowledgements A previous version of this paper has been presented at the Centre Cournot, Paris, the 4th International Conference on Computational and Financial Econometrics, London, and at the BI Business School, Oslo, at the Monetary Authority in Hong Kong, an international conference in Guangzhou, China, the 2011 SNDE conference in Washington, DC, and workshops at Humboldt University and Giessen University, Germany. We want to thank the audience of those conferences and workshops for valuable comments. We are also grateful for encouraging communications with Markus Brunnermeier. Willi Semmler would like to thank the Center of Economic

[60] Note that most of the components of the financial stress index are some measure of credit spread.

[61] In earlier literature on Keynesian macrodynamics this has been called "corridor stability": small shocks have only small effects but larger shocks could be destabilizing and have large effects.

Risk and Wolfgang Haerdle at Humboldt University for a great hospitality in the Winter Semester 2010/2011.

Appendix A: The Numerical Solution of the Model

We have used the dynamic programming method by Gruene and Semmler (2004) to study the dynamics of the stochastic version of debt-asset ratio with consumption and growth rate of assets as controls. The dynamic programming method can explore the local and global dynamics by using a coarse grid for a larger region and then employing grid refinement for a smaller region. As Brunnermeier and Sannikov (2014) correctly state, the dynamics should not be studied by first or second order Taylor approximations to solve for the local dynamics, since this will not capture the global instabilities of the model, in particular below the steady state. Instead we use dynamic programming, which can provide us with the truly global dynamics in a larger region of the state space without losing much accuracy (see Becker et al. 2007). In contrast, local linearization, as has been argued there, does not give sufficient information on the global dynamics.

Hence, before going into the model discussion, we start by briefly describing this dynamic programming algorithm and the mechanism by which it enables us to numerically solve our dynamic model variants. The adaptive discretization of the state space feature of the dynamic programming algorithm leads to high numerical accuracy with moderate use of memory. In particular, the algorithm is applied to discounted infinite horizon dynamic decision problems of the type introduced in Section 4. In our model variants we have to numerically compute $V(x)$:

$$V(x) = \max_u \int_0^\infty e^{-rt} f(x,u)dt$$

$$\text{s.t.} \quad \dot{x} = g(x,u), \quad x(0) = x_0 \in \mathbb{R}^1$$

where u represents the decision variable, and x a vector of state variables. Note that the time index t, as used in Section 4.3 of the paper, could be one of the state variables.

In the first step, the continuous time optimal decision problem has to be replaced by a first order discrete time approximation given by

$$V_h(x) = \max_{u \in U} J_h(x,u)$$

where $J_h(x,u) = h\sum_{i=0}^{\infty}(1-\theta h) f(x_h(i), u_i)$, and x_h is defined by the discrete dynamics

$$x_h(0) = x, \qquad x_h(i+1) = x_h(i) + hg\big(x_h(i), u_i\big)$$

and $h > 0$ is the discretization time step. Note that U denotes the set of discrete control sequences $u = (u_1, u_2, \ldots)$ for $u_i \in U$.

The value function is the unique solution of a discrete Hamilton-Jacobi-Bellman equation such as

$$V_h(x) = \max_{u \in U} \{ hf(x,u) + (1 - \theta h) V_h(x_h(1)) \} \tag{22}$$

where $x_h(1) = x + hg(x,u)$ denotes the discrete solution corresponding to the control and initial value x after one time step h. Using the operator

$$T_h(V_h)(x) = \max_{u \in U} \{ hf(x,u_o) + (1 - \theta h) V_h(x_h(1)) \}$$

the second step of the algorithm now approximates the solution on a grid Γ covering a compact subset of the state space, i.e. a compact interval $[0, K]$ in our setup. Denoting the nodes of Γ by x^i with $i = 1, \ldots, P$, we are now looking for an approximation V_h^Γ satisfying

$$V_h^\Gamma(x^i) = T_h(V_h^\Gamma)(x^i) \tag{23}$$

for each node x^i of the grid, where the value of V_h^Γ for points x which are not grid points (these are needed for the evaluation of T_h) is determined by linear interpolation. We refer to Gruene and Semmler (2004) for the description of iterative methods for the solution of (23). This procedure allows then the numerical computation of approximately optimal trajectories.

In order to distribute the nodes of the grid efficiently, we make use of an a posteriori error estimation. For each cell C_l of the grid Γ we compute

$$\eta_l := \max_{k \in c_l} \left| T_h(V_h^\Gamma)(k) - V_h^\Gamma(k) \right|$$

More precisely, we approximate this value by evaluating the right hand side in a number of test points. It can be shown that the error estimators η_l give upper and lower bounds for the real error (i.e., the difference between V_j and V_h^Γ) and hence serve as an indicator for a possible local refinement of the grid Γ. It should be noted that this adaptive refinement of the grid is particularly effective for computing steep value functions, non-differential value functions and models with multiple equilibria, see Gruene et al. (2004) and Gruene and Semmler (2004). These are all cases where local linearizations are not sufficiently informative.

Appendix B: Periodic Components in Credit Spreads, Bank Failures, Leveraging, and Bonus Payments

We take the BAA bond yield as a proxy of time varying credit cost that includes a risk premium. We apply the *Fast Fourier Transform* to Moody's Seasoned BAA Cor-

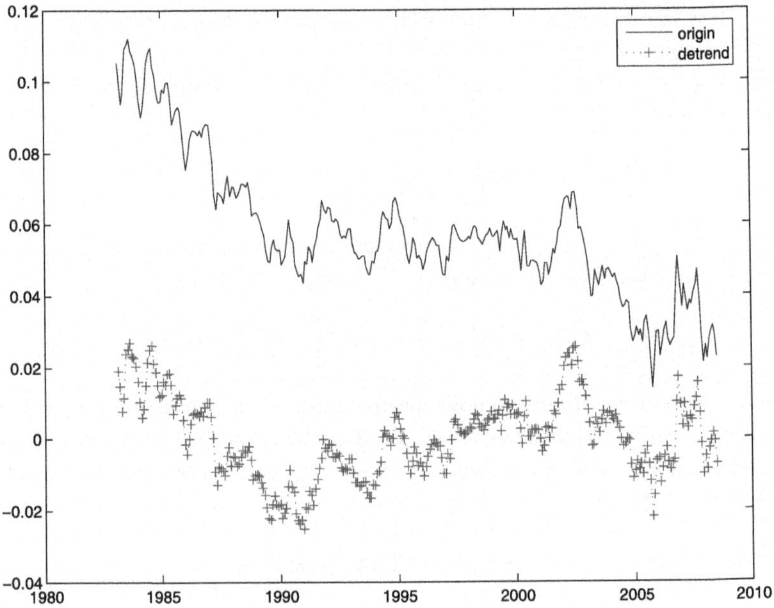

Fig. 12 Original and de-trended real BAA yields

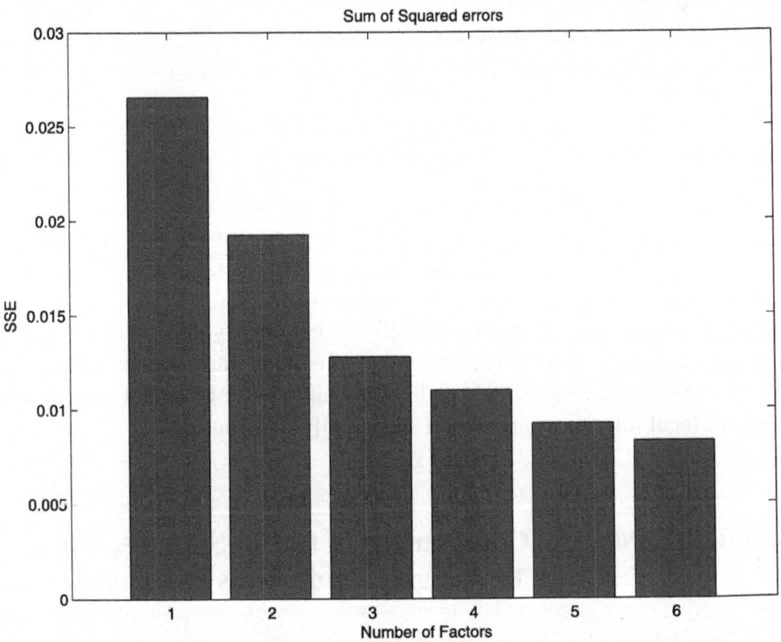

Fig. 13 Sum of the squared errors

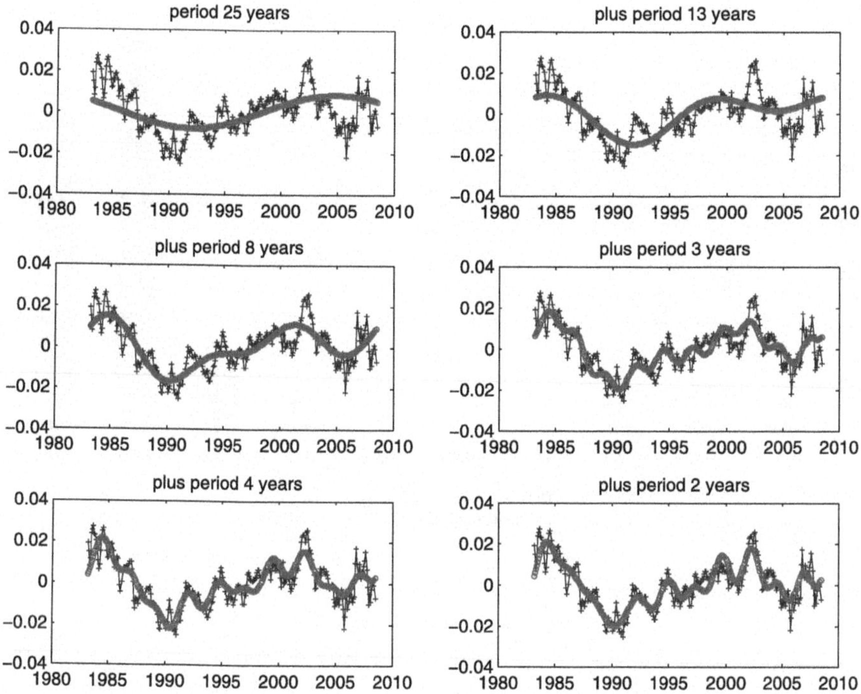

Fig. 14 Harmonic fit of the BAA yields

porate Bond Yield. The time period is from February 1983 to June 2008 at monthly frequency (305 data points). Data are: (A) Moody's BAA corporate bond yield from St Louis Fed,[62] and the inflation rate: (B) the CPI (seasonal adjusted) consumer price index of all urban areas from Bureau of Labor Statistics of U.S. Department of Labor.[63] The realized real bond yield is then: (A)–(B).

First we de-trend the real BAA yields

$$\text{Detrend rb} = \text{Original rb} - \big(-0.0022(t - t_0) + 0.0862\big) \tag{24}$$

and illustrate it in Figure 12.

Next we apply the FFT on the de-trended real BAA Yields and obtain the loading/power of periods, which helps us to select the first few harmonic components of the fit. The harmonic fit is implemented, the coefficients estimated (reported in the table below), and the results are then illustrated in Figures 13–14. The empirical estimates are based on linear regressions based on the trigonometric functions, which means we fit the time series x_t using the sin/cos functions of the given period. The

[62] See http://research.stlouisfed.org/fred2/series/BAA.

[63] See http://www.bls.gov/cpi/home.htm.

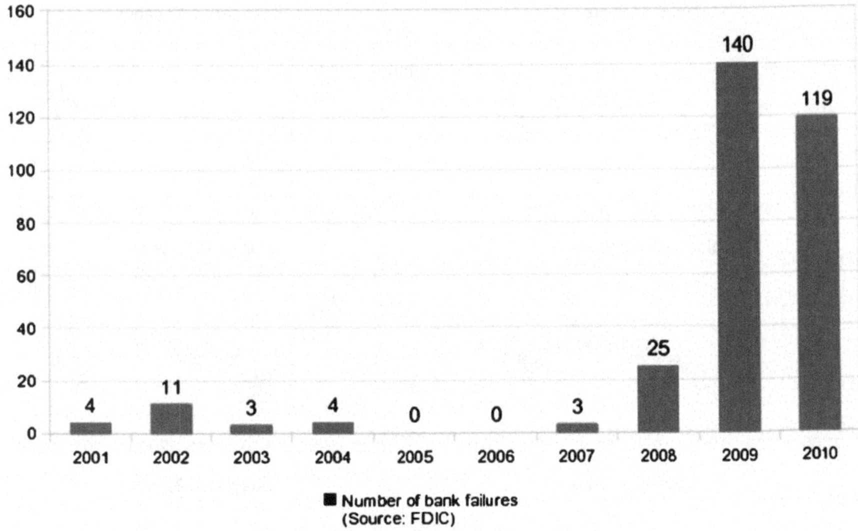

Fig. 15 Bank failures

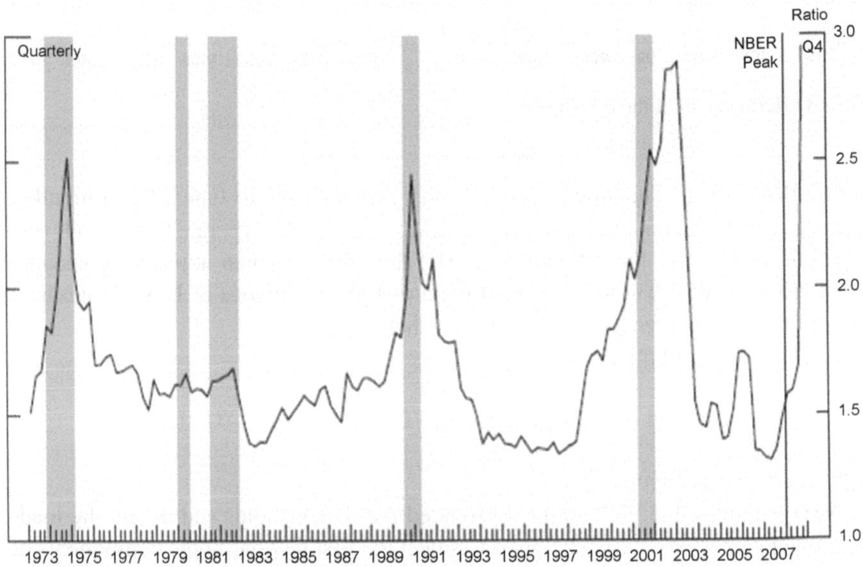

Fig. 16 Leveraging (source: Gilchrist et al. 2009)

linear regression formula is given by

$$x_t = \sum_{i=1}^{n} \left(a_i \sin\left(\frac{2\pi}{\tau_i}(t - t_0) \right) + b_i \cos\left(\frac{2\pi}{\tau_i}(t - t_0) \right) \right) \tag{25}$$

Table 2 Coefficients of the harmonic fit of the real bond yield in Eq. (25)

i	1	2	3	4	5	6
τ_i (month)	305.0	152.5	101.7	30.5	43.6	27.7
a_i	−0.0066	0.0062	0.0063	0.0007	0.00222	0.0025
b_i	0.0049	0.0031	0.0016	−0.0033	−0.0026	−0.0004

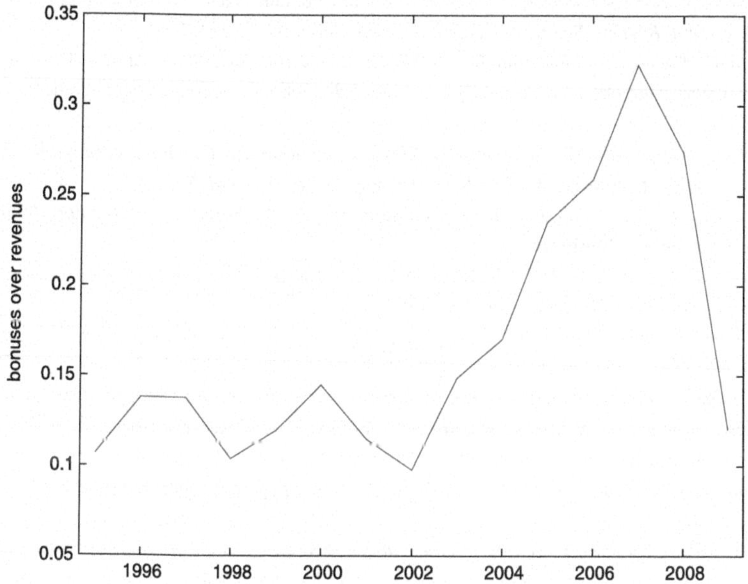

Fig. 17 Ratio of bonus payments to revenue (source: DiNapoli and Bleiwas 2010)

Figure 13 gives the sum of squared errors of the harmonic fit for the BAA corporate bond yields, Table 2 the estimated coefficients of the harmonic fit, and Figure 14 the periodic components.

References

Adrian, T., & Shin, H. S. (2009). Money, liquidity, and monetary policy. *American Economic Review, American Economic Association*, *99*(2), 600–605.

Adrian, T., Moench, A., & Shin, H. S. (2010). *Macro risk premium and intermediary balance sheet quantities* (Staff Report no 428). Federal Reserve Bank of New York.

Becker, S., Gruene, L., & Semmler, W. (2007). Comparing second order approximation with dynamic programming. *Computational Economics*, *30*(1), 65–91.

Bernanke, B., Gertler, M., & Gilchrist, S. (1999). The financial accelerator in a quantitative business cycle framework. In J. Taylor & M. Woodford (Eds.), *NBER working paper*: No *6455*. *Handbook of macroeconomics*. Amsterdam: North-Holland.

Brunnermeier, M. K., & Sannikov, Y. (2014, forthcoming). *A macroeconomic model with the financial sector*. *American Economic Review*.

Brunnermeier, M., & Pedersen, L. (2009). Market liquidity and funding liquidity. *The Review of Financial Studies*, *22*(6), 2210–2238.

Caballero, R. J., & Krishnamurthy, A. (2009). Global Imbalances and Financial Fragility. *American Economic Review, American Economic Association*, *99*(2), 584–588.

Carlstrom, C., Fuerst, T., & Paustian, M. (2009). Optimal monetary policy in a model with agency costs. In *Fed conference on monetary policy and the financial market*, Washington, DC, June 2009.

Chan, W.-S., Wong, A. C. S., & Tong, H. (2004). Some nonlinear threshold autoregressive time series models for actuarial use. *North American Actuarial Journal*, *8*, 37–61.

Chiarella, C., Flaschel, P., Franke, R., & Semmler, W. (2009). *Financial market and the macro economy*. London: Routledge.

Christensen, I., & Dib, A. (2008). The financial accelerator in an estimated new Keynesian model. *Review of Economic Dynamics*, *11*, 155–178.

Curdia, V., & Woodford, M. (2009a). *Credit frictions and optimal monetary policy*. BIS Working Paper, No. 278.

Curdia, V., & Woodford, M. (2009b). *Credit spreads and monetary policy*. Paper presented at the Conference on Financial Market and Monetary Policy, Federal Reserve Board, June 5–6.

Del Negro, M., Eggersson, G., Ferrero, A., & Kiyotaki, N. (2010). *The great escape: a quantitative evaluation of the Fed's non-standard policies* (Working paper). Federal Reserve Bank of New York.

DiNapoli, T. P., & Bleiwas, K. B. (2010). *Securities industries in New York city* (Report 10-2011).

Ehrmann, M., Ellison, M., & Valla, N. (2003). Regime-dependent impulse response functions in a Markov-switching vector autoregressive model. *Economics Letters*, *78*, 295–299.

Ernst, E., Mittnik, S., & Semmler, W. (2010). *Interaction of labor and credit market frictions: a theoretical and empirical analysis*. Paper prepared for the Winter Meeting of the Econometric Society, Atlanta, January 2010.

Geanakoplos, J. (2010). *The Leverage cycle* (Cowles Foundation Discussion Paper No. 1715). Yale University—Cowles Foundation.

Geanakoplos, J., & Farmer, J. (2009). The virtues and vices of equilibrium and the future of financial economics. *Complexity*, *14*(3), 11–38 [CFDP 1647, CFP 1274].

Geneva Report on the World Economy 11 (2009). *Fundamental principles of financial regulations*. ICMB.

Gertler, M., & Karadi, P. (2011). A model of unconventional monetary policy. *Journal of Monetary Economics*, *58*(1), 17–34.

Gilchrist, S., & Zagrajsek, S. (2012). Credit spreads and business cycle fluctuations. *American Economic Review 102*, 1692–1720.

Gilchrist, S., Ortiz, A., & Zagrajsek, S. (2009). *Credit risk and the macroeconomy: evidence from an estimated DSGE model*. Manuscript, Boston University.

Gorton, G. (2009). *Information, liquidity, and the (ongoing) panic of 2007*. Prepared for the Federal Reserve Bank of Kansas City, Jackson Hole Conference, August 2008.

Gorton, G. B. (2010). *Slapped by the invisible hand: the panic of 2007*. Oxford: Oxford University Press.
Greenwald, B., & Stiglitz, J. (1993). Financial market imperfections and business cycles. *Quarterly Journal of Economics*, *108*, 77–114.
Gruene, L., & Semmler, W. (2004). Using dynamic programming with adaptive grid scheme for optimal control problems in economics. *Journal of Economic Dynamics & Control*, *28*, 2427–2456.
Gruene, L., & Semmler, W. (2005). Default risk, asset pricing, and debt control. *Journal of Financial Econometrics*, *3*(1), 79–106.
Gruene, L., Semmler, W., & Sieveking, M. (2004). Creditworthiness and threshold in a credit market model with multiple equilibria. *Economic Theory*, *25*(2), 287–315.
Gruene, L., Semmler, W., & Bernard, L. (2007). Firm value, diversified capital assets and credit risk. *Journal of Credit Risk*, *3*(4), 81–113.
Hakkio, C. S., & Keeton, W. R. (2009). *Financial stress: what is it, how can it be measured, and why does it matter*? Federal Reserve Bank of Kansas City.
Hall, E. R. (2010). *Forward looking decision making*. Princeton: Princeton University Press.
Hamilton, J. D. (1989). A new approach to the economic analysis of nonstationary time series and the business cycle. *Econometrica*, *57*, 357–384.
Hatzius, J., Hooper, P., Mishkin, F., Schoenholtz, K., & Watson, M. (2010). *Financial condition indexes: a fresh look after the financial crisis*. www.Princeton.edu/~mwatson.
He, Z., & Krishnamurthy, A. (2008). *Intermediary asset pricing*. Manuscript, University of Chicago.
Hsiao, C.-Y., & Semmler, W. (2009). *Harmonic fitting for long run economic data*. Manuscript. www.newschool.edu/nssr/cem (there financial market).
Illing, M., & Liu, Y. (2006). Measuring financial stress in a developed country: an application to Canada. *Journal of Financial Stability*, *2*(3), 243–265.
Kindleberger, C., & Aliber, R. (2005). *Manias, panics, and crashes: a history of financial crises* (5th ed.). New York: Wiley.
Koop, G., Pesaran, M., & Potter, S. (1996). Impulse responses in nonlinear multivariate models. *Journal of Econometrics*, *74*, 119–147.
Lux, T. (2009). *Mass psychology in action: identification of social interaction effects in the German stock market* (Working paper, No: 1514). Kieler Institut for the World Economy.
Merton, R. C. (1974). On the pricing of corporate debt. The risk structure of the interest rate. *Journal of Finance*, *2*, 449–470.
Minsky, H. P. (1975). *John Maynard Keynes*. New York: Columbia University Press.
Minsky, H. P. (1982). *Can it happen again?* Armonk: ME Sharpe.
Mittnik, S., & Semmler, W. (2012). Regime dependence of the fiscal multiplier. *Journal of Economic Behavior & Organization*, *83*(3), 502–522.
Mittnik, S., & Semmler, W. (2013). The real consequences of financial stress. *Journal of Economic Dynamics & Control*, *37*(8), 1479–1499.
Mittnik, S., & Zadrozny, P. (1993). Asymptotic distributions of impulse responses, step responses, and variance decompositions of estimated linear dynamic models. *Econometrica*, *61*, 857–887.
Reinhart, C., & Rogoff, K. S. (2009). *This time is different: eight centuries of financial folly*. Princeton: Princeton University Press.
Semmler, W., & Bernard, L. (2012). Boom-bust cycles: leveraging, complex securities, and asset prices. *Journal of Economic Behavior & Organization*, *81*(2), 442–465.
Semmler, W., & Lucas, B. (2009). Boom-bust cycles: leveraging, complex securities, and asset prices. *Journal of Economic Behavior and Organization*, *82*(2), 442–465 (2012) (revised version). www.newschool.edu/nssr/cem.
Semmler, W., Gruene, L., & Oehrlein, L. C. (2009). Dynamic consumption and portfolio decisions with time varying asset returns. *Journal of Wealth Management*, *12*(3), 21–47.
Stein, J. (2012). *Stochastic optimal control and U.S. financial debt crisis*. Heidelberg: Springer.

Tong, H. (1978). On a threshold model. In C. H. Chen (Ed.), *Pattern recognition and signal processing*. Amsterdam: Kluwer.
Tong, H. (1983). *Threshold models in non-linear time series analysis*. New York: Springer.
Tsay, R. S. (1998). Testing and modelling multivariate threshold models. *Journal of the American Statistical Association, 93*, 1188–1202.

U.S. Business Cycles, Monetary Policy and the External Finance Premium

Enrique Martínez-García

Abstract I investigate a model of the U.S. economy with nominal rigidities and a financial accelerator mechanism à la Bernanke et al. (Handbook of Macroeconomics, Vol. 1, Elsevier Sci., Chap. 21, pp. 1341–1393, 1999). I calculate total factor productivity (TFP) and monetary policy deviations for the U.S. and quantitatively explore the ability of the model to account for the cyclical patterns of U.S. per capita real private GDP (excluding government), U.S. per capita real private investment, U.S. per capita real private consumption, the share of hours worked per capita, (year-over-year) inflation and the quarterly interest rate spread between the Baa corporate bond yield and the 20-year Treasury bill rate during the Great Moderation. I show that the magnitude and cyclicality of the external finance premium depend nonlinearly on the degree of price stickiness (or lack thereof) in the Bernanke et al. (Handbook of Macroeconomics, Vol. 1, Elsevier Sci., Chap. 21, pp. 1341–1393, 1999) model and on the specification of both the target Taylor (Carnegie-Rochester Conf. Ser. Pub. Policy 39:195–214, 1993) rate for policy and the exogenous monetary shock process. The strong countercyclicality of the external finance premium (the interest rate spread) induces substitution away from consumption and into investment in periods where output grows above its long-run trend as the premium tends to fall below its steady state and financing investment becomes temporarily "cheaper". The less frequently prices change in this environment, the more accentuated the fluctuations of the external finance premium are and the more dominant they become on the dynamics of investment, hours worked and output. However, these features—the countercyclicality and large volatility of the spread—are counterfactual and appear to be a key impediment limiting the ability of the model to account for the U.S. data over the Great Moderation period.

E. Martínez-García (✉)
Federal Reserve Bank of Dallas, 2200 N. Pearl Street, Dallas, TX 75201, USA
e-mail: enrique.martinez-garcia@dal.frb.org
url: http://www.dallasfed.org/research/economists/martinez-garcia.cfm

F. Schleer-van Gellecom (ed.), *Advances in Non-linear Economic Modeling*, Dynamic Modeling and Econometrics in Economics and Finance 17,
DOI 10.1007/978-3-642-42039-9_2,

1 Introduction

The 2007 recession has led to renewed concern about the role of the financial system among researchers and policymakers alike. The 'credit crunch' in the U.S. has focused attention back on the determinants of lending and the impact of financing conditions on the transmission mechanism for monetary policy. However, the standard variants of the New Keynesian framework that had become dominant for the analysis of monetary policy since the 1990s (see, e.g., Woodford 2003 and Galí 2008) typically abstract from financial frictions. Evidence from past banking crises and the 2007 downturn suggests—or, at least, has re-invigorated the view—that the role of the financial channel may be important in the propagation and amplification of shocks.

The role of monetary policy rules and their interaction with financial frictions has become also an issue of first-order importance in academic and policy circles. Indeed, the monetary authorities' reaction—both in the U.S. and other major industrialized countries—has been unusual during the current episode and very aggressive relative to their prior experience over the past 25 years of the so-called Great Moderation. In this context, the role of monetary policy is once again being hotly contested. A heated debate about the scope of monetary policy and the contribution to business cycles of deviations from well-established policy rules such as Taylor (1993)'s rule has ensued, and it is likely to continue for a long time.

To provide a quantitative analysis of the issues raised by these ongoing policy debates, I focus my attention on the nexus between monetary policy and financial frictions. In particular, I ask how one can evaluate the macroeconomic performance of monetary policy in an environment where policymakers understand that the nominal short-term interest rate they control—net of inflation—is not equal to the marginal lending rates that determine the cost of borrowing for economic agents—in other words, in economic environments where there is a non-trivial spread between the actual cost of borrowing and the real risk-free rate.

In a conventional New Keynesian model with no financial frictions, the transmission mechanism for monetary policy is rather stylized. Borrowing and lending has no impact on the monetary transmission mechanism and, consequently, no real effects. In a world with financial frictions, the implications of the Modigliani-Miller theorem no longer hold and the capital structure of firms and other economic agents becomes important, so the financial-side of the model can no longer be ignored.[1]

To investigate the economic consequences of financial frictions, I draw on the well-known financial accelerator model of Bernanke et al. (1999) where interest rate spreads are tied to the aggregate characteristics of the borrowers (mor precisely, to the borrowers' leverage ratio). This model offers a tractable framework for integrating financial frictions into an otherwise standard New Keynesian general equilibrium model with nominal rigidities. Moreover, the model has the appealing feature

[1]The Modigliani-Miller theorem, derived from the seminal work of Modigliani and Miller (1958), is also known as the capital structure irrelevance principle. The theorem indicates that, lacking some specific frictions or taxes, the value of the firm does not depend on whether the firm is financed by issuing equity (from their net worth) or debt (or simply taking on loans).

relative to other models of financial frictions that: (a) defaults and spreads (the external finance premium) occur endogenously in equilibrium, and (b) asset prices (the price of capital) feed into the spreads linking the two together endogenously.

I find that the economy has a stronger financial mechanism when the model incorporates standard New Keynesian features such as monopolistic competition and price stickiness. I emphasize that the financial accelerator by itself has only mild effects unless it interacts with frictions such as the type of nominal rigidities favored in the New Keynesian literature. I also illustrate that the financial accelerator model can have a significant amplification effect when it interacts with different specifications of the policy rule and with the addition of monetary policy shocks. However, these results are very sensitive to: (a) the degree of price stickiness assumed under Calvo price-setting, (b) the specification of the systematic part of the monetary policy rule, and (c) the interpretation one assigns to the exogenous and discretionary component of monetary policy.

Furthermore, I also show that a stronger financial accelerator mechanism does not necessarily mean that the model of Bernanke et al. (1999) is better suited to explain the path of endogenous variables like real per capita private output (excluding government), real per capita investment, real per capita consumption, the share of hours worked per capita, the year-over-year inflation rate or even the quarterly interest rate spread between the Baa corporate bond yield and the 20-year Treasury bill rate since the onset of the Great Moderation. In fact, a plain vanilla Real Business Cycle (RBC) model parameterized in a way consistent with that of the financial accelerator model—or a variant of it augmented with the financial friction, but no nominal rigidities—produce simulations of the endogenous variables that correlate more strongly with the actual data than the full-fledge financial accelerator model does.

I have several additions to the literature. First, I consistently and thoroughly examine the U.S. data and provide a coherent mapping between the data and the model. I also explicitly consider the possibility that there was a level shift in the data after the 2007 recession in establishing the mapping of the data into the model. The consistency between the way in which the model is laid down to account for the business cycle fluctuations and how the data itself is measured and detrended (or expressed in deviations from a long-run mean or target) is crucial in helping evaluate the strength and weaknesses of the model.

Second, I quantitatively investigate the ability of the financial accelerator model of Bernanke et al. (1999) to explain the cyclical fluctuations in the U.S. data. Although this is not the first paper to investigate the financial accelerator model's performance (see, e.g., the estimation in Meier and Müller 2006), it is the first paper to my knowledge that does it by the simulation method taking as given the realizations of the detrended Solow residual and the monetary policy deviations straight from the data—rather than estimating them based on imposing ex ante the structure of the model on the observable variables. While both approaches are complementary, I argue that the exercise I conduct in this paper is useful for the purpose of evaluating the model and accounting for the cyclical features of the data without having to worry (among other things) that misspecification may be biasing the estimates of the structural parameters. Moreover, it is also quite useful as a tool to inspect the financial accelerator mechanism and understand how it operates.

Third, I also aim to provide insight about the first-order effects of the interaction between financial frictions and nominal rigidities in the model of Bernanke et al. (1999). To do so, I adopt a simple first-order perturbation method to characterize the short-run dynamics of the financial accelerator model as Bernanke et al. (1999) did too. First-order approximations to the equilibrium conditions can be very useful to track fluctuations around the steady state arising from small perturbations of exogenous shocks, but might be quite inaccurate when the shocks are fairly large or the economy is far away from its long-run steady state. When I take account of the non-stationarity in the U.S. data and calculate the realization of the TFP and monetary shocks driving the business cycle, it is reassuring that I do not see a strong case to back the idea that fluctuations have been unusually pronounced during most of the period since the mid-1980s—although in the case of the monetary shocks the question may be far less settled.

While the short-run dynamics of the model are indeed linear in the variables under the first-order approximation that I have adopted, the coefficients are highly nonlinear functions of the structural parameters of the model. I contend that these nonlinearities in the coefficients are important to understand the interaction between nominal rigidities and financial frictions. This nonlinear interaction, in turn, can have large effects on the path the endogenous variables take in response to a given realization of the shocks—I find the degree of price stickiness, in particular, to be crucial for the amplification of fluctuations in the external finance premium and on investment.

My paper proceeds as follows: Section 2 outlines the Bernanke et al. (1999) financial accelerator and several nested variants that abstract from all frictions (the RBC model), that abstract from nominal rigidities (the FA model), and that eliminate the financial friction (the DNK model). I continue in Section 3 with a discussion of the parameterization of the model and the derivation of the shock realizations, and then I present the quantitative findings in Section 4. Section 5 provides some discussion and concludes.

2 The Financial Accelerator Model

One framework incorporating a financial accelerator in general equilibrium that has been extensively used in the literature is Bernanke et al. (1999)'s model of financial intermediation with 'costly state verification'. Costly monitoring of the realized return on capital of the defaulting borrowers and an endogenous probability of default result in increased borrowing costs on loans over the risk-free rate and introduce time-variation on the loan rates over the business cycle. The *external finance premium*—the spread of the loan rate over the risk-free rate—makes investment and capital accumulation more expensive. This, in turn, intensifies the impact and can even alter the propagation of a given shock. The model of Bernanke et al. (1999), however, includes other distortions—in particular, it includes standard New Keynesian frictions such as monopolistic competition and nominal price rigidities.

I adopt the model of Bernanke et al. (1999) for its tractability and intuitive economic appeal. Also, because financial intermediation plays a key role in funding investment—a connection that I want to explore further in light of the investment collapse observed in the U.S. data during the 2007 recession.[2] The model shares an important characteristic with the framework of collateral borrowing constraints articulated by Kiyotaki and Moore (1997) in that asset price movements serve to reinforce credit market imperfections. Fluctuations on the value of capital contribute directly to volatility in the leverage of the borrowers. This feature is missing in the Carlstrom and Fuerst (1997) framework which also builds on the idea of 'costly state verification', as noted by Gomes et al. (2003). Another difference between the Carlstrom and Fuerst (1997) and Bernanke et al. (1999) environments is that financial intermediation is intratemporal in the former and intertemporal in the latter.[3]

The model of Bernanke et al. (1999) is populated by households and entrepreneurs, a variety of firm types (capital producers, wholesale producers, and retailers) as well as financial intermediaries (banks) and a central bank entrusted with the conduct of monetary policy. Households own all capital producing firms, retailers and banks. Capital producers determine a relative price for investment goods, and are subject to technological constraints in how they can transform final output into productive capital that can be used to produce wholesale output.

Retailers are separated from wholesale producers in order to introduce differentiation in the wholesale goods, and add nominal rigidities into the model. Wholesale producers are formed and operated by entrepreneurs. The capital returns they generate tomorrow with today's allocation of capital are paid net of borrowing costs as dividends back to the entrepreneurs if there is no default. Capital returns on wholesalers are subject to idiosyncratic shocks that affect the revenue stream for the entrepreneurs who own them. Therefore, entrepreneurs are exposed to bankruptcy risk on the wholesale firms which occurs whenever capital returns fall short of the required loan repayment. In that case, the entrepreneurs lose the capital returns and the undepreciated stock of capital on the defaulting wholesalers.

The financial system intermediates between the households and the entrepreneurs. Banks are risk-neutral firms facilitating loans to the risk-neutral entrepreneurs who borrow to fund the stock of capital they need for wholesale production. Entrepreneurs are more impatient than households, dying out at an exogenous rate, and that motivates them to borrow. Entrepreneurs' deaths also prevents them from accumulating enough net worth (internal funds) to be able to self-finance their capital holdings every period.

[2]The literature has investigated other roles of financial intermediation: for instance, funding the wage bill instead of the capital bill (see, e.g., Carlstrom and Fuerst 2001). The financial accelerator model of Bernanke et al. (1999) has the potential to amplify the effects of a shock, but by constraining capital accumulation, it can affect the propagation of shocks as well.

[3]Faia and Monacelli (2007) and Walentin (2005) provide a comparative analysis of the Bernanke et al. (1999) and Carlstrom and Fuerst (1997) models.

Capital returns are determined by the marginal product of capital and the capital gains on the value of the assets (the capital), but also by the realization of an idiosyncratic shock which is observable to the entrepreneurs but not to the financial intermediaries. Banks can only determine the realization of the idiosyncratic shock and, therefore, the true returns to capital after paying a non-zero monitoring or verification cost. Loan contracts cannot be made conditional on the realization of the idiosyncratic shock because they are unobserved by the banks. However, the design of the loans is meant to reduce the costs associated with this asymmetry of information between the entrepreneurs who own the wholesale firms and the banks.

Financial intermediaries offer one-period deposits available to households promising the real risk-free rate and use the funds they are able to raise to make one-period loans available to the entrepreneurs. The implied loan rate charges a spread over the real risk-free rate—the external finance premium—for banks to cover the costs of monitoring the defaulting entrepreneurs and any shortfall on loan repayment that may occur. All entrepreneurs face the same borrowing costs. Ex post there is always a fraction of wholesale producers with low draws from the idiosyncratic shock that do not generate enough revenue from their capital for the entrepreneurs to meet the loan repayment, causing them to default.

Ex ante the banks know the distribution of the idiosyncratic shock and can determine the probability of default and its associated costs under the terms of the loan—even if banks do not know which entrepreneurs will end up defaulting next period, they know how many defaults to expect. Banks are perfectly competitive so they structure their loans to cover solely the costs of default (as they face no other costs), and make no profits on the loans. The expected default rates priced into the loan rates are always confirmed ex post in equilibrium. Banks supply whatever loan amount is desired by the entrepreneurs under the terms of the offered loan, and accept any amount that households wish to deposit at the prevailing real risk-free rate. As a result, ex post banks always break even and distribute zero-profits in every period to the households who own them.

Finally, a central bank is added which sets monetary policy in terms of a nominal short-term interest rate. Monetary policy is non-neutral in the short run, irrespective of the capital structure of the entrepreneurs or the functioning of the loan market. Monetary policy non-neutrality arises as in the standard New Keynesian framework simply because of nominal rigidities on prices. I modify the model of Bernanke et al. (1999) to include a more standard monetary policy rule à la Taylor (1993) to characterize the perceived monetary policy regime over the Great Moderation period. The model is, otherwise, the same one derived in Bernanke et al. (1999) in log-linear form with only minor simplifications in the timing of pricing decisions and the role of entrepreneurs' consumption and government consumption shocks.

The contribution of this paper is not predicated on any theoretical improvement upon what is already a well-established framework for understanding financial distortions, but it is primarily a quantitative one. For a conventional parameterization of the model, I provide a careful quantitative evaluation of the ability (of lack thereof) of this financial accelerator channel to answer questions on the role of monetary policy over the U.S. business cycle, on the cyclical factors behind the Great Moderation, and on the financial aspects of the 2007 recession.

Log-linearized Equilibrium Conditions of the Financial Accelerator Model Since the model of Bernanke et al. (1999) is quite well-known, I refrain from a detailed discussion of its first principles. This section describes the log-linearized equilibrium conditions of the model that I use and a frictionless variant—the RBC counterpart—to make the presentation as compact as possible. As a notational convention, all variables identified with lower-case letters and a caret on top are expressed in logs and as deviations relative to their steady state values. Since the model abstracts from population growth and accounts only for stationary cyclical fluctuations, the endogenous variables are matched whenever appropriate to do so with observed time series expressed in per capita terms and detrended (or demeaned). Further discussion on the mapping between the data and the model can be found in Appendices A and B.

On the demand-side, households are infinitely-lived and maximize their lifetime discounted utility, which is additively separable in consumption and leisure in each period. Aggregate consumption evolves according to a standard Euler equation,

$$\widehat{c}_t \approx \mathbb{E}_t[\widehat{c}_{t+1}] - \sigma \widehat{r}_{t+1}, \tag{1}$$

where $\widehat{c}_t$ denotes real aggregate consumption, and $\widehat{r}_{t+1}$ is the Fisherian real interest rate. This consumption Euler equation is fairly standard and implies that the financial frictions do not directly affect the consumption-savings decision of the households. Financial intermediaries pay the same real risk-free rate on deposits. The intertemporal elasticity of substitution, $\sigma > 0$, regulates the sensitivity of the consumption-savings decision to the Fisherian real interest rate.

The Fisherian real interest rate is defined as the one-period nominal (risk-free) interest rate minus the expected inflation over the next quarter, i.e.,

$$\widehat{r}_{t+1} \equiv \widehat{i}_{t+1} - \mathbb{E}_t[\widehat{\pi}_{t+1}], \tag{2}$$

where $\widehat{\pi}_t \equiv \widehat{p}_t - \widehat{p}_{t-1}$ is the inflation rate, and $\widehat{p}_t$ is the consumer price index (CPI). Nominal (uncontingent) one-period bonds are traded in zero net supply and guarantee a nominal risk-free rate of $\widehat{i}_{t+1}$ paid at time $t+1$ but set at time t. Here, $\mathbb{E}_t[\cdot]$ denotes the expectations operator conditional on information available up to time t.

The first-order condition on labor supply from the households' problem can be expressed as follows,

$$\widehat{w}_t - \widehat{p}_t \approx \frac{1}{\sigma}\widehat{c}_t + \frac{1}{\varphi}\widehat{h}_t, \tag{3}$$

where $\widehat{h}_t$ represents aggregate household labor, and $\widehat{w}_t$ is the competitive nominal wage. The Frisch elasticity of labor supply, $\varphi \equiv \eta(\frac{1-H}{H}) > 0$, indicates the sensitivity of the supply of labor to changes in real wages, ceteris paribus. The parameter η corresponds to the inverse of the coefficient of relative risk aversion on leisure, and H defines the share of hours worked in steady state.[4]

[4]Total hours worked H_t and hours spent in leisurely activities L_t are normalized to add up to one (i.e., $H_t + L_t = 1$). If consumption and leisure are additively separable as assumed by Bernanke

On the supply-side, there are retailers, capital producers, wholesale producers (owned and operated by the entrepreneurs), and financial intermediaries. I implicitly assume that the only input required in the production of retail varieties is the wholesale good. Retailers acquire wholesale output, costlessly differentiate the wholesale goods into retailer-specific varieties, and sell their varieties for either consumption or investment. Preferences are defined over all the retail varieties, but not directly over the wholesale goods which are only utilized as inputs in the production of retail varieties.

Each retailer has monopolistic power in its own variety and chooses its price to maximize the expected discounted value of its current and future profits, subject to a downward-sloping demand constraint. Price stickiness is modeled à la Calvo (1983), so in each period only a fraction $0 < 1 - \alpha < 1$ of the retailers gets to re-optimize prices.[5] The CPI inflation dynamics resulting from aggregating over all retail prices are given by the following forward-looking Phillips curve,

$$\widehat{\pi}_t \approx \beta \mathbb{E}_t[\widehat{\pi}_{t+1}] + \left(\frac{(1-\alpha\beta)(1-\alpha)}{\alpha} \right) \widehat{mc}_t, \tag{4}$$

where I define the real marginal cost as $\widehat{mc}_t \equiv (\widehat{p}_t^w - \widehat{p}_t)$ and denote the wholesale output price as $\widehat{p}_t^w$. The intertemporal discount factor of the households is $0 < \beta < 1$. Under flexible prices, the retailers intermediate the exchanges in the market for wholesale goods charging a mark-up over marginal costs but have no discernible impact on the short-run dynamics (i.e., $\widehat{mc}_t = 0$) since the monopolistic competition mark-up is time-invariant. The mark-up, however, still distorts the steady state allocation relative to the case under perfect competition.

In keeping with the precedent of Bernanke and Woodford (1997), Bernanke et al. (1999) assume that prices are set prior to the realization of any aggregate time t shock. The timing in Bernanke et al. (1999) distorts the equilibrium beyond what the monopolistic competition mark-up and Calvo (1983) price stickiness already do. In turn, I adopt the convention that prices are set after observing the realized shocks at time t as in Woodford (2003). The model solution then approximates the case where prices equal a mark-up over marginal costs in the limit when only an arbitrarily small fraction of firms $\alpha \to 0$ cannot re-optimize. This facilitates the comparison between the financial accelerator model and the frictionless model that I investigate in the paper.

Capital accumulation evolves according to a standard law of motion,

$$\widehat{k}_{t+1} \approx (1-\delta)\widehat{k}_t + \delta \widehat{x}_t, \tag{5}$$

---

et al. (1999), and I define the per-period preferences over leisure generically as $V(L_t)$, then it follows that in steady state $\eta^{-1} \equiv -\frac{LV''(L)}{V'(L)}$.

[5]The retailers add a 'brand' name to the wholesale good which introduces differentiation across varieties and, consequently, retailers gain monopolistic power to charge a mark-up in their prices. The retailers are not price-takers under this market structure.

where $\widehat{k}_t$ denotes the stock of capital available at time t and $\widehat{x}_t$ stands for real investment in the same period. The depreciation rate of physical capital is given by $0 < \delta < 1$. The capital goods producers use the same aggregate of retail varieties that households consume in the production of new capital. To be consistent with the convention of Bernanke et al. (1999), I also assume that entrepreneurs buy all capital they need from the capital goods producers—the period before production takes place—and then sell the depreciated capital stock back to them after being used for the production of wholesale goods.

Capital goods producers face increasing marginal adjustment costs in the production of new capital, modelled in the form of an increasing and concave adjustment cost which is a function of the investment-to-capital ratio.[6] The technological constraint on capital goods producers implies that the investment-to-capital ratio $(\widehat{x}_t - \widehat{k}_t)$ is tied to the shadow value of an additional unit of capital (or Tobin's q) in units of consumption, $\widehat{q}_t$, by the following relationship,

$$\widehat{q}_t \approx \chi(\widehat{x}_t - \widehat{k}_t). \tag{6}$$

The degree of concavity of the cost function around its steady state, $\chi \geq 0$, regulates the sensitivity of the investment-to-capital ratio to fluctuations in Tobin's q. Without adjustment costs (i.e., if $\chi = 0$), Tobin's q becomes time-invariant, i.e.,

$$\widehat{q}_t \approx 0, \tag{7}$$

and the investment-to-capital ratio is unconstrained. However, without adjustment costs the financial accelerator mechanism in Bernanke et al. (1999) would lose the characteristic that asset price movements serve to reinforce loan market imperfections.

The wholesale firms employ homogenous labor supplied by both households and entrepreneurs as well as capital in order to produce wholesale output. Entrepreneurs' labor is differentiated from that of the households. All factor markets are perfectly competitive, and each wholesale producer relies on the same Cobb-Douglas technology in capital and in labor from households and entrepreneurs. Aggregate wholesale output can be expressed as follows,

$$\widehat{y}_t \approx \widehat{a}_t + \psi\widehat{k}_t + (1 - \psi - \varrho)\widehat{h}_t, \tag{8}$$

where $\widehat{y}_t$ denotes wholesale output, and $\widehat{a}_t$ is an aggregate productivity (TFP) shock. The capital share in the production function is $0 < \psi < 1$, while the entrepreneurs' labor share is $0 \leq \varrho < 1$ and the households' labor share is $0 < 1 - \psi - \varrho < 1$.[7] Entrepreneurs' labor is assumed to be inelastically supplied and time-invariant, so

[6] As in Bernanke et al. (1999), profits of the capital goods producers are of second-order importance and, therefore, omitted. For more details, see footnote 13 in p. 1357.

[7] As in Bernanke et al. (1999), the entrepreneurs' labor share is chosen to be small enough that this modification of the standard production function does not have a significant direct effect on the aggregate dynamics of the model.

it drops out of the log-linearized production function in (8). The TFP shock follows an $AR(1)$ process of the following form,

$$\widehat{a}_t = \rho_a \widehat{a}_{t-1} + \varepsilon_t^a,\ \varepsilon_t^a \sim N\left(0, \sigma_a^2\right), \tag{9}$$

where ε_t^a is a zero mean, uncorrelated and normally-distributed innovation. The parameter $-1 < \rho_a < 1$ determines the persistence of the TFP shock and σ_a its volatility.

The competitive real wage paid to households is equal to their marginal product, i.e.,

$$\widehat{w}_t - \widehat{p}_t \approx \widehat{mc}_t + (\widehat{y}_t - \widehat{h}_t). \tag{10}$$

Entrepreneurs' real wages—which differ from those of the households—are not needed to characterize the short-run dynamics of the model, though. Combining the labor supply equation for households in (3) with the households' labor demand in (10), I derive a households' labor market equilibrium condition in the following terms,

$$\widehat{mc}_t + (\widehat{y}_t - \widehat{h}_t) - \frac{1}{\sigma}\widehat{c}_t \approx \frac{1}{\varphi}\widehat{h}_t. \tag{11}$$

This condition suffices to describe the real marginal costs faced by the retailers, without having to keep track of any real wages explicitly.

Entrepreneurs operating the wholesale firms buy the capital stock every period from the capital goods producers at a price determined by Tobin's q, using both internal funds (that is, their own net worth) and external loans from the financial intermediaries. After production takes place the next period, the depreciated stock of capital is sold back to the capital goods producers. Accordingly,

$$\widehat{r}_t^k \approx (1-\epsilon)\left(\widehat{mc}_t + (\widehat{y}_t - \widehat{k}_t)\right) + \epsilon \widehat{q}_t - \widehat{q}_{t-1}, \tag{12}$$

where the aggregate real return on capital, $\widehat{r}_t^k$, is equal to a weighted combination of the marginal product of capital, $\widehat{mc}_t + (\widehat{y}_t - \widehat{k}_t)$, and the re-sale value of the depreciated capital stock (as captured by Tobin's q), $\widehat{q}_t$, minus the cost of acquiring the stock of capital from the capital goods producers in the previous period, $\widehat{q}_{t-1}$.

The composite coefficient in the definition of the returns to capital in (12) is characterized as $\epsilon \equiv (\frac{1-\delta}{\upsilon(\gamma_n^{-1})\beta^{-1}})$. This composite depends on the gross steady state ratio between the cost of external funding for entrepreneurs and the real risk-free rate $\upsilon(\gamma_n^{-1}) \equiv \frac{R^k}{R} \geq 1$. Moreover, $\upsilon(\gamma_n^{-1})$ is a function of the steady state gearing or leverage ratio of the entrepreneurs, $\gamma_n^{-1} \equiv \frac{K}{N}$, that is the ratio of total assets—the stock of capital K—over the total real net worth—equity N of the entrepreneurs. Tobin's q is equal to 1 in steady state and, therefore, K corresponds to both the stock of capital as well as its value in units of consumption.

Following the logic of the 'costly state verification' framework embedded in Bernanke et al. (1999), the returns to capital of each wholesale producer are subject to idiosyncratic (independent and identically-distributed) shocks that are observable to the entrepreneurs but costly to monitor for the financial intermediaries. The idiosyncratic shocks are realized only after capital is acquired for wholesale production and external loans for funding have been secured. Therefore, such idiosyncratic shocks have a direct impact on the capital returns that entrepreneurs obtain from allocating capital to wholesale production, but do not affect the allocation of capital itself to each wholesale producer.

Financial intermediaries raise funds from households by offering deposits that pay the real risk-free rate, $\widehat{r}_{t+1}$, and make loans in real terms to entrepreneurs to finance their capital stock. On one hand, the return on deposits for households is guaranteed and inflation-protected. On the other hand, entrepreneurs can default on their loan contract obligations, and financial intermediaries can find out about their true capital returns (that is, learn about the realization of the idiosyncratic shock) only after paying a monitoring or verification cost. The bank lenders solely monitor the entrepreneurs who default, pay the verification costs when default occurs, and seize all income revenues obtained from the allocation of capital and the remaining assets (capital) of the defaulting entrepreneurs.[8]

In equilibrium, the financial intermediaries—which are assumed to be risk-neutral—price into their loan contracts the probability and costs of default, so an endogenous spread arises between the cost at which banks fund themselves through deposits from households (the real risk-free rate) and the real cost of external financing through loans faced by the entrepreneurs. The entrepreneurs—who are also assumed to be risk-neutral—borrow up to the point where the expected real return to capital equals the real cost of external funding through loans, i.e.,

$$\mathbb{E}_t\left[\widehat{r}^k_{t+1}\right] \approx \widehat{r}_{t+1} + \vartheta\left(\widehat{q}_t + \widehat{k}_{t+1} - \widehat{n}_{t+1}\right). \tag{13}$$

As shown in Bernanke et al. (1999), the external financing premium or spread over the real risk-free rate demanded by the financial intermediaries, $\widehat{sp}_t \equiv \mathbb{E}_t[\widehat{r}^k_{t+1}] - \widehat{r}_{t+1}$, is a function of the leverage ratio of the entrepreneurs in any given period, $\widehat{q}_t + \widehat{k}_{t+1} - \widehat{n}_{t+1}$, where $\widehat{n}_{t+1}$ denotes the net worth (or equity) of the entrepreneurs at the end of time t and $\widehat{q}_t + \widehat{k}_{t+1}$ denotes the total value of their assets (the value of their outstanding stock of capital) also at the end of time t.

The composite coefficient in (13) is characterized as $\vartheta \equiv \left(\frac{\upsilon'(\gamma_n^{-1})\gamma_n^{-1}}{\upsilon(\gamma_n^{-1})}\right)$ where the parameter $\upsilon'(\gamma_n^{-1}) \equiv \frac{\partial \upsilon(\gamma_n^{-1})}{\partial \gamma_n^{-1}} \geq 0$ is the first derivative of the external financing premium with respect to the entrepreneurs' leverage ratio γ_n^{-1} in steady state. Then, the composite coefficient ϑ can be interpreted as the elasticity of the external financing premium with respect to the entrepreneurs' leverage ratio evaluated in steady state. The lower the entrepreneurs' leverage in steady state (i.e., the closer $\gamma_n^{-1} \equiv \frac{K}{N}$ is to one), the lower the associated costs of default (and the smaller the elasticity ϑ) will be.

---

[8]Loan contracts are enforced under limited liability, so the bank cannot appropriate more than the value of the collateral assets (capital) and earned capital income of the defaulting entrepreneurs. Default takes place before the entrepreneurs earn any labor income.

The balance sheet of the entrepreneurs requires the real value of the stock of capital to be equal to real net worth (equity) plus the real amount in borrowed external funds (loans), i.e.,

$$\widehat{q}_t + \widehat{k}_{t+1} \approx \gamma_n \widehat{n}_{t+1} + (1-\gamma_n)\widehat{l}_{t+1}, \tag{14}$$

where $\widehat{l}_{t+1}$ denotes the total loans in real terms provided by the financial intermediaries to fund the stock of capital, $\widehat{k}_{t+1}$, valued at $\widehat{q}_t$ per unit of capital at time t. As a result, the leverage or gearing ratio of the entrepreneurs is simply proportional to the entrepreneurs' debt-to-equity ratio, i.e.,

$$\widehat{q}_t + \widehat{k}_{t+1} - \widehat{n}_{t+1} \approx (1-\gamma_n)(\widehat{l}_{t+1} - \widehat{n}_{t+1}). \tag{15}$$

Hence, the more indebted the entrepreneurs become—or the least equity they have at stake—in any given period, the more leveraged they are and the costlier it gets for entrepreneurs to fund their desired stock of capital with bank loans given the capital demand in (13).

Banks are perfectly competitive and real deposits held by households must be equal to the total loanable funds in real terms supplied to the entrepreneurs in every period t, i.e.,

$$\widehat{l}_t \approx \widehat{d}_t, \tag{16}$$

where $\widehat{d}_t$ represents the real value of the households' deposits. Given the simplicity of the balance sheet of the banks, it can be said that the model of Bernanke et al. (1999) is silent about the bank lending channel and in turn places all the emphasis on the borrowers-side. Hence, the external finance premium is unaffected by the characteristics of the lenders, and only depends on the characteristics of the borrowers (more specifically, on the leverage of the entrepreneurs). I leave for future research the extension of the model to incorporate an economically-relevant bank lending channel.

The aggregate real net worth of the entrepreneurs accumulates according to the following law of motion,

$$\begin{aligned}\widehat{n}_{t+1} \approx{}& (\zeta\beta^{-1}\gamma_n^{-1})\left(\widehat{r}_t^k - \widehat{r}_t\right) + \widehat{r}_t + \widehat{n}_t + \cdots \\ & \left(\upsilon(\gamma_n^{-1}) - 1\right)\gamma_n^{-1}\left(\widehat{r}_t^k + \widehat{q}_{t-1} + \widehat{k}_t\right) + \cdots \\ & \frac{\varrho}{\psi}\left(\upsilon(\gamma_n^{-1})\beta^{-1} - (1-\delta)\right)\gamma_n^{-1}\widehat{y}_t + \widehat{mc}_t,\end{aligned} \tag{17}$$

where $0 < \zeta < 1$ is interpreted as a *survival rate* for entrepreneurs in the same spirit as Bernanke et al. (1999). Households' consumption and savings are governed by the standard consumption Euler equation described in (1), but the entrepreneurs' consumption $\widehat{c}_t^e$ is simply proportional to their net worth $\widehat{n}_{t+1}$, i.e.,

$$\widehat{c}_t^e \approx \widehat{n}_{t+1}, \tag{18}$$

plus a term of second-order importance that drops out from the log-linearized model.

Equation (17) indicates that the real net worth (or equity) of the entrepreneurs, $\widehat{n}_{t+1}$, accumulates over the previous period real net worth, $\widehat{n}_t$, at the real risk-free rate, $\widehat{r}_t$, plus a retained share of the capital returns net of borrowing costs which is proportional to the real capital return relative to the real risk-free rate, $\widehat{r}_t^k - \widehat{r}_t$. The fraction of net real capital returns retained is a function of the steady state gearing or leverage ratio γ_n^{-1}, the steady state real interest rate β^{-1}, and the survival rate of the entrepreneurs ζ. The law of motion for net worth in (17) also includes a variety of additional terms of lesser importance under standard parameterizations—partly related to entrepreneurial labor income.

Entrepreneurs are risk-neutral and discount the future at the same rate β as households. The assumption that a fraction of entrepreneurs $(1 - \zeta)$ dies out in every period and gets replaced by the same proportion of new entrepreneurs without any net worth of their own—but with some labor income—introduces entry and exit in the model. In that case, the effective discount rate for entrepreneurs $\beta\zeta < \beta$ is lower than that of households. Entrepreneurs, who are more impatient as a result, borrow to fund the acquisition of capital; households save the loanable funds through riskless deposits with the risk-neutral financial intermediaries.

Entrepreneurs have an incentive to borrow, but also to delay consumption and accumulate net worth (equity) in order to retain more of the high returns on capital that can be obtained using internal funds. This is because the opportunity cost of internal funds is given by the risk-free rate $\widehat{r}_t$ which is lower than the implied loan rates from the financial intermediaries. The assumption that a fraction of entrepreneurs $(1 - \zeta)$ dies out in every period, therefore, is also meant to preclude entrepreneurs from becoming fully self-financing over the long-run since that would eliminate the need for external finance through banks and kill the financial accelerator channel.

The resource constraint can be approximated as follows,

$$\widehat{y}_t \approx \gamma_c \widehat{c}_t + \gamma_x \widehat{x}_t + \gamma_{c^e} \widehat{c}_t^e, \tag{19}$$

where $0 < \gamma_c < 1$ denotes the households' consumption share in steady state, $0 < \gamma_x < 1$ is the investment share, and $0 < \gamma_{c^e} < 1$ is the entrepreneurs' consumption share. By construction, it must be the case that $\gamma_c \equiv 1 - \gamma_x - \gamma_{c^e}$. The investment share is a composite coefficient of the structural parameters of the model given by $\gamma_x \equiv \delta \frac{K}{Y} = \delta \left(\frac{\psi}{\mu(\upsilon(\gamma_n^{-1})\beta^{-1} - (1-\delta))} \right)$ where $\mu \equiv \frac{\theta}{\theta - 1} > 1$ is the monopolistic competition mark-up and $\theta > 1$ is the elasticity of substitution across retail varieties. Monopolistic competition distorts the dynamics of the model through the resource constraint in (19) because the mark-up lowers the long-run investment share and increase the share of consumption. Similarly, the investment share is also distorted by the gross steady state ratio between the cost of external funding for entrepreneurs and the real risk-free rate $\upsilon(\gamma_n^{-1}) \equiv \frac{R^k}{R}$. The higher the ratio between these two rates, the lower the investment share will be.

The entrepreneurs have been largely modeled as in Bernanke et al. (1999), but I depart from them in one respect: instead of assuming that dying entrepreneurs consume all their entire net worth and disappear, I assume that they consume only an arbitrarily small fraction as they exit the economy while the rest is transferred to the households. This does not change the entrepreneurs' consumption relationship with net worth described in (18), but it affects the entrepreneurs' consumption share in steady state γ_{c^e} and the resource constraint in (19). The steady state share γ_{c^e} under this alternative assumption is chosen to be very small such that the entrepreneurs' consumption does not have a significant direct effect on the model dynamics.

As discussed in Christiano et al. (2003) and Meier and Müller (2006), this assumption suffices to ensure the objective function of the entrepreneurs is well-defined. It also has the desirable feature that entrepreneurs' consumption—which is an artifact of the heterogeneity across agents needed to introduce borrowing and lending—is almost negligible and, therefore, that total consumption is essentially pined down by the households' consumption and governed by the standard Euler equation from the households' maximization problem.

The resource constraint in (19) abstracts from the consideration of the resources devoted to monitoring costs, as those ought to be negligible on the dynamics of the model under standard parameterizations according to Bernanke et al. (1999). It also equates final aggregate output of all varieties for consumption and investment purposes with the wholesale output that is used as the sole input in the production of each retail variety.

In Bernanke et al. (1999) government consumption is modeled as an exogenous shock which detracts resources from the resource constraint. I simplify the financial accelerator model by excluding government consumption entirely. I contend that eliminating government consumption shocks does not fundamentally alter the financial accelerator mechanism developed in Bernanke et al. (1999) or the dynamics of the model in response to monetary and TFP shocks since fiscal policy is not fleshed out beyond the exogenous impact of this government shock on aggregate demand. In turn, I focus my investigation primarily on the traditional main driver of the business cycle (aggregate TFP) and on the connection between lending and monetary policy.[9] I leave the investigation of the role of fiscal policy and its interplay with loan market imperfections for future research.

Another important departure from the original model set-up comes from replacing the monetary policy rule of Bernanke et al. (1999) with a more standard specification. In line with most of the current literature, I assume that the central bank follows a conventional Taylor (1993)-type reaction function under a dual mandate that adjusts the short-term nominal rate, $\widehat{i}_t$, to respond to fluctuations in inflation, $\widetilde{\pi}_t$, and some real economic activity measure of the business cycle, $\widetilde{y}_t$. Thus, monetary policy is determined by the following general expression,

$$\widehat{i}_{t+1}^{AR} = \rho_i \widehat{i}_t^{AR} + (1 - \rho_i)[\phi_\pi \widetilde{\pi}_t + \phi_y \widetilde{y}_t] + \widehat{m}_t, \quad (20)$$

where the policy parameters $\phi_\pi \geq 1$ and $\phi_y \geq 0$ regulate the sensitivity of the policy rule to inflation and output fluctuations, and $0 \leq \rho_i < 1$ is the interest rate smoothing parameter. I use the annualized short-term interest rate as the relevant policy

[9] To make the data consistent with the model, however, output is measured as private market output (excluding government compensation of employees).

instrument, $\widehat{i}_t^{AR}$, i.e.,

$$\widehat{i}_{t+1}^{AR} \approx 4\widehat{i}_{t+1}. \tag{21}$$

The monetary policy shock, $\widehat{m}_t$, follows an $AR(1)$ process that can be represented as,

$$\widehat{m}_t = \rho_m \widehat{m}_{t-1} + \varepsilon_t^m, \quad \varepsilon_t^m \sim N\left(0, \sigma_m^2\right), \tag{22}$$

where ε_t^m is a zero mean, uncorrelated and normally-distributed innovation. The parameter $-1 < \rho_m < 1$ determines the persistence of the policy shock and the parameter $\sigma_m \geq 0$ its volatility. I assume that monetary and TFP shocks are uncorrelated.

In keeping with Taylor (1993)'s original prescription, I consider a specification where the inflation rate is measured over the previous four quarters, $(\widehat{p}_t - \widehat{p}_{t-4})$, and real economic activity over the business cycle is tracked with output in deviations from its steady state, $\widehat{y}_t$, i.e.,

$$\widetilde{y}_t \approx \widehat{y}_t, \tag{23}$$

$$\widetilde{\pi}_t \approx (\widehat{p}_t - \widehat{p}_{t-4}) \approx \widehat{\pi}_t + \widehat{\pi}_{t-1} + \widehat{\pi}_{t-2} + \widehat{\pi}_{t-3}, \tag{24}$$

I also experiment with an alternative specification of the policy rule in which $(\widehat{p}_t - \widehat{p}_{t-4})$ is replaced with the annualized quarter-over-quarter rate, $\widehat{\pi}_t^{AR}$, i.e.,

$$\widetilde{\pi}_t \approx \widehat{\pi}_t^{AR} \approx 4\widehat{\pi}_t. \tag{25}$$

The inflation rate in (25) is consistent with how the Taylor rule is specified in most quantitative and theoretical models, but is not the preferred measure of inflation in Taylor (1993).[10] Another alternative conception of the monetary policy reaction function that I do consider here respond to deviations of output from its potential, $\widehat{x}_t$, i.e.,

$$\widetilde{y}_t \approx \widehat{x}_t, \tag{27}$$

rather than to deviations of output from its long-run steady state (i.e., $\widehat{y}_t$). The output gap $\widehat{x}_t \equiv \widehat{y}_t - \widehat{y}_t^F$ measures the deviations of output $\widehat{y}_t$ from potential $\widehat{y}_t^F$ where the potential is defined as the output level that would prevail in the frictionless model (abstracting from monopolistic competition, nominal rigidities and the financial frictions under 'costly state verification').

Nested Models Without Nominal Rigidities and/or Financial Frictions The financial accelerator mechanism developed in Bernanke et al. (1999) is integrated

[10] The rule of Bernanke et al. (1999) characterizes monetary policy in the following form,

$$\widehat{i}_{t+1} = \rho_i \widehat{i}_t + (1 - \rho_i)\psi_\pi \widehat{\pi}_t + \widehat{m}_t. \tag{26}$$

This feedback rule assumes monetary policy is inertial and the inflation rate is quarter-over-quarter, but that policymakers do not respond to output at all (i.e., $\psi_y = 0$).

into an otherwise standard New Keynesian model that features nominal rigidities—that is, price stickiness and monopolistic competition—as well. The combination of both frictions constitutes my benchmark—which I refer to as the BGG model. In investigating the amplification and propagation effects of the financial accelerator mechanism over the business cycle, I need to establish a reference for what would have happened otherwise without these two frictions. The frictionless allocation abstracting from nominal rigidities and financial frictions—which reduces the BGG model to a standard Real Business Cycle (RBC) economy—offers a natural point of reference for the assessment.

Up to a first-order approximation, the dynamics of the RBC model without frictions differ from those of the financial accelerator model only in the specification of a small subset of the log-linearized equilibrium conditions described before. Hence, the RBC variant of the model can be easily nested within the framework of Bernanke et al. (1999).

Moreover, the financial accelerator also nests other economically-relevant variants that strip down either financial frictions or nominal frictions alone. Abstracting from each friction separately conveys useful information to quantitatively asses the contribution of each friction and the interaction between them in the set-up of Bernanke et al. (1999). The specification variant that eliminates solely the financial friction reduces the BGG model to a Dynamic New Keynesian (DNK) one. In turn, the specification that assumes flexible prices and perfect competition—without nominal rigidities—can be interpreted as an RBC model augmented with financial frictions. I refer to this latter variant of the BGG model as the Financial Accelerator (FA) model.

The Phillips curve equation in (4)—which emerges under Calvo price stickiness —and the resource constraint in (19) are two of the equilibrium conditions that need to be modified under the assumption of flexible prices and perfect competition. The allocation abstracting from nominal rigidities and monopolistic competition mark-ups can be obtained simply assuming that: (a) the Phillips curve in (4) is replaced with a formula that equates real marginal costs $\widehat{mc}_t$ to zero since under flexible prices and perfect competition the price charged by retailers must be equal to its marginal costs; and (b) the monopolistic competition (gross) mark-up is set to 1 (i.e., $\mu = 1$) in the resource constraint in (19) given the assumption of perfect competition. The changes postulated in (a) and (b) are needed for the RBC and FA variants of the model, as they both abstract from nominal rigidities.

Equation (13), which determines the optimal capital allocation, is another one of the equilibrium conditions that needs to be changed whenever state-contingent loans can be used to diversify away all idiosyncratic risks under the additional assumption of perfect information among borrowers and lenders. In that case, the allocation abstracting from financial distortions and inefficiencies can be obtained assuming that: (c) the gross external finance premium in steady state is set to 1 (i.e., $\upsilon(\gamma_n^{-1}) = 1$) in Eqs. (12) and (13) which implies that the borrowing cost is equal to the opportunity cost (the cost of internal funds) given by the real risk-free rate; and (d) the elasticity of the external finance premium relative to the entrepreneurs' leverage ratio evaluated in steady state is set to 0 (i.e., $\upsilon'(\gamma_n^{-1}) = 0$ or $\vartheta = 0$) which eliminates the spread between real borrowing rates and the real risk-free rate in Eq. (13). The changes required under the terms of (c) and (d) are necessary to implement the frictionless allocation of the RBC model in addition to (a) and (b). Conditions (c) and (d) are also needed in the standard DNK model set-up.

Assumptions (a) and (b) eliminate the standard New Keynesian distortions, while assumptions (c) and (d) ensure that it becomes efficient and optimal to accumulate capital to the point where the expected real return on capital equals the real risk-free rate. However, the role of the entrepreneurs' must also be reconsidered in the frictionless RBC and in the DNK cases as it becomes negligible for the aggregate dynamics. Entrepreneurs' consumption and labor income are already negligible by construction.[11] Absent financial frictions, entrepreneurs' aggregate characteristics do not matter for the determination of the investment path either. The leverage of the entrepreneurs (the borrowers) and their net worth (equity)—which is a state variable given by Eq. (17)—become irrelevant to set the borrowing costs and, therefore, the demand for capital. Hence, entrepreneurs' can be dropped without much loss of generality whenever the financial friction is eliminated.

Why does the model of Bernanke et al. (1999) incorporate entrepreneurs anyway? The financial accelerator model distinguishes between two types of economic agents, households and entrepreneurs. Entrepreneurs are risk-neutral agents which decide on the capital to be accumulated for the purposes of wholesale production and on how to finance that stock of capital with a combination of internal funds (net worth or equity) and external borrowing. The households are savers originating the external funds that are intermediated by the banks and eventually borrowed by the entrepreneurs. These two types of agents characterize the borrowers and savers of the economy, respectively.

Absent any financial distortions, the funding costs between internal and external sources must be equalized and given by the real risk-free rate. The predictions of the Modigliani-Miller theorem in a sense are restored and how the capital stock is funded should not matter for the aggregate dynamics of the economy. Therefore, the distinction between savers and borrowers becomes irrelevant for the allocation when the capital structure is undetermined—after all, funding from internal or external sources costs basically the same and should not affect the capital demand or any other economic decision.

Given the negligible impact of the entrepreneurs, the frictionless allocation of the RBC model and the DNK set-up can be approximated under the additional simplifying assumption that: (e) entrepreneurs can be ignored entirely by imposing $\varrho = 0$ and $\gamma_{c^e} = 0$ in order to derive the first-best allocation in the RBC case or the standard DNK solution. The simplification introduced in (e), which abstracts from entrepreneurs altogether, is of little quantitative significance to describe the dynamics of either variant of the model, but it has the advantage of reducing the number of

[11]The labor share of entrepreneurs in the production function is small by assumption (guarantees the entrepreneurs only a small income stream in every period). The steady state consumption share of the entrepreneurs is small by assumption too.

state variables since tracking the entrepreneurs' net worth as in Eq. (17) is no longer needed.

These modifications and simplifications of the financial accelerator model of Bernanke et al. (1999) suffice to characterize an approximation to the frictionless RBC allocation with flexible prices, perfect competition and no-financial accelerator. This approximation of the frictionless model defines the notion of potential for the economy as it abstracts from all frictions. Together with the DNK and FA variants, it also provides the basis on which to assess the contribution to account for the U.S. business cycle of the financial distortion and the New Keynesian frictions (monopolistic competition and nominal rigidities) embedded in the Bernanke et al. (1999) model.

3 Model Parameterization

3.1 Structural Parameters

In this section I describe the choice of the parameter values summarized in Table 1. The values for the taste and technology parameters that I use are fairly standard in the literature, and keep the model comparable to that of Bernanke et al. (1999) also in its parameterization. I assume that the discount factor, β, equals 0.99, which implies an annualized real rate of return of 4 %. The intertemporal elasticity of substitution, σ, and the preference parameter on leisure, η, are both equal to 1. Given that the utility function is assumed to be additively separable in consumption and leisure, the parameterization of σ and η ensures that preferences on both consumption and leisure are logarithmic and, therefore, that the model is consistent with a balanced growth path. The Frisch elasticity of labor supply, $\varphi \equiv \eta(\frac{1-H}{H})$, is determined by the share of hours worked in steady state, H, and the preference parameter η. Given that $\eta = 1$, I fix the share of hours worked, H, to be 0.25 in order to match the Frisch elasticity of labor supply, φ, of 3 preferred by Bernanke et al. (1999).[12]

The capital share, ψ, is set to 0.35 and the share of entrepreneurial labor, ϱ, is kept small at 0.01 as in Bernanke et al. (1999). I maintain the capital share, but set the entrepreneurial labor share to 0 abstracting from the entrepreneurs altogether whenever financial frictions are excluded. As a result, the households' labor share, $1 - \psi - \varrho$, is 0.64 in the financial accelerator BGG model and 0.65 in the DNK and RBC cases. The quarterly depreciation rate, δ, is set to 0.025, which implies an annualized depreciation rate of approximately 10 %. The elasticity of Tobin's q with respect to the investment-to-capital ratio, given by the coefficient χ, is taken to be 0.25.

[12]The share of hours worked is broadly consistent with the U.S. data. The average hours worked relative to hours available per quarter in the U.S. for the period between 1971:III and 2007:IV is 0.2664. The average for the Great Moderation between 1984:I and 2007:IV is similar at 0.2771. For more details on the dataset, see Appendices A and B.

Table 1 Benchmark model parameterization

Parameters	BGG model		FA model	DNK model	RBC model
Intertemporal discount factor	$\beta = 0.99$	Bernanke et al. (1999)	0.99	0.99	0.99
Elasticity of intertemporal substitution	$\sigma = 1$	Bernanke et al. (1999)	1	1	1
Inverse coefficient of risk aversion on leisure	$\eta = 1$	Bernanke et al. (1999)	1	1	1
Steady state share of household hours worked	$H = 0.25$	Bernanke et al. (1999)	0.25	0.25	0.25
Capital income share	$\psi = 0.35$	Bernanke et al. (1999)	0.35	0.35	0.35
Entrepreneurs' labor share	$\varrho = 0.01$	Bernanke et al. (1999)	0.01	0[a]	0[a]
Depreciation rate	$\delta = 0.025$	Bernanke et al. (1999)	0.025	0.025	0.025
Adjustment cost parameter	$\chi = 0.25$	Bernanke et al. (1999)	0.25	0.25	0.25
Calvo price stickiness	$\alpha = 0.75$	Bernanke et al. (1999)	–	0.75	–
Elasticity of substitution across retail varieties	$\theta = 10$	Basu (1996)	–	10	–
Entrepreneurs' "Survival Rate"	$\zeta = 0.9728$	Bernanke et al. (1999)	0.9728	–	–
Entrepreneurs' (inverse) leverage ratio	$\gamma_n = 0.5$	Bernanke et al. (1999)	0.5	–	–
Entrepreneurs' consumption share	$\gamma_{c^e} = 0.01$	Assumption	0.01	0[a]	0[a]
Steady state external finance premium	$\upsilon(\gamma_n^{-1}) = 1.003951$	Data (1984:I–2007:IV)	1.003951	1	1
Steady state slope of external finance premium	$\upsilon'(\gamma_n^{-1}) = 0.0337$	Meier and Müller (2006)	0.0337	0	0
Taylor rule parameters					
Policy inertia	$\rho_i = 0$	Taylor (1993)	0	0	0
Sensitivity to inflation	$\phi_\pi = 1.5$	Taylor (1993)	1.5	1.5	1.5
Sensitivity to output	$\phi_y = 0.5$	Taylor (1993)	0.5	0.5	0.5

Table 1 (Continued)

Parameters	BGG model		FA model	DNK model	RBC model
Shock process parameters					
Persistence of the TFP shock	$\rho_a = 0.8789$	Data (1984:I–2007:IV)	0.8789	0.8789	0.8789
Volatility of the TFP shock	$\sigma_a = \exp(-0.3967)$	Data (1984:I–2007:IV)	exp(−0.3967)	exp(−0.3967)	exp(−0.3967)
Persistence of the monetary policy shock	$\rho_m = 0.8753$	Data (1984:I–2007:IV)	0.8753	0.8753	0.8753
Volatility of the monetary policy shock	$\sigma_m = \exp(-0.4660)$	Data (1984:I–2007:IV)	exp(−0.4660)	exp(−0.4660)	exp(−0.4660)

This table defines the benchmark parameterization of the BGG model as well as the FA, DNK and RBC model variants used in my simulations
[a]This superscript indicates a parameter value change that is used to eliminate entrepreneurs from the model and to simplify its set-up whenever financial frictions are excluded

The Calvo price stickiness parameter, α, is assumed to be 0.75. This parameter value implies that the average price duration is 4 quarters. The (inverse of the) leverage or gearing ratio of the entrepreneurs, $\gamma_n \equiv \frac{N}{K}$, is set at 0.5 and the entrepreneurs' quarterly survival rate in each quarter, ζ, is chosen to be 0.9728. All the parameter choices so far are taken directly from Bernanke et al. (1999), but for the remaining structural parameters I use additional sources to select their values. The elasticity of substitution across varieties, $\theta > 1$, is set to 10. This parameter characterizes the (gross) price mark-up $\mu \equiv \frac{\theta}{\theta-1} > 1$ and its value is consistent with a plausible net mark-up of 11 % (documented in the U.S. data, for instance, by Basu 1996). Notice here that the structural parameters α, γ_n, ζ and θ do not affect the aggregate dynamics of the frictionless RBC economy and that only a subset of them are needed for the parameterization of the FA and DNK variants.

I choose a tiny share of 0.01 for the steady state entrepreneurial consumption, γ_{c^e}, in the financial accelerator model and set this share to 0 in the absence of financial frictions. This modification of the Bernanke et al. (1999) set-up ensures that consumption is essentially determined by the households' Euler equation and that entrepreneurs' consumption is negligible for the dynamics of the model, as discussed before. I set the monetary policy inertia, ρ_i, to 0, the response of the monetary policy rule to fluctuations in inflation, ϕ_π, to 1.5 and the response to fluctuations in output, ϕ_y, to 0.5 to be consistent with the policy recommendation of Taylor (1993). Although the proposal for $(\rho_i, \phi_\pi, \phi_y)$ in Taylor (1993) was based on a reaction function fitted with year-over-year inflation and detrended output, I impose the same parameter values in (20) in all cases—even when the policy rule reacts to annualized quarter-over-quarter inflation and/or output gap measures (which is closer to how this policy rule is often specified in the literature for quantitative and theoretical work).

The steady state external finance premium, $\upsilon(\gamma_n^{-1}) \equiv \frac{R^k}{R}$, is set to 1.003951 in the financial accelerator model, which corresponds to the average quarterly ratio between the Baa corporate yield and the 20-year Treasury yield during the Great Moderation period from 1984:I until 2007:IV (see Appendices A and B for further discussion on how it is calculated). This ratio is consistent with a spread, $R^k - R \equiv (\frac{R^k}{R} - 1)R$, of approximately 160 basis points at an annualized rate given that $R = \beta^{-1} = \frac{1}{0.99}$.[13] This is a bit smaller than the 200 basis points of the historical average spread between the prime lending rate and the six-month Treasury bill rate that Bernanke et al. (1999) used to parameterize their model, but I believe it offers a cleaner measure of the risks modeled. Absent the financial friction, the steady state external finance premium $\upsilon(\gamma_n^{-1})$ is simply set to 1 and—accordingly—the spread $R^k - R$ becomes equal to 0.

Meier and Müller (2006) estimated a similar financial accelerator model and reported plausible values for the composite coefficient $\vartheta \equiv \left(\frac{\upsilon'(\gamma_n^{-1})\gamma_n^{-1}}{\upsilon(\gamma_n^{-1})}\right)$ around 0.0672, which is close to the value implied by the parameterization of Bernanke

[13] Note that I multiply $(\frac{R^k}{R} - 1)R$ by 400 in order to express the spread on an annual basis and in percentages.

et al. (1999). I adopt the value suggested by the estimates of Meier and Müller (2006) which implies that a 1 % increase in the leverage ratio, $(\frac{\Delta\gamma_n^{-1}}{\gamma_n^{-1}})$, is on average associated with a 6.72 basis points increase in the interest rate ratio, $(\frac{\Delta\frac{R^k}{R}}{\frac{R^k}{R}})$. Therefore, given my parameterization of the entrepreneurs' leverage ratio, $\gamma_n^{-1} = 2$, and the external finance premium, $\upsilon(\gamma_n^{-1}) = 1.003951$, the slope coefficient $\upsilon'(\gamma_n^{-1}) \equiv \frac{\partial\upsilon(\gamma_n^{-1})}{\partial\gamma_n^{-1}} = \vartheta\frac{\upsilon(\gamma_n^{-1})}{\gamma_n^{-1}}$ is set equal to 0.0337 in order to match the value of 0.0672 of the composite coefficient ϑ. In the frictionless RBC case and the DNK case, I set the steady state slope of the external finance premium, $\upsilon'(\gamma_n^{-1})$, to 0 in order to bring ϑ to 0 as well shutting down the financial frictions of the model.

It is worthwhile to consider here the implications of the parameterization on the long-run allocation of expenditures. The steady investment share in the model of Bernanke et al. (1999), $\gamma_x \equiv \delta(\frac{\psi}{\mu(\upsilon(\gamma_n^{-1})\beta^{-1}-(1-\delta))})$, is a composite coefficient of structural parameters that is distorted by monopolistic competition and by the long-run external finance premium. In the frictionless RBC steady state, the investment share is simply given by $\delta(\frac{\psi}{\beta^{-1}-(1-\delta)})$ which takes the value of 0.25 under the parameterization I adopt here. The monopolistic competition mark-up, $\mu \equiv \frac{\theta}{\theta-1}$, is a function of the elasticity of substitution across retail varieties, θ, which does not appear anywhere else in the log-linearized equilibrium conditions. Imposing a plausible mark-up of approximately 11 % alone reduces the steady state investment share to 0.22 for the DNK case.

The investment share in steady state is also affected by the size of the external finance premium in steady state, $\upsilon(\gamma_n^{-1}) \equiv \frac{R^k}{R}$. This distortion does not only affect the steady state investment share, because it also enters in the elasticity of the external finance premium to changes in the leverage of the borrowers, $\vartheta \equiv (\frac{\upsilon'(\gamma_n^{-1})\gamma_n^{-1}}{\upsilon(\gamma_n^{-1})})$, as well as in the weight of capital gains in the returns to capital, $\epsilon \equiv (\frac{1-\delta}{\upsilon(\gamma_n^{-1})\beta^{-1}})$. In any case, the combined effect of the monopolistic competition mark-up and the external finance premium reduces the investment share in the BGG model to just 0.20, which implies a very significant shift away from investment over the long-run.[14]

Finally, I assume a share of entrepreneurial consumption, γ_{c^e}, of 0.01 for the financial accelerator model—as a modeling simplification relative to the Bernanke et al. (1999) set-up—and of 0 in the absence of financial frictions. As a result, the households' consumption share, γ_c, becomes equal to 0.79 in the model variants with financial frictions and 0.80 otherwise. I do not incorporate government consumption in the model, as noted earlier, so the consumption and investment shares are related to their counterparts in the data based on real private output (excluding government compensation).

[14]The quarterly share of real investment over private real output is broadly consistent with the U.S. data. The average quarterly share in the U.S. for the period between 1971:III and 2007:IV is 0.1757. The average for the Great Moderation between 1984:I and 2007:IV is similar at 0.1719. For more details on the dataset, see Appendices A and B.

3.2 Shock Processes and Macro Observables

I parameterize the financial accelerator model to be consistent with the existing literature and comparable with Bernanke et al. (1999). The parameters that characterize the monetary policy regime and the shock processes depart somewhat from those of Bernanke et al. (1999) to conform with the long-run features of the data observed during the Great Moderation. To be more precise, the policy specification is set as in Taylor (1993) to describe the prevailing monetary policy regime. The features of the TFP and the monetary shock processes are estimated from actual data on the Solow residual and the deviations between the Federal Funds rate and the Taylor (1993)'s prescribed policy rates during the Great Moderation period (as detailed in Appendices A and B).

The observable data that the model tries to explain is detrended—or demeaned, as the case might be. The estimates of the trend or the level of these macro variables are based on data for the Great Moderation period, which I project forwards but also backwards to get longer time series to work with. Before anything else, of course, I need to clarify what I consider to be the time span of the Great Moderation. While different authors date the start at different times, most authors agree that the major decline in macro volatility began in 1984. McConnell and Pérez-Quirós (2000) estimate a break date of 1984:I using quarterly real output growth data between 1953:II and 1999:II. As it has become common practice to follow the dating of McConnell and Pérez-Quirós (2000), I also adopt 1984:I as the starting quarter of the Great Moderation for the purposes of this paper.

Given that the policy framework in the model is geared towards describing the post-Bretton Woods era, I do not attempt to assess in this paper the path of the macro series prior to 1971:III.[15] Moreover, I also abstract from discussing in great depth the structural changes that took place during the 1970s. In turn, I focus solely on the period since the onset of the Great Moderation. This avoids the structural breaks found in the data prior to the 1980s as well as the two consecutive recessions of the early 1980s, so it makes more straightforward the mapping of the data into the model. The Great Moderation period since 1984:I is largely characterized by stable trends, except in the aftermath of the 2007 recession. In fact, that is the only break that I consider here. I investigate the 2007 recession as a break with the Great Moderation allowing explicitly for the possibility of a level—but not a growth—shift on the long-run path along which the U.S. economy evolved. For my analysis, I compute detrended (or demeaned) variables that incorporate and ignore that level shift.

I settle with 2007:IV as the end of the Great Moderation period in order to ensure that my estimation results of the underlying trends of the data and, more generally,

[15]The dollar became a fiat currency after the U.S. unilaterally terminated convertibility of the U.S. dollar to gold on August 15, 1971, ending the Bretton Woods international monetary system that had prevailed since the end of World War II. Floating exchange rates and increasing capital account openness characterize the post-Bretton Woods period in the U.S., in a major break with the prevailing monetary policy regime under Bretton Woods.

the parameterization of the model would not be driven by a few observations during and after the 2007 recession. I consider the possibility of a level shift occurring after the 2007 recession with the implication of increasing the size of the interest rate spreads, while lowering the share of hours worked and the levels of real private output (excluding government compensation), real private investment and real private consumption. The break itself is dated where the fall in real private output (excluding government compensation) is the highest in percentage terms—in the dataset described in Appendices A and B, by this metric, the level shift occurs in 2009:II.

Table 2 summarizes the empirical estimates of the Solow residual and monetary shock processes, as well as the detrending (or demeaning) of the observable macro variables of interest—the data includes real private output (excluding government compensation), real private investment, real private consumption, and the share of hours worked all expressed in per capita terms, as well as year-over-year consumption (of nondurables and services) inflation and the spread between the Baa corporate and the 20-year Treasury yield. Each specification is set in state space form and estimated by Maximum Likelihood with data for the Great Moderation period between 1984:I and 2007:IV.

Subsequently, I fix the coefficients of the specification at their estimated values for [1984:I, 2007:IV] and add a recession dummy that takes the value of 1 from 2009:II onwards to take account of the possibility of a level shift. I expand the estimation sample to go up to 2012:I and estimate the coefficient on the recession dummy to determine both the size and the significance (if any) of the break. This estimation strategy implemented in two stages preserves the estimates obtained with data prior to the 2007 recession as such, while allowing me to incorporate the data available up to 2012:I in order to test the hypothesis that a level shift may have occurred in the aftermath of the 2007 recession. Table 2 summarizes the empirical evidence for such structural shift whenever it is statistically significant in the data and reports the size of the break based on the currently available data.

Macro Observables Figure 1 illustrates the behavior of the share of hours worked, consumer (nondurables and services) price inflation in year-over-year rates and the quarterly interest rate spread around a constant level. I use historical estimates of the mean during the Great Moderation period for the share of hours worked and the interest rate spread, while the inflation level is set to the implicit monetary policy target of 2 percent sought over the past three decades and assumed by Taylor (1993). I maintain the description of the monetary policy framework invariant in my current analysis even after the 2007 recession, so the inflation target of 2 percent is unchanged before and after 2009:II. In turn, I allow for the possibility of a level shift in both hours worked as well as the quarterly interest rate spread that is shown in Table 2 to be statistically significant. The interest rate spread went up by around 10 basis points on average after 2009:II, while the share of hours worked in logs declined on average around 5.78 percent.

Figure 2 illustrates the path of the time series for real private output in logs, real private investment in logs and real private consumption in logs along a linear trend. In linearly detrending these macro time series, however, I impose a priori only a

Table 2 Summary of maximum likelihood estimates

Dependent variable	Regression estimates	Sample
Endogenous macro aggregates		
$\overline{h}_t \equiv \ln(h_t)100$	$\overline{h}_t = \underset{(0.344779)}{-128.3841} + u_t^h,\ \ln\sigma(u_t^h) = \underset{(0.076884)}{1.104195}$	1984:I–2007:IV
	$\overline{h}_t = -128.3841 - \underset{(2.827576)}{7.4254401}_{(t\geq 2009:\text{II})} + u_t^h,\ \ln\sigma(u_t^h) = 1.104195$	
$\overline{\pi}_t \equiv (\frac{P_t - P_{t-4}}{P_{t-4}})100$	$\overline{\pi}_t = 2 + u_t^\pi,\ \ln\sigma(u_t^\pi) = \underset{(0.081194)}{0.394633}$	1984:I–2012:I
$spread_t \equiv (\rho_t^{Baa\ Corp} - \rho_t^{20\ yr\ T\text{-}Bill})100$	$spread_t = \underset{(0.010858)}{0.395125} + u_t^{spread},\ \ln\sigma(u_t^{spread}) = \underset{(0.088871)}{-2.377567}$	1984:I–2007:IV
	$spread_t = 0.395125 + \underset{(0.043423)}{0.1045791}_{(t\geq 2009:\text{II})} + u_t^{spread},\ \ln\sigma(u_t^{spread}) = -2.377567$	
$\rho_t^j \equiv 0.25\ln(1+\frac{R_t^j}{100})$, $j \in$ {Baa Corporate, 20 yr Treasury}		1984:I–2007:IV
$\overline{y}_t \equiv \ln(y_t)100$	$\begin{pmatrix} 100\frac{c_t}{y_t} \\ 100\frac{x_t}{y_t} \end{pmatrix} = \begin{pmatrix} \underset{(0.138189)}{63.91049} \\ \underset{(0.139952)}{17.16505} \end{pmatrix} + \begin{pmatrix} u_t^{c\_share} \\ u_t^{x\_share} \end{pmatrix}$,	1984:I–2012:I
	$\begin{pmatrix} \overline{y}_t \\ \overline{c}_t \\ \overline{x}_t \end{pmatrix} = 100\ln(\underset{(110.5673)}{10934.29}) + 100\begin{pmatrix} 0 \\ \ln(\frac{1}{100}(\underset{(0.138189)}{63.91049})) \\ \ln(\frac{1}{100}(\underset{(0.139952)}{17.16505})) \end{pmatrix} + \underset{(0.019412)}{0.431212}t + \begin{pmatrix} u_t^y \\ u_t^c \\ u_t^x \end{pmatrix}$,	
	$\begin{pmatrix} \ln\sigma(u_t^{c\_share}) = \underset{(0.277974)}{0.612087},\ \ln\sigma(u_t^{x\_share}) = \underset{(0.257966)}{0.168934}, \\ \ln\sigma(u_t^y) = \underset{(0.516894)}{1.047145},\ \ln\sigma(u_t^c) = \underset{(0.328568)}{2.124132},\ \ln\sigma(u_t^x) = \underset{(0.135221)}{0.269371} \end{pmatrix}$	

Table 2 (Continued)

Dependent variable	Regression estimates	Sample
$\overline{c}_t \equiv \ln(c_t)100$	$\begin{pmatrix} 100\frac{c_t}{y_t} \\ 100\frac{x_t}{y_t} \end{pmatrix} = \begin{pmatrix} 63.91049 + \underset{(2.370044)}{4.9547861}_{(t\geq 2009:\text{II})} \\ 17.16505 \underset{(0.668740)}{-4.219118}\, 1_{(t\geq 2009:\text{II})} \end{pmatrix} + \begin{pmatrix} u_t^{c\_share} \\ u_t^{x\_share} \end{pmatrix},$	1984:I–2007:IV
	$\begin{pmatrix} \overline{y}_t \\ \overline{c}_t \\ \overline{x}_t \end{pmatrix} = 100\ln(10934.29) - \underset{(3.730605)}{16.126611}_{(t\geq 2009:\text{II})} + 0.431212t + \cdots$	
$\overline{x}_t \equiv \ln(x_t)100$	$100\begin{pmatrix} 0 \\ \ln(\frac{1}{100}(63.91049 + \underset{(2.370044)}{4.9547861}_{(t\geq 2009:\text{II})})) \\ \ln(\frac{1}{100}(17.16505 \underset{(0.668740)}{-4.219118}\, 1_{(t\geq 2009:\text{II})})) \end{pmatrix} + \begin{pmatrix} u_t^y \\ u_t^c \\ u_t^x \end{pmatrix},$	1984:I–2012:I
	$\begin{pmatrix} \ln\sigma(u_t^{c\_share}) = 0.612087, \ln\sigma(u_t^{x\_share}) = 0.168934, \\ \ln\sigma(u_t^y) = 1.047145, \ln\sigma(u_t^c) = 2.124132, \ln\sigma(u_t^x) = 0.269371 \end{pmatrix}$	
Exogenous shock processes		
$\overline{s}_t \equiv (\ln(\frac{S_t}{S_0}))100$	$\overline{s}_t = \underset{(1.589296)}{55.73225} + \underset{(0.026276)}{0.185077}t + \widehat{a}_t$	1984:I–2007:IV
	$\widehat{a}_t = \underset{(0.055162)}{0.878870}\widehat{a}_{t-1} + \varepsilon_t^a, \ln\sigma(\varepsilon_t^a) = \underset{(0.059299)}{-0.396746}$	
$i_t^{AR} \equiv$ federal funds rate (effective)	$i_t^{AR} = 2 + \overline{\pi}_t + 0.5(\overline{\pi}_t - 2) + 0.5\widehat{y}_t + \widehat{m}_t$	1984:I–2007:IV
	$\widehat{m}_t = \underset{(0.040514)}{0.875284}\widehat{m}_{t-1} + \varepsilon_t^m, \ln\sigma(\varepsilon_t^m) = \underset{(0.062270)}{-0.465995}$	

This table reports the Maximum Likelihood estimates with standard errors between parentheses. The E-views 8 codes to generate the estimates are available from the author upon request

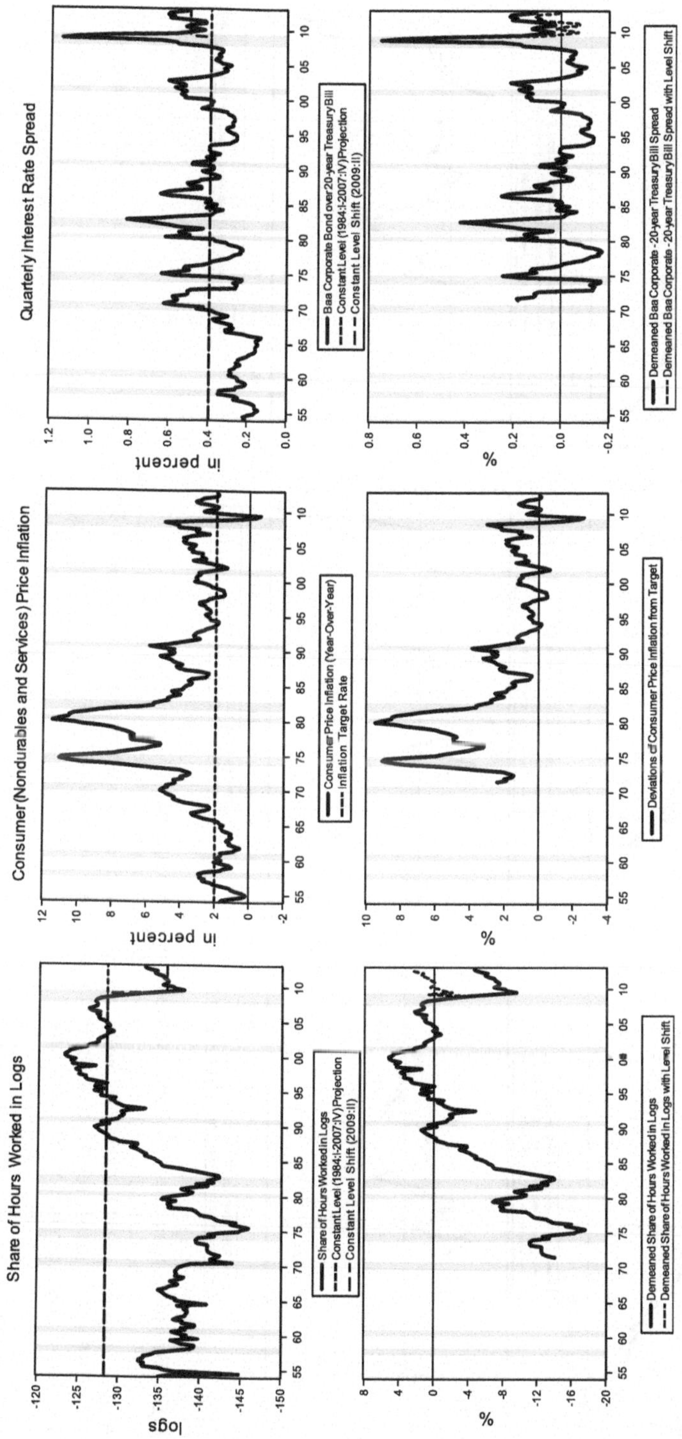

Fig. 1 Deviations from constant level on hours worked, inflation and interest rate spreads. This graph plots the share of hours worked, the consumer (nondurables and services) price inflation in year-over-year terms, and the Baa corporate spread over the 20-year Treasury Bill in deviations. I include a constant level estimated over the Great Moderation period (1984:I–2007:IV) projected backwards and forwards allowing for a level shift on 2009:II (except for inflation), and the corresponding variables in deviations since 1971:III. For more details, see Appendix B 'U.S. Dataset' and the estimates reported in Table 2. The *shaded areas* represent NBER recessions

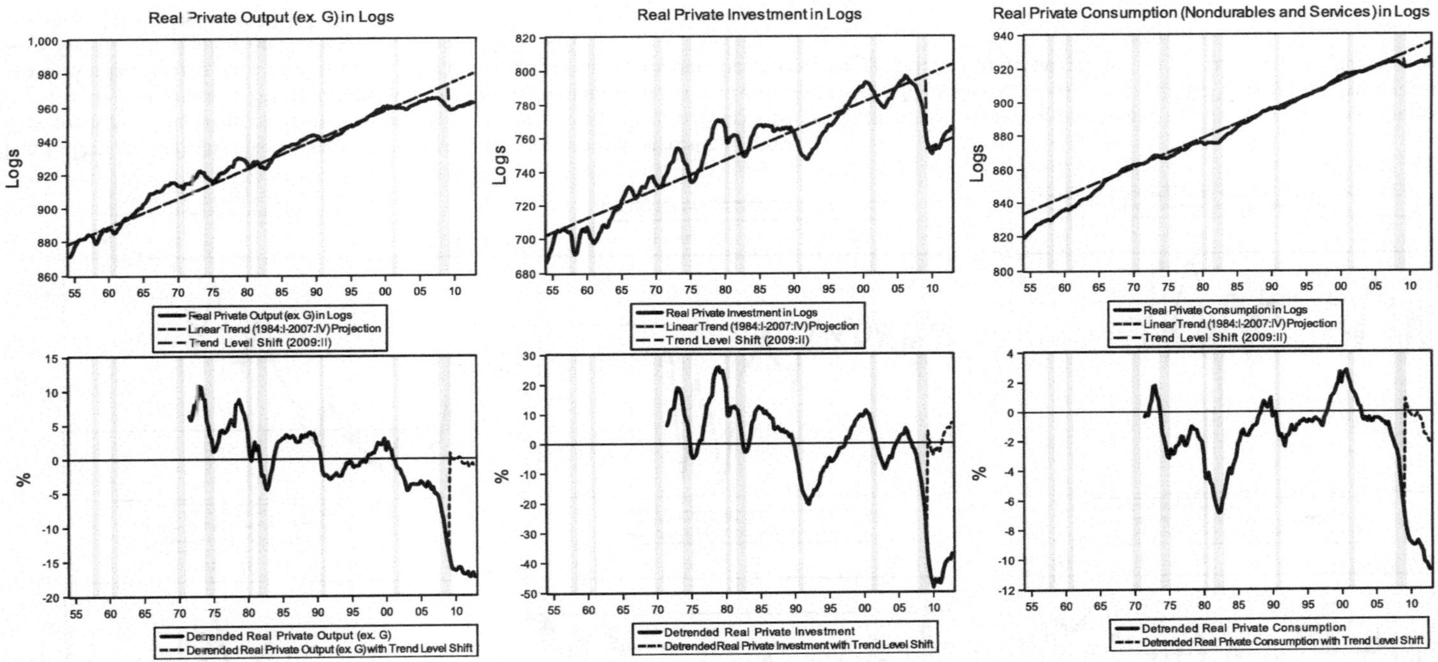

Fig. 2 Deviations from trend on output, investment and consumption.
This graph plots the cetrended real private output (ex. G), real private investment and real private consumption. I include the log-linear trend over the Great Moderation period (1984:I–2007:IV) projected backwards and forwards allowing for a level shift on 2009:II, and the corresponding detrended variables since 1971:III. For more details, see Appendix B 'U.S. Dataset' and the estimates reported in Table 2.
The *shaded areas* represent NBER recessions

minimal set of theoretical constraints with the aim to limit the violence done to the data. In particular, I only require the following two model-consistent features to be satisfied by the specification: First, the financial accelerator model that I use is consistent with a balanced growth path in which consumption and investment grow at the same rate as output, so I assume a time trend that is linear and has a common slope on all three variables (as described in Table 2 and in Appendices A and B).

Second, any structural change that affects the steady state of the financial accelerator model can result in a level shift in output that is consistent with an economy growing along the new path at the same constant rate as before but from a different level. Structural changes could also produce shifts in the steady state consumption and investment shares, though. I, therefore, impose the constraint that intercept of the linear time trend be consistent with the historical shares for consumption and investment observed during the Great Moderation period. I allow in the specification for the possibility that the level shift after the 2007 recession resulted in a decline in the level of output (the intercept) as well as resulted in a change in the long-run shares of consumption and investment.

The evidence reported in Table 2 seems to be consistent with a statistically-significant level shift in the trend specification for real private output, consumption and investment after 2009:II. As seen in Figure 2, the break is basically matched by the decline in real private output and appears to be largely permanent. A simultaneous downward shift of the long-run investment share of around 4 percentage points absorbed by a similar increase in the consumption share accounts for the large decline observed in investment and the more moderate downward shift on consumption.[16]

Adjusting for the estimated level shift and the perceived decline in the investment share, real private output and consumption remain below their new long-run path while real private investment bounced-back above trend by 2012:I. The evidence does not necessarily suggest that there is a break in trend growth in the data in the aftermath of the 2007 recession. In turn, ignoring the possibility of a level shift altogether would produce deviations from the Great Moderation trend that are unprecedented—for the post-Bretton Woods period since 1971:III.

Solow Residual Since the model abstracts from population growth, the Solow residuals are computed from data on the stock of capital, the share of hours worked, and private output expressed in per capita terms. For exact details on the calculation of the U.S. Solow residual, see the data description in Appendices A and B. I extract the relevant features of the stationary shock process $\widehat{a}_t$ used in the model taking into

[16]The basic intuition of the permanent income hypothesis implies that consumption, as a fraction of the permanent income of households, moves in sink with permanent income changes. In the context of the model I use here that same logic implies that consumption shifts should follow from permanent changes in output. However, that is not the full story, as consumption declines depend on the long-run consumption and investment shares which also appear to have shifted around with the break.

account that what is observed is the Solow residual, S_t, and this measure includes a trend component that arises from labor-augmenting growth. Hence, I specify a deterministic linear time trend for S_t with autoregressive residuals to recover $\widehat{a}_t$ and to estimate the persistence and volatility of this stationary process. I cast the model for the Solow residual into state space form and estimate it by Maximum Likelihood for the period of the Great Moderation between 1984:I and 2007:IV.

The estimates of the stationary part of the Solow residual in logs are fitted to an $AR(1)$ process which characterizes the TFP shock dynamics of the model for $\widehat{a}_t$ described in Eq. (9). These features of the TFP shock process are common knowledge and economic agents factor that information in forming their own expectations. The persistence of TFP given by ρ_a is, therefore, set at the estimated value of 0.878870. Similarly, the volatility of the shock σ_a is equal to $\exp(-0.396746) = 0.6725$. Figure 3 illustrates the linear time trend and the stationary components of the actual series for the Solow residual in logs, S_t.

I consider the possibility of a level shift in the path of S_t after 2009:II as well. However, the p-value on the coefficient of the recession dummy that indicates the possibility of this level shift comes at 0.8577, so I cannot reject the null hypothesis that there is no such shift in the trend of the observed Solow residuals. The lack of evidence of a level shift in S_t suggests that the fall documented in real private output (excluding consumption) as well as on other macro variables does not follow from a level shift in productivity, but must be the result of structural changes that affect the steady state of the financial accelerator model. One distinctive possibility that would be consistent with the model and the data showing higher interest rate spreads and lower investment shares since the 2007 recession is this: external funding costs may have become significantly higher in the aftermath of the recession, making investment costlier, and therefore reducing the long-run capital-to-labor ratio and the level of economic activity.

The evidence gathered in the data also shows that trend growth has been noticeably higher for real private output than for the Solow residual. However, this fact can also be accounted by theory—this could be the case because output growth itself does not have to grow at the same rate as the Solow residual by the contribution of other growth factors (e.g., a trend decline in the relative price of capital goods) as discussed in Appendices A and B.

Monetary Policy Shock I define the monetary policy rule in the spirit of Taylor (1993), where the monetary policy instrument is the (effective) Federal Funds rate in percent per annum. As in Taylor (1993), the central bank reacts to the percentage inflation rate over the previous four quarters and to the percent deviation of real GDP from a log-linear trend (where the trend of real private output is estimated with data for the Great Moderation period only). I also maintain the parametric assumptions of Taylor (1993) implying that the response to fluctuations in inflation, ϕ_π, is 1.5, the response to fluctuations in detrended output, ϕ_y, is 0.5, and the interest rate smoothing parameter, ρ_i, is set to 0. All the sources on U.S. monetary policy rates are described in Appendices A and B.

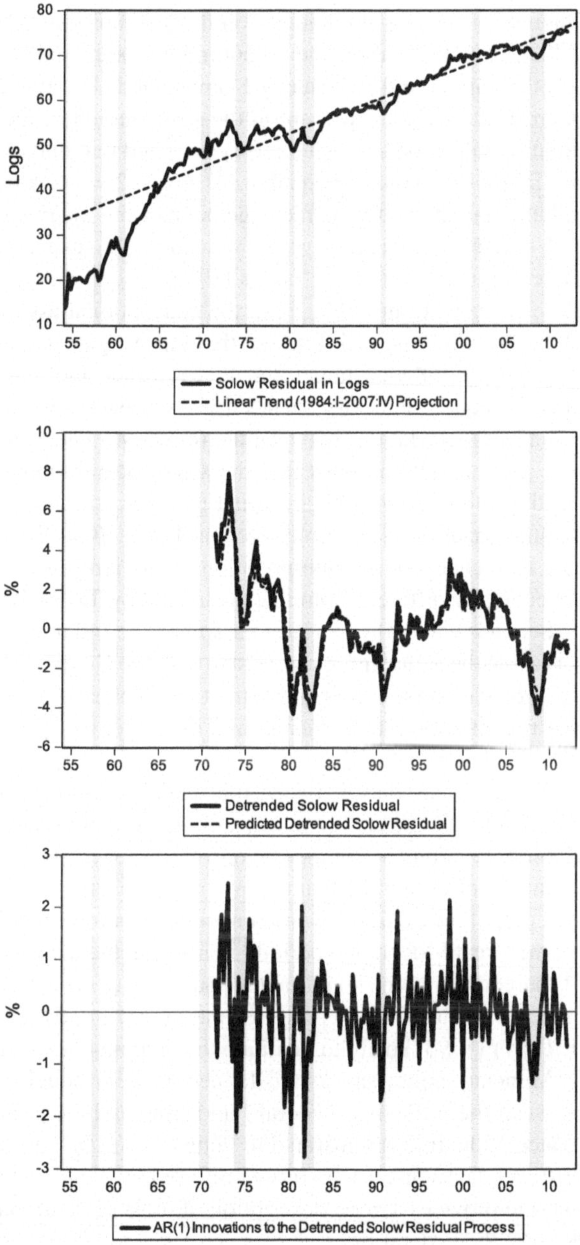

Fig. 3 Solow residuals: trend and stationary components. This graph plots the U.S. Solow residual, detrended and fitted to follow an $AR(1)$ process. I include the log-linear trend over the Great Moderation period (1984:I–2007:IV) projected backwards and forwards, the corresponding detrended variables and the innovations of the $AR(1)$ since 1971:III. For more details, see Appendix B 'U.S. Dataset' and the estimates reported in Table 2.
The *shaded areas* represent NBER recessions

The Taylor (1993) implied annualized rates (in percentages), i_{t+1}^{AR}, are calculated with the following mathematical formula,

$$
\begin{aligned}
i_t^{AR} &= 2 + \overline{\pi}_t + 0.5(\overline{\pi}_t - 2) + 0.5\widehat{y}_t + \widehat{m}_t \\
&= 4 + 1.5(\overline{\pi}_t - 2) + 0.5\widehat{y}_t + \widehat{m}_t, \qquad (28)
\end{aligned}
$$

where $\overline{\pi}_t$ is the rate of inflation over the previous four quarters in percentages, and $\widehat{y}_t$ is the detrended real private output in logs expressed in percentages. Taylor (1993) sets the implicit inflation rate at 2 percent and also adds a long-run (annualized) real interest rate r^{AR} of 2 percent in the specification of the rule in (28). Hence, if the inflation rate is on target (i.e., if $\overline{\pi}_t = 2$) and real output is on trend (i.e., if $\widehat{y}_t = 0$), the Taylor rate would be equal to $i_t^{AR} \equiv 2 + 2 = 4$—two percentage points from the inflation target and two percentage points from the real rate.

I derive the monetary policy deviations $\widehat{m}_t$ using the formula in (28) and the same parameterization as Taylor (1993) to calculate the Taylor rates, but at least three caveats are in order: First, the conventional assumption underlying the class of models with nominal rigidities that I investigate here is that the long-run inflation rate and the inflation target are 0. Given that, the real and nominal interest rates must be equal along the balanced growth path—assuming that the unconditional mean of the deviations between the (effective) Fed Funds rate and the Taylor rates is 0 as well. This implies that the steady state nominal interest rate i^{AR} and the steady state real rate r^{AR} are equal to 4 percent annualized by consistency with a parameterization of the time discount factor, β, at 0.99. In other words, while the rule is unchanged, the interpretation of the long-run inflation and interest rates is conceptually different from that postulated by Taylor (1993) in his empirical work.

Second, while Taylor (1993) assumes the inflation target to be 2 percent, I observe that the actual inflation average over the Great Moderation period is 3.14 %. To treat the data on inflation and extract the cyclical component, I assume nonetheless that the inflation rate moves around the target of 2 percent set by Taylor (1993)—instead of demeaning the data.

Third, consistency between the model definitions and the data is maintained throughout the paper. For instance, I define real private output to be real GDP excluding government compensation of employees to be consistent with the model definition of output. I calculate my own measure of the log-linear trend of real private output (excluding government compensation) in order to fit the estimated trend for the Great Moderation period.[17] I also calculate the relevant inflation rate in terms of the consumption (nondurables and services) price deflator for the same reason. Hence, I depart from the preferred measures of real GDP and the GDP deflator used in Taylor (1993) solely to facilitate the mapping between the data and the model.

Monetary policy shocks are defined by the residual $\widehat{m}_t$, implied by the deviations between the (effective) Federal Funds rate and the policy rule in (28). The performance of the rule is illustrated in Figure 4. As can be seen, even though I have used different data sources than those preferred by Taylor (1993), the long-held view that the rule provides a good description of most of chairman Greenspan's tenure at the helm of the Federal Reserve between 1987 and 2002 remains unchanged.

A number of further qualifications need to be made regarding the conduct of monetary policy during the Great Moderation period and about the interpretation of the monetary shocks derived in this way. First, I am surely missing some transitional

[17]Taylor (1993) estimates trend real GDP with a shorter sample from 1984:I till 1992:III.

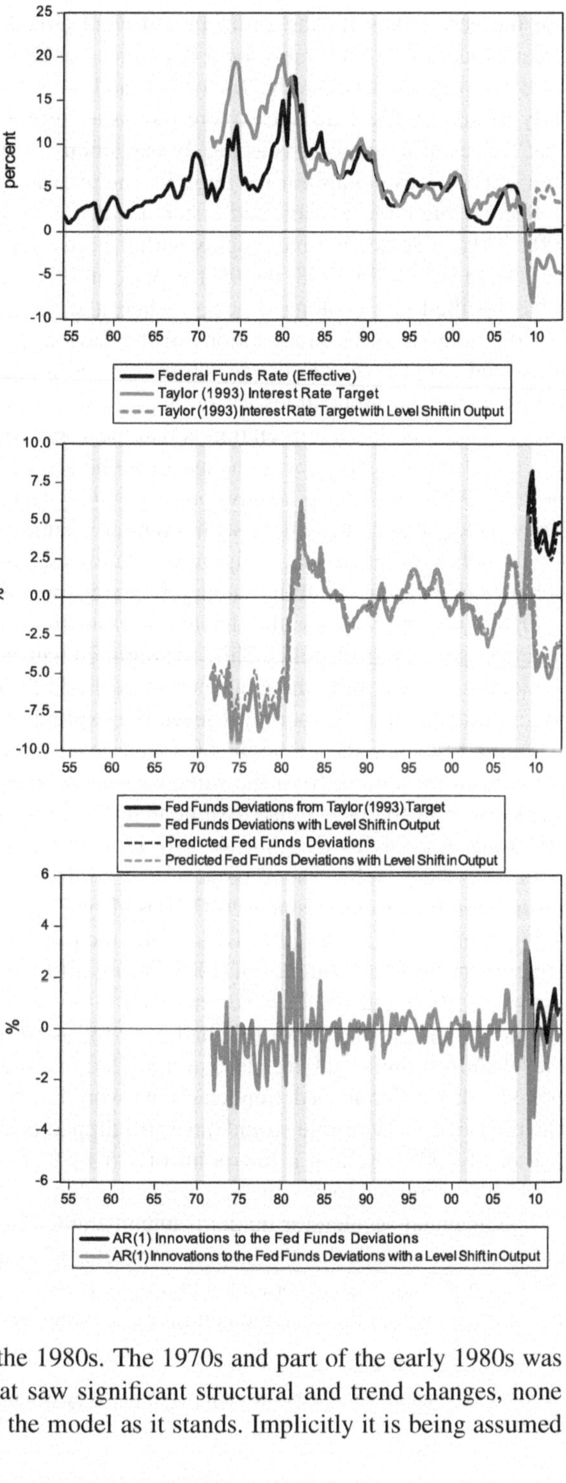

Fig. 4 U.S. monetary policy under the Taylor (1993) rule. This graph plots the Federal Funds rate (effective, annualized), as the policy instrument of reference. It also includes the Taylor rule rates based on the Taylor (1993) specification of the monetary policy rule reacting to changes in the year-over-year inflation rate and detrended output. The Taylor rule residuals are treated as exogenous and defined as the difference between the Federal Funds rate effective and the Taylor rule rates. The Taylor rule residuals are fitted to an $AR(1)$ process over the period of the Great Moderation (1984:I–2007:IV) and then projected backwards and forwards. I also include the corresponding predicted Taylor residuals and the innovations of the $AR(1)$ since 1971:III. For more details, see Appendix B 'U.S. Dataset'. The *shaded areas* represent NBER recessions

dynamics in the first half of the 1980s. The 1970s and part of the early 1980s was a convulse period of time that saw significant structural and trend changes, none of which is fully captured by the model as it stands. Implicitly it is being assumed that the new trends for the entire period were already known at the onset of the Great Moderation. The transitional dynamics could, perhaps, account for some of the discrepancies between the Taylor rule and the (effective) Fed Funds rate in the early-to-mid-1980s. I do not explore the issue further in the paper and, therefore, treat the resulting deviations as purely exogenous monetary shocks—in any event, their impact does not appear too large for the relevant period after 1984:I.

Second, there is a sizeable and systematic downward deviation from the rule after 2002. This coincided in time roughly with the aftermath of the Asian Crisis of 1997, the LTCM bailout in 1998, the 9/11 events, and the subsequent recession of 2001. It has resulted in a prolonged period where the Federal Funds rate has been kept too low relative to the prescriptions of the Taylor (1993) rule. This fact has been noted and extensively discussed before, but the model laid down here allows me to investigate its implications for the U.S. business cycle in a general equilibrium setting. One possible interpretation is that these systematic deviations of the policy rule could be indicative of a change in monetary policy regime that occurred in the late 1990s, leading to an environment with systematically lower interest rates. Many factors can contribute to such a regime change, for instance, a change in the weight policymakers assign to fighting inflation and promoting sustainable growth, a change in the long-run inflation target, or a change in the long-run real rates.

Distinguishing whether the deviations from the rule are exogenous after 2002 or reflect some sort of policy shift (or regime change) is probably one of the key challenges to determine the contribution to the U.S. business cycle that monetary policy has had over this period. I leave the exploration of alternative explanations for future research, and I treat the observed deviations under the Taylor rule in (28) as realizations coming from the same exogenous process for the monetary policy shocks as prior to 2002. I also assume that economic agents did not perceive those deviations as implying a regime shift for monetary policy.

Finally, there is the crucial issue of how to handle monetary policy at the zero-lower bound, especially since 2007. Based on my dataset for the U.S. economy and my characterization of the Taylor rule, the prescribed rate should have become negative in the fourth quarter of 2008 hitting a low point of $-8.11\ \%$ in the third quarter of 2009 and would have remained in negative territory for the rest of my sample. In turn, if the Taylor rule had been followed recognizing the possibility of a level shift in output taking place around 2009:II, the prescribed path would have looked rather different as detrended output would look very different (see Figure 2). The prescribed Taylor rate would have still dipped below the zero-line in the fourth quarter of 2008 reaching a low point of $-5.13\ \%$ in the first quarter of 2009 but would have returned to positive territory after that.

The financial accelerator model is unconstrained in the setting of the policy rate and, therefore, entails that no agent incorporates in its decision-making the practical fact that nominal rates are bounded below by zero.[18] This is an issue that cannot be disregarded even in the case the central bank would have recognized the possibility

[18]Unless some unorthodox measures are put in place by central banks that I am not considering here either.

of a level shift in output from very early on. In any event, I will leave the exploration of the zero-lower bound for further research. Instead, the deviations between the unconstrained Taylor rate and the constrained (effective) Federal Funds rate are merely treated as realizations of the same exogenous monetary policy shock process.

With all those caveats in mind about what constitutes a monetary policy shock, I fit the series of Taylor rule deviations to an $AR(1)$ process. The persistence of the monetary shock process given by ρ_m is, therefore, set at the estimated value of 0.875284. Similarly, the volatility of the monetary shock σ_m is equal to $\exp(-0.465995) = 0.6275$. This estimated process characterizes the dynamics of the monetary policy shock described in (22). I maintain the conventional assumption that all agents know about these shock dynamics and factor them into their decision-making process in forming their expectations.

4 Simulation and Quantitative Findings

In this paper I investigate the strengths and weaknesses of the financial accelerator mechanism of Bernanke et al. (1999) to account for the business cycle fluctuations observed in the U.S. data during the Great Moderation period and the 2007 recession. I focus my attention primarily on real private output per capita, real private consumption per capita, real private investment per capita, the share of hours worked per capita, and (year-over-year) inflation, since the path of these variables often provides a useful gauge of the model's overall performance and the effectiveness of monetary policy. I also track the quarterly interest rate spreads as a key indicator of the financial mechanism that the model is trying to describe.

Then, I ask the following questions from the Bernanke et al. (1999) framework: (a) to what extent does the financial accelerator model replicate the path followed by the macro variables of interest during the Great Moderation?; and (b) to what extent can a first-order approximation of the financial accelerator model such as the one proposed in the paper account for the unusual path that the U.S. economy has taken since the 2007 recession? In other words, is the financial accelerator model of Bernanke et al. (1999) a good benchmark to interpret the Great Moderation and the 2007 recession?

Given some initial conditions, the linearized equilibrium equations and the stochastic shock processes described in Section 2 constitute a fully specified linear rational expectations model. To answer my own questions about the model, I first derive the policy functions implied by the linear rational expectations model laid out in Section 2.[19] I use those policy functions to map the realizations of the detrended U.S. Solow residual in logs (the TFP shock) and the U.S. monetary policy deviations presented in Section 3.2—and also discussed in Appendices A and B—into measures of the cyclical behavior of (per capita) real private output, consumption

[19] All the policy functions used in my simulations are derived using the software package Dynare. The parameterization satisfies the Blanchard-Kahn conditions, so a solution exists and is unique.

and investment, the share of hours worked per capita, year-over-year inflation and quarterly interest rate spreads (the external finance premium). I then compare the model simulations against the U.S. data also presented in Section 3.2.

To initialize each simulation, I assume that the economy is growing at (or near) its balanced growth path at the starting quarter. Prior to the Great Moderation period, 1983:IV stands out as a quarter where actual real private output per capita is approximately equal to its potential (as implied by the log-linear trend estimated for the period 1984:I–2007:IV). Hence, I take that quarter to be the initial period in all the simulations and set the exogenous state variables to match the values for the detrended Solow residual in logs and the monetary policy deviation for that quarter. In turn, all endogenous state variables of the model are set to zero in 1983:IV. For every subsequent quarter, the state variables are simulated using the realizations of the detrended Solow residual in logs and the monetary policy deviations obtained from the U.S. data. I only report the simulated endogenous series for the relevant period since 1984:I onwards.

I run a number of policy experiments and counterfactual simulations intended to gauge the strength of the financial accelerator mechanism, the contribution of TFP versus monetary shocks over the business cycle, and the sensitivity of the predictions to some key modelling assumptions. In order to test the robustness of the results, I specifically explore changes to the benchmark model that have been suggested already in the literature. More concretely, I investigate the role of the inflation rate measure (year-over-year versus quarter-over-quarter rates) and output fluctuations (detrended output versus the output gap) to which monetary policy reacts, the degree of nominal rigidities, and the sensitivity of the external finance premium to monetary shocks.

4.1 Model Comparison and Assessment

To establish a clear point of reference, I evaluate the framework laid-out here during the Great Moderation—as this is the time period that my parameterization is meant to characterize. I simulate the financial accelerator model under the Bernanke et al. (1999) set-up specification—the benchmark BGG model—together with three nested variants—that include the financial accelerator model without nominal rigidities (FA), the standard New Keynesian model without financial frictions (DNK), and the standard Real Business Cycle (RBC) model without nominal rigidities or financial frictions. I compare all of those simulations against the observed data along three conventional dimensions: in their ability to match the standard business cycle moments documented in the data (Table 3), on the evidence of comovement between the simulations and the data (Tables 4A and 4B), and in their contribution to account for the movements of the data (illustrated by Figure 5 including the post-2007 recession period).

All simulations evaluated here are derived under the assumption that the central bank's monetary policy during the Great Moderation can be well-approximated by the Taylor (1993) rule introduced in its general form in Eq. (20). Subsequently, I will assess simple departures from this particular conception about the conduct of monetary policy that illustrate the importance of policy over the business cycle in the presence of different frictions. I consider the role of monetary shocks as drivers of business cycles (Figure 6), the economic significance of price stickiness—which introduces monetary non-neutrality into the model—and its interaction with the financial friction (Figure 7), and experiment with different measurements of inflation and output fluctuations in the policy reaction function (Figure 8).

Tables 3, 4A and 4B are constructed with actual and simulated data from 1984:I until 2007:IV which excludes entirely the 2007 recession and its aftermath—it also excludes the period of nominal policy rates at the zero-lower band that followed. As indicated earlier, the simulations take as given the Taylor (1993) specification introduced in Eq. (20) and the parameterization in Table 1. The rule seems to track fairly well the path of the (effective) Federal Funds rate during most of the Great Moderation period since the mid-1980s until around 2002 and, therefore, can be thought of as a referent for the prevailing monetary policy regime for the period of interest in this study. Tables 3, 4A and 4B, therefore, provide insight on the key question of how well does the BGG model or any of its nested variants do in explaining the Great Moderation era under a conventional characterization of the policy regime.

In Table 3, the reported moments are all unconditional—for the simulated data these unconditional moments are all derived under the exact same realization of the TFP shock and the monetary policy shock derived earlier. I review standard business cycle moments (such as standard deviations, autocorrelations and correlations) for the main macro variables of interest—output, investment, consumption, hours worked, inflation and the external finance premium. The standard deviations summarize the volatility inherent in the actual and simulated data. This metric reveals important differences across models and shortcomings in accounting for the volatility observed during the Great Moderation. The BGG model or the FA variant without nominal rigidities tend to provide a better match for the standard deviations found in the data during the Great Moderation period, but it is notable that the patterns of volatility are altered by the combination of frictions—nominal rigidities and financial frictions—in rather complex ways.

The frictionless model, that represents the standard RBC way of interpreting business cycles, deviates from the data primarily because it severely undershoots the volatility of hours worked (0.426 versus 3.033 in the data) and investment (3.466 versus 8.540 in the data). This means that output is somewhat smoother than the data too, while the volatility of consumption is pretty much right on target. The RBC model is subject to both TFP and monetary policy shocks. However, the monetary shocks only have an impact on the nominal variables and all real variable are driven by the realization of the TFP shock because monetary policy is neutral in this case. Naturally, the specification of the systematic part of the Taylor rule itself only has consequences for the nominal variables as well. The resulting volatility of inflation overshoots that found in the data (2.041 versus 0.955 in the data). Absent any financial frictions, the RBC model is silent about the external finance premium.

Relative to the frictionless equilibrium—of the RBC model—the New Keynesian model (DNK) with monopolistic competition and sticky prices introduces monetary

Table 3 Business cycle moments: simulated vs. empirical

	Data	BGG model	FA model	DNK model	RBC model
Std. deviations					
$\sigma(\widehat{y}_t)$	2.856	1.565	1.955	1.521	1.806
$\sigma(\widehat{x}_t)$	8.540	12.373	4.630	3.008	3.466
$\sigma(\widehat{c}_t)$	1.367	2.078	1.204	1.110	1.284
$\sigma(\widehat{h}_t)$	3.033	3.052	0.602	1.012	0.426
$\sigma(\widehat{p}_t - \widehat{p}_{t-4})$	0.955	1.447	2.038	1.529	2.041
$\sigma(\widehat{sp}_t)$	0.093	0.549	0.069	–	–
Autocorrelation					
$\rho(\widehat{y}_t, \widehat{y}_{t-1})$	0.951	0.840	0.885	0.895	0.880
$\rho(\widehat{x}_t, \widehat{x}_{t-1})$	0.973	0.904	0.863	0.865	0.861
$\rho(\widehat{c}_t, \widehat{c}_{t-1})$	0.910	0.940	0.913	0.917	0.900
$\rho(\widehat{h}_t, \widehat{h}_{t-1})$	0.914	0.831	0.851	0.452	0.848
$\rho(\widehat{p}_t - \widehat{p}_{t-4}, \widehat{p}_{t-1} - \widehat{p}_{t-5})$	0.888	0.965	0.898	0.954	0.896
$\rho(\widehat{sp}_t, \widehat{sp}_{t-1})$	0.890	0.959	0.852	–	–
Correlations					
$\sigma(\widehat{y}_t, \widehat{x}_t)$	0.628	0.804	0.986	0.988	0.987
$\sigma(\widehat{y}_t, \widehat{c}_t)$	0.192	−0.325	0.983	0.992	0.989
$\sigma(\widehat{y}_t, \widehat{h}_t)$	−0.181	0.721	0.962	0.291	0.944
$\sigma(\widehat{y}_t, \widehat{p}_t - \widehat{p}_{t-4})$	0.173	0.333	−0.704	−0.523	−0.699
$\sigma(\widehat{y}_t, \widehat{sp}_t)$	0.210	−0.727	−0.986	–	–

These moments are based on the Taylor (1993) specification of the monetary policy rule reacting to changes in the year-over-year inflation rate and detrended output. The moments are calculated for detrended real private output (ex. Government compensation) per capita, detrended real private investment per capita, real private consumption per capita, demeaned hours worked per capita, cyclical inflation—computed as the deviation from a 2 percent target—and the demeaned quarterly interest rate spread between the Baa corporate bond yield and the 20-year Treasury bill rate. The full sample covers the period between 1984:I and 2007:IV

This table reports the theoretical moments for each series given my parameterization. All statistics on simulations are computed after each series is H-P filtered (smoothing parameter = 1600). I use Matlab R2012a (7.14.0.739) and Dynare v4.3.2 for the stochastic simulation

non-neutrality. Hence, that means monetary shocks now contribute to the fluctuations of all real variables—output, investment, consumption and hours worked. Based on data for the Great Moderation period, the DNK model generates somewhat lower volatility on all variables (nominal and real) except on hours worked where the volatility jumps from 0.426 in the RBC case to 1.012 in the DNK model. The distortion that nominal rigidities introduce in the dynamics of the economy causes mainly a static response in hours worked, while investment becomes somewhat smoother as capital accumulation is favored more to distribute the impact of the TFP and monetary shocks intertemporally. Hence, the DNK model does not re-

Table 4A Time series correlations: simulated and empirical data

	Data	BGG model	FA model	DNK model	RBC model
Real private output (ex. G) per capita, $\widehat{y}_t$					
Data	1	−0.216	0.117	0.096	0.141
BGG model		1	−0.164	0.151	−0.157
FA model			1	0.895	0.998
DNK model				1	0.888
RBC model					1
Real private investment per capita, $\widehat{x}_t$					
Data	1	−0.510	0.477	0.369	0.460
BGG model		1	−0.546	-0.134	−0.558
FA model			1	0.835	0.998
DNK model				1	0.827
RBC model					1
Real private consumption per capita, $\widehat{c}_t$					
Data	1	0.322	0.500	0.545	0.512
BGG model		1	0.573	0.459	0.641
FA model			1	0.936	0.995
DNK model				1	0.930
RBC model					1

These correlations are based on the Taylor (1993) specification of the monetary policy rule reacting to changes in the year-over-year inflation rate and detrended output. The correlations correspond to detrended real private output (ex. Government compensation) per capita, real private investment per capita and real private consumption per capita for the entire period between 1984:I and 2007:IV This table reports the theoretical moments for each series given my parameterization. All statistics on simulations are computed after each series is H-P filtered (smoothing parameter = 1600). I use Matlab R2012a (7.14.0.739) and Dynare v4.3.2 for the stochastic simulation

solve the two big "misses" of the RBC model—it only seems to improve relative to the low volatility of hours worked of the RBC model, but at the expense of worsening the predicted volatility of investment. Absent any financial frictions, the DNK model is also silent about the external finance premium.

The model with financial frictions alone (FA) preserves the monetary non-neutrality of the standard RBC model, so all real variables respond solely to the realization of the TFP shock and only nominal variables are affected by the realization of the monetary shock. As in the RBC case, different specifications of the systematic part of the Taylor rule that respond to inflation and output fluctuations will only affect the path of the nominal variables. However, relative to the frictionless equilibrium—of the RBC model—the FA model introduces an interest rate spread (the external finance premium) between the borrowing costs and the real risk-free rate that distorts the investment decisions. The result is that the FA model matches fairly well the demeaned quarterly interest rate spread between the Baa corporate

Table 4B Time series correlations: simulated and empirical data

	Data	BGG model	FA model	DNK model	RBC model
Share of hours worked per capita, $\widehat{h}_t$					
Data	1	−0.189	0.235	0.015	0.227
BGG model		1	−0.611	0.751	−0.632
FA model			1	−0.194	0.992
DNK model				1	−0.219
RBC model					1
Inflation (year-over-year rate), $\widehat{p}_t - \widehat{p}_{t-4}$					
Data	1	0.666	0.703	0.687	0.721
BGG model		1	0.690	0.710	0.735
FA model			1	0.980	0.997
DNK model				1	0.980
RBC model					1
Quarterly interest rate spread (external finance premium), $\widehat{sp}_t$					
Data	1	0.093	−0.058	–	–
BGG model		1	−0.712	–	–
FA model			1	–	–
DNK model				1	–
RBC model					1

These correlations are based on the Taylor (1993) specification of the monetary policy rule reacting to changes in the year-over-year inflation rate and detrended output. The correlations correspond to demeaned hours worked per capita, year-over-year inflation and the quarterly interest rate spread (or external finance premium) for the entire period between 1984:I and 2007:IV

This table reports the theoretical moments for each series given my parameterization. All statistics on simulations are computed after each series is H-P filtered (smoothing parameter = 1600). I use Matlab R2012a (7.14.0.739) and Dynare v4.3.2 for the stochastic simulation

bond yield and the 20-year Treasury bill rate (0.069 versus 0.093 in the data) and increases the investment volatility from 3.466 in the RBC model to 4.630 in the FA case. The increased investment volatility relative to the RBC case brings with it an increase in the volatility of hours worked and output, but—in spite of that—the volatility of investment, hours, and output still remains too low when compared against the actual data.

The BGG model of Bernanke et al. (1999) combines the two broad types of frictions highlighted by the FA model (financial frictions) and the DNK model (monopolistic competition and price stickiness). However, the interaction of both frictions often is more than just the sum of their separate effects. Nominal rigidities breakdown with the monetary neutrality of the frictionless RBC model, while financial frictions imply that borrowing costs—or the opportunity cost of investment—is no longer equal to the real risk-free rate as in the frictionless RBC case. Moreover, with nominal rigidities in the BGG model, monetary non-neutrality implies that the spread—the external finance premium—responds to the realization of the monetary shock and not just to the realization of the TFP shock. In other words, the external finance premium and ultimately the path of investment itself will be amplified by the realization of the monetary shock process that describes the exogenous deviations from the policy rate target set according to Taylor (1993)'s rule.[20] The amplification of the external finance premium will, in turn, crucially depend on the importance of the demand distortion caused by the degree of price stickiness, as I discuss in a later subsection.

The volatility magnification attained by the BGG model with a combination of monetary and TFP shocks under the benchmark parameterization cannot be more dramatic as it rises the volatility of the external finance premium from 0.069 in the FA case to 0.549 in BGG—well above the 0.093 seen in the data—and the volatility of investment from 4.630 in the FA case to 12.373 in BGG—also above the 8.540 observed in the data. Consumption and hours worked become somewhat more volatile in the BGG model than in the FA case, while output and inflation are less so. In other words, while variants of the model with financial frictions such as the BGG and FA ones tend to generate volatility patterns closer to those observed in the data for the Great Moderation period, I interpret from the results reported in Table 3 that breaking away from monetary neutrality by incorporating nominal rigidities alters the financial accelerator mechanism in a fundamental way by allowing monetary shocks to have real effects and to influence the external finance premium by distorting the demand allocation.[21] The resulting amplification of the spread (the external finance premium) generates a large increase in the volatility of investment as well.

The first-order autocorrelations also reported in Table 3 offer one simple measure of the persistence of fluctuations in the actual and simulated data. The evidence I have collected reveals that most variants of the model generate very persistent dynamics that are not far from those actually observed in the data. I find interesting that the persistence of hours worked under the DNK model is 0.452 while it is 0.831 or above in all other cases. However, aside from this noticeable difference, there is no strong support in Table 3 for the view that the addition of nominal rigidities and/or financial frictions changes the propagation of shocks—which are themselves quite persistent, as noted before—endogenously generated by the frictionless RBC model in any significant or systematic way.

The correlations between output and all other macro variables of interest—investment, consumption, hours worked, inflation and the external finance premium—provide a sense of the cyclicality of the actual and simulated data. Unlike the

[20] I explore this issue further in Figure 6 and in the next subsection where I consider explicitly the impact that monetary shocks have on the dynamics of the BGG model.

[21] In the BGG model under the Taylor rule specified in (20), the decomposition of the variance of the spread is 35.48 from TFP innovations and 64.52 from monetary policy innovations. Needless to say, both the contribution of monetary shocks to the external finance premium as well as the impact that monetary shocks have on other variables will depend on the systematic part of the rule too. I explore that issue a little closer in the remainder of the paper.

situation described with the first-order autocorrelations, the results I report in Table 3 suggest that the combination of financial frictions and nominal rigidities results in major differences relative to the cyclical patterns of the frictionless RBC model or even relative to the variants that include solely nominal rigidities (DNK) or only financial frictions (FA). The RBC model under monetary neutrality generates a strong positive comovement between output and all other real variables—investment at 0.987 versus 0.628 in the data, consumption at 0.989 versus 0.192, and hours worked at 0.944 versus −0.181. These strong correlations are at odds with the data, but the RBC model also delivers a countercyclical inflation of −0.699 that is far away from the mild procyclicality observed in the data (with a positive correlation between output and inflation of 0.173). The negative correlation between output and inflation is to be expected whenever inflation is primarily driven by real or supply-side shocks (TFP shocks in the framework that I am investigating here), but is otherwise inconsistent with the empirical evidence and therefore indicative that the contribution of TFP and monetary shocks is not well-captured in the frictionless RBC setting.

Breaking from the monetary neutrality of the RBC model with the addition of nominal rigidities only seems to weaken both the procyclicality of the hours worked—0.291 in the DNK model versus 0.944 in the RBC model and −0.173 in the actual data—and the counter-cyclical behavior of inflation. Adding financial frictions as in the FA model preserves the monetary neutrality of the RBC specification, and does very little to change the cyclicality of investment, consumption, hours worked and inflation. However, the FA model introduces an interest rate spread (the external finance premium) between the borrowing costs and the real risk-free rate that is solely driven by TFP shocks, as discussed before. In this case, a very strong negative correlation emerges between output and the external finance premium (−0.986 in the FA model versus 0.210 in the data). In other words, while the model implies that "low" spreads ought to be expected in "good" times and vice versa, the data suggests that spreads have been mildly pro-cyclical. This counterfactual evidence is precisely one of the major challenges that the financial accelerator mechanism embedded in the framework of Bernanke et al. (1999) confronts—the strong countercyclicality of the external finance premium is at odds with the empirical evidence.

Combining both frictions—the nominal rigidities of the DNK model and the financial friction of the FA model—into the BGG framework, however, significantly alters the cyclicality patterns that arise from the RBC model and from any of the variants that incorporate just one of the frictions at a time. The external finance premium is now influenced by monetary shocks since the nominal rigidities incorporated by the BGG model break the monetary neutrality of the RBC and FA models. However, the external finance premium remains strongly countercyclical although less so than in the FA case—at −0.727 in the BGG case versus −0.986 in the FA case and 0.210 in the data. Hours worked and output also remain far from what I observe in the data, but improve relative to what the FA model can deliver by weakening their strong procyclicality bias to some extent.

The two major shifts are to be found on inflation and consumption. In regards to inflation, the BGG reverts the countercyclicality found in the RBC, DNK and FA cases—at 0.333 versus 0.173 in the data. This suggests that the contribution of monetary shocks to the dynamics of inflation overwhelms the strength of the TFP shocks unlike what happened in other variants of the model. In regards to consumption, the BGG model also reverts the procyclicality of all other model variants with a correlation of −0.325 versus 0.983 in the FA case and 0.192 in the data. This suggests an interesting reading of the model predictions: whenever monetary shocks have real effects as in the BGG model, the financial accelerator mechanism gets accentuated and this, in turn, has a substitution effect.

In "good" times, the external finance premium tends to be "low" and there is a strong incentive to postpone consumption for later in order to invest more now—taking advantage of the fact that it is relatively less expensive to borrow for investment purposes. Similarly, in "bad" times, the external finance premium tends to be "high" and there is a strong incentive to consume now and invest later. However, while the pattern is well understood, it is nonetheless counterfactual for the Great Moderation period. The implication of all these results is that the BGG model is an incomplete framework with which to account for the business cycle features in the U.S. since the mid-1980s. More research still is needed to understand the role of other frictions and shocks, and to quantify their contribution to the cyclical fluctuations during this period.

I explore in Table 3 the performance of each of the models discussed in this paper by investigating their strengths and weaknesses in attempting to match key features of the data (that is, in trying to match key business cycle moments). Tables 4A and 4B illustrate now the evidence of comovement between the simulations across all model specifications and the data for each one of the macro variables of interest. The goal of this exercise is to provide a simple metric to assess the ability of the simulated data to track the path of the actual data or the path of data simulated by other models—in other words, the exercise provides some insight into the similarities across model simulations and with respect to the actual data.

The conclusions inferred from Tables 4A and 4B reinforce the perception that the BGG model set-up of Bernanke et al. (1999) dramatically alters the dynamics of the frictionless RBC model (even those of simpler models with either nominal rigidities or financial frictions alone), but remains inadequately prepared to account for the Great Moderation period. The first notable observation here is that the correlation between the simulated data from the RBC model, the DNK model with nominal rigidities and the FA model with financial frictions is very high for all variables except for hours worked. This is consistent with the evidence reported in Table 3 that shows the business cycle moments from the DNK model differ from those of the frictionless model (RBC) or those of the model augmented with financial frictions (FA) largely in regards to hours worked.

Absent other frictions in the model, the impact of the distortion caused by the presence of nominal rigidities (monopolistic competition and price stickiness) is absorbed primarily by the intratemporal margin provided by hours worked. For the Great Moderation period, the correlation of simulated hours worked between the DNK and RBC model stands at −0.219 while the correlation between the DNK and FA models is −0.194. However, the correlation between the simulated data from the DNK model and actual data is almost negligible at 0.015, while the correlations of the RBC and FA simulations with the actual data still have some information content at 0.227 and 0.235 respectively. This, in turn, translates into a weaker correlation between simulated and actual data on output for the DNK model than for the alternative RBC and FA variants.

A second notable observation is that the RBC, DNK and FA model simulations track better the data on inflation than on consumption and investment. These three models do worst matching the path of the actual data on hours worked and output, with the DNK model providing less information than the other two models for these variables. At its best, the correlation between actual and simulated output is merely 0.141 for the RBC model. Moreover, the FA model with financial frictions shows also a very poor result in tracking the external finance premium—as the correlation with the actual data is merely −0.058. However, the BGG model does generally worst than any of the other three models considered here and in most cases it generates a radically different path that moves in the opposite direction as the observed data. This happens, for instance, with consumption (where the correlation with the data is −0.510), with output (where the correlation stands at −0.216) and with hours worked (at −0.189). The correlation between the external finance premium in the BGG model and the data becomes positive, but it is still pretty low at 0.093 and strongly negatively correlated with the external finance premium simulated by the FA model at −0.712.

In other words, the accentuation of the financial distortion mechanism by the combined effect of TFP and monetary shocks when nominal rigidities are added to the BGG model to break away from monetary neutrality produces a large deviation from the dynamics to be expected from the frictionless RBC model or from the FA and DNK variants. This departure amplifies the volatility of the model (especially on investment and the external finance premium), and it introduces a strong motive to substitute away from consumption and into investment. This pattern, however, does not fit well with the evidence during the Great Moderation period as the data in Tables 4A and 4B clearly indicate.

Figure 5 plots the actual data for each one of the macro variables of interest—output, investment, consumption, hours worked, inflation and the external finance premium—together with the simulations from each one of the model variants I have considered in the paper. Aside from illustrating some of the distinctive features that I already noted based on the evidence reported in Tables 3, 4A and 4B for the Great Moderation period between 1984:I and 2007:IV, it also includes observations and simulations for the period of the 2007 recession and its aftermath up to 2012:I. The actual data is plotted based on the trends from the Great Moderation period ending in 2007:IV (black solid line) and also adjusted to account for the possibility of a level shift in 2009:II discussed earlier (black dashed line). Similarly, the simulated series are reported using a realization of the monetary shock inferred under the assumption that output continued evolving along the same trend that prevailed prior to the 2007 recession (black solid line) and using a realization of the monetary shock that is derived from the assumption that output experienced a level shift after 2009:II (black dashed line).

It is worth pointing out that the pattern of investment and hours worked generated by the BGG model (gray line) during the entire period is clearly dominated by the

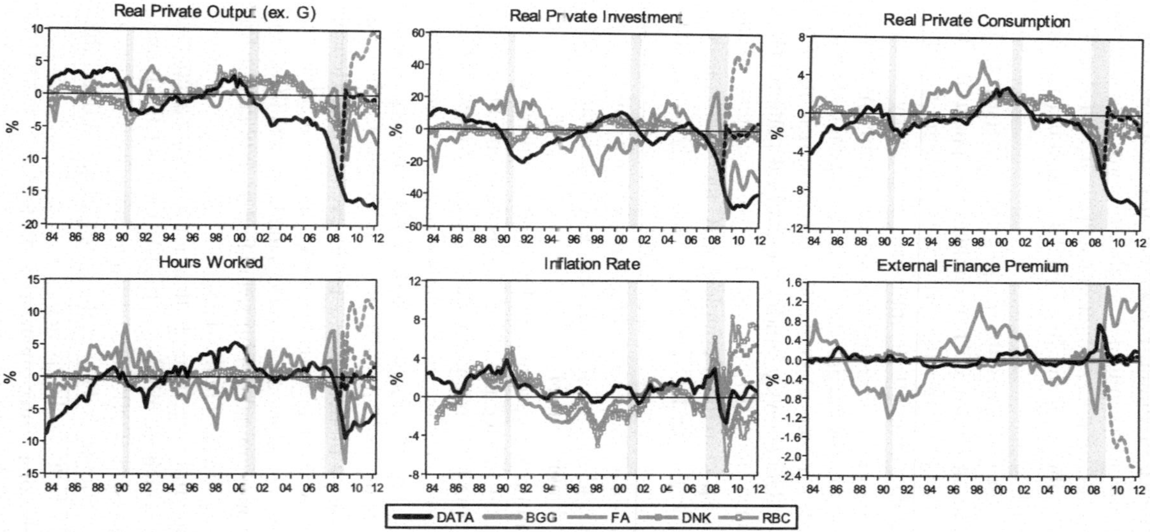

Fig. 5 Comparison across model variants: simulations vs. data.
This *graph plots in black* the detrended per capita real private output (ex. G) in logs, the detrended per capita real private investment in logs, the detrended per capita real private consumption in logs, the demeaned share of hours worked per capita in logs, the cyclical deviations of consumer (nondurables and services) price inflation in year-over-year terms relative to a 2 percent inflation target, and the demeaned Baa corporate quarterly spread over the 20-year Treasury Bill. I estimate the trends and means in the data for the Great Moderation period (1984:I–2007:IV), allowing for a level shift on 2009:II (except for the inflation variable). The *black solid line* represents the variables in deviations with respect to their Great Moderation trends, while the *black dashed line* indicates the variables in deviations accounding for a level shift in 2009:II. The *shaded areas* represent NBER recessions.
This *graph plots in solid gray* the simulation of the BGG model with nominal rigidities (price stickiness and monopolistic competition) and financial frictions, in *solid gray with circle markers* the FA model with flexible prices and perfect competition but with financial frictions, in *solid gray with square filled markers* the Dynamic New Keynesian (DNK) model without financial frictions, and in *solid gray with square outlined markers* the standard Real Business Cycle (RBC) model. The *dashed line* represents the simulation of the corresponding model whenever the monetary shocks are derived under a Taylor rule which responds to a measure of output that has incorporated the adjustment for a level shift in economic activity occuring in 2009:II. For more details on the derivation of the realization of the monetary shock, see Appendix B 'U.S. Dataset'. I use Matlab R2012a (7.14.0.739) and Dynare v4.3.2 for the stochastic simulation

long and deep swings in the external finance premium. These swings are strongly counter-cyclical—unlike the actual data—and much more sizeable than anything I observe in the quarterly interest rate spread between the Baa corporate bond yield and the 20-year Treasury bill rate. There is an evident comovement between these three simulated series (external finance premium, investment and hours worked) showing that whenever the external finance premium is "high", both investment and hours worked are "low". This, however, produces counterfactual predictions for all three variables that further limit the ability of the BGG model to account for the observations in the data, as noted earlier.

As expected, the RBC, DNK and FA variants of the model show a similar pattern among them when displayed together in Figure 5. Their simulations appear also to be quite different from the BGG simulation. In summary, the key margin that the financial accelerator model of Bernanke et al. (1999) introduces is given by the external finance premium which influences the investment path in the model. Without nominal rigidities as in the FA variant, the magnitude of the endogenous spread generated is somewhat lower than that observed in the data but the spread itself is strongly counter-cyclical while the data shows a mildly pro-cyclical pattern. The distortion that this margin adds to investment is rather small, so the discrepancies in the path of the real variables with respect to the frictionless RBC model are of second order importance and with respect to the DNK model they are only significant for hours worked.

With nominal rigidities as in the BGG model, monetary shocks have real effects and price stickiness distorts the demand allocation. Under the benchmark parameterization, this results in endogenous fluctuations of the external finance premium that are almost 10 times as volatile as in the FA case and still strongly counter-cyclical. The large movements of the external finance premium in this case end up dominating the evolution of investment (and, by extension, that of hours worked) while favoring a substitution away from consumption and towards investment. These features are all intimately connected, but generate simulated data that is largely at odds with the observed data—so generally the BGG model path is a worst one for the data than the path implied by the alternative models (RBC, FA or even DNK). Hence, a successful model of the business cycle that builds upon the financial mechanism of Bernanke et al. (1999) first and foremost needs to address the large amplification of the external finance premium under nominal rigidities and find an explanation for the mild procyclicality of the spread observed in the data.

One argument that has been forcefully discussed both among policymakers and scholars is that financial frictions—which are missing in the frictionless RBC framework or in the standard New Keynesian (DNK) model—may have been a crucial factor in the 2007 recession and its aftermath. One of the most popular models that accounts for the role of financial frictions explicitly in a general equilibrium setting is the Bernanke et al. (1999) model that I investigate here. In looking at the data after 2007 through this particular lens I explicitly take into account the possibility that a level shift may have occurred in the aftermath of the recession. I distinguish between actual or simulated series that are adjusted (dashed lines) and those that are not adjusted for such a level shift (solid line) in Figure 5 (as well as in Figures 6, 7 and 8 subsequently).

In any event, the BGG model turns out to display very large movements of the endogenous external finance premium in the post-2007 period that produce fluctuations in hours worked and investment that are hard to reconcile with the data. The model does well in tracking the consumption series adjusted for a level shift, but that is something that can also be attained with any of the alternative specifications. Hence, the shortcomings of the BGG model that make it insufficient to account for the business cycle fluctuations during the Great Moderation period also limit the insight that the framework can provide for the observed macro data in the U.S. since 2007.

These findings give some perspective and set the stage for a further exploration of the role of monetary policy and monetary policy shocks in the financial accelerator model of Bernanke et al. (1999).

4.2 Claim 1: The Role of Monetary Shocks

The benchmark financial accelerator model assumes price stickiness implying an average duration of each price spell of 4 quarters (i.e., $\alpha = 0.75$) and a parameterization of the sensitivity of the external finance premium implying that ceteris paribus a one percent increase in the leverage of borrowers raises the cost of external finance by almost 7 basis points per quarter (i.e., $\vartheta \equiv (\frac{\upsilon'(\gamma_n^{-1})\gamma_n^{-1}}{\upsilon(\gamma_n^{-1})}) = 0.0672$). The model is also endowed with a Taylor (1993)-type monetary policy rule as described in Eq. (20). All these features are viewed as consistent over the Great Moderation period with the empirical evidence available on the monetary policy regime and the features of the nominal rigidities and financial frictions.

In the experiment plotted in Figure 6, I look at the simulation of the benchmark financial accelerator model—the BGG model developed by Bernanke et al. (1999)—and compare it against the RBC model and a variant of the BGG model driven exclusively by TFP shocks in order to gauge the role played by monetary shocks in this environment. All three models are compared against the data to evaluate the strength of the quantitative findings.

In the frictionless RBC case, monetary neutrality ensures that all real variables are entirely driven by the realization of the TFP shock (gray line with square outlined markers). The BGG model incorporates nominal rigidities that introduce monetary non-neutrality, as I discussed earlier. The variant of the BGG model driven solely by TFP shocks (solid gray line with triangle markers) shows how the propagation of TFP differs relative to the RBC case whenever nominal rigidities and financial frictions are incorporated. The BGG model (solid gray line), in turn, introduces monetary shocks which have real effects as well. The BGG model, therefore, illustrates how the addition of monetary shocks further alters the dynamics of the macro variables of interest.

The monetary policy rule in all cases is described by Eq. (20) under the standard parameterization reported in Table 1. However, this requires some qualifications in the case of the BGG model driven exclusively by TFP shocks. In this particular

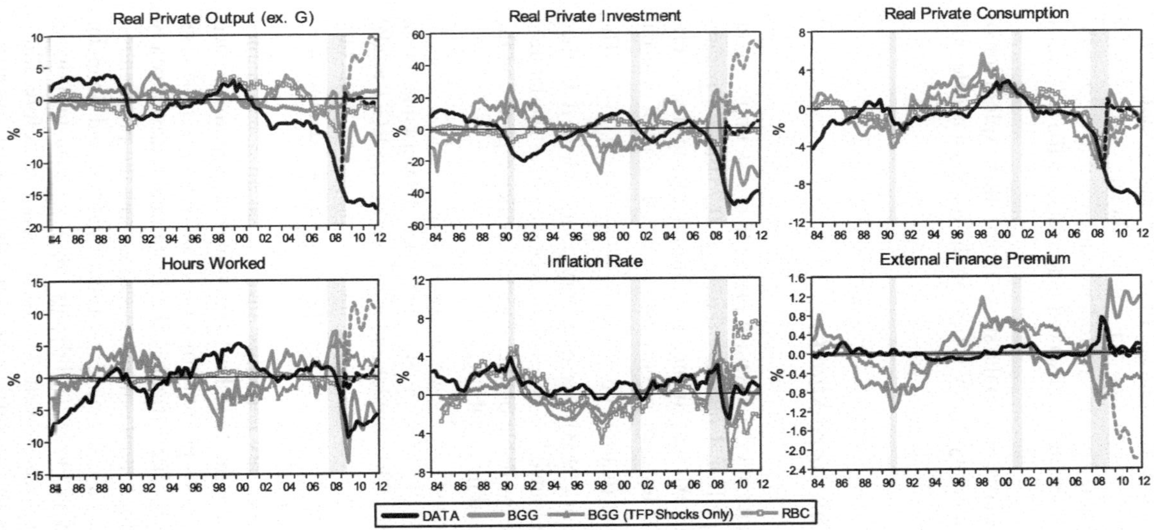

Fig. 6 Claim 1: the role of monetary shocks.
This *graph plots in black* the detrended per capita real private output (ex. G) in logs, the detrended per capita real private investment in logs, the detrended per capita real private consumption in logs, the demeaned share of hours worked per capita in logs, the cyclical deviations of consumer (nondurables and services) price inflation in year-over-year terms relative to a 2 percent inflation target, and the demeaned Baa corporate quarterly spread over the 20-year Treasury Bill. I estimate the trends and means in the data for the Great Moderation period (1984:I–2007:IV), allowing for a level shift on 2009:II (except for the inflation variable). The *black solid line* represents the variables in deviations with respect to their Great Moderation trends, while the *black dashed line* indicates the variables in deviations accounding for a level shift in 2009:II. The *shaded areas* represent NBER recessions.
This *graph plots in solid gray* the simulation of the BGG model with nominal rigidities (price stickiness and monopolistic competition) and financial frictions, in *solid gray with triangle markers* the same BGG model but driven solely by the realization of the TFP shock rather than by a combination of TFP and monetary shocks, and in *solid gray with square outlined markers* the standard Real Business Cycle (RBC) model. The *dashed line* represents the simulation of the corresponding model whenever the monetary shocks are derived under a Taylor rule which responds to a measure of output that has incorporated the adjustment for a level shift in economic activity occuring in 2009:II. For more details on the derivation of the realization of the monetary shock, see Appendix B 'U.S. Dataset'. I use Matlab R2012a (7.14.0.739) and Dynare v4.3.2 for the stochastic simulation

variant, the specification of monetary policy involves the joint assumption that the central bank never deviates from the given Taylor rule and that all agents know (and believe) that the central bank is not going to deviate from that policy rule. Therefore, the comparison of the BGG model with and without monetary policy shocks provides further insight on the business cycle contributions from a combination of monetary policy deviations and TFP shocks beyond what can be accounted for with TFP shocks alone. To be precise, this does not isolate the effect of the monetary policy shocks but it shows how different would the endogenous dynamics of the model be including monetary shocks compared to the case where business cycles are entirely driven by TFP shocks.

It can be argued on the basis of the findings that I report here that the financial accelerator mechanism together with nominal rigidities has a strong amplification effect over the business cycle that is accentuated when I combine monetary and TFP shocks. Interestingly, I find that during the entire Great Moderation period and even in the 2007 recession, the external finance premium simulated by the model solely driven by TFP shocks tends to be more volatile and counter-cyclical than the actual spreads observed in the data—although it also noticeably differs from the endogenous spread under the BGG specification that combines both monetary and TFP shocks. It is interesting to see how smooth the simulated output series is relative to the standard BGG case with both monetary and TFP shocks, and nonetheless how different the path is relative to the standard frictionless RBC case.

By and large, the external finance premium in the variant of the BGG model without monetary shocks follows the same path as in the standard BGG model that includes monetary shocks as well. The two notable exceptions correspond to periods where monetary policy appears to have largely deviated from the Taylor rule prescription: the period of low interest rates between 2002 and 2006 that is often regarded as the build-up period for the 2007 recession and the period of interest rates at the zero-lower band since 2009:I.

In the early part of the 1980s, monetary policy deviations are larger in size—with the Federal Funds rate above the Taylor-implied rate—which neither variant of the model considered here is well-suited to account for. The deviations of monetary policy that I detect in the data after 1987 (as seen in Figure 4) are rather modest in size during most of chairman Greenspan's tenure at the Fed until 2002. In spite of that, the differences between the BGG model and the BGG model without monetary shocks are of first-order importance during those years (between 1987 and 2002).

The period of low interest rates between 2002 and 2006 is well-known for the size of the policy deviations—with the Federal Funds rate well below the Taylor-implied rate. It is precisely during this time that there is strong evidence of divergence between the BGG model and the BGG variant driven by TFP shocks alone—where the difference translates into a "high" spread (positive in deviations from its mean) if I look at the BGG model with TFP shocks alone and a "low" spread (negative in deviations from its mean) in the standard BGG case.

The other period of low interest rates corresponds with the time of policy rates set at the zero-lower band (since 2009). I explicitly allow for the possibility of a level shift in output in the calculation of the monetary policy deviations since 2009:II, but this is only of relevance for the simulation of the BGG model when it includes monetary shocks as the BGG model with TFP shocks alone would be unaffected for obvious reasons. A significant discrepancy in the derived external finance premium and, by extension, on the dynamics of the economy emerges during this period nonetheless.

Neither of the instances of large monetary policy deviations reviewed here, though, suggests that combining monetary and TFP shocks helps improve the ability of the BGG model to capture the patterns observed in the data. These results highlight some of the inherent weaknesses of the financial accelerator model of Bernanke et al. (1999), but also indicate that neither variant of the model is capable of successfully explaining the turn of events during the 2007 recession.

4.3 Claim 2: The Role of Nominal Rigidities

As monetary non-neutrality in the standard New Keynesian (DNK) model rests on the assumption of nominal rigidities, the question arises as to how important is this feature for the strength of the financial accelerator mechanism in the BGG set-up. The simulations reported in this sub-section are all based on the same Taylor (1993) specification of monetary policy described in Eq. (20). Here, I compare the BGG model (gray line)—where the average duration of a pricing spell is of 4 quarters—against a BGG variant with lower nominal rigidities (gray line with cross markers)—where the average duration of a pricing spell is set at 2 quarters—and against the FA model specification (gray line with square filled markers) that abstracts entirely from nominal rigidities (that is, from price stickiness and monopolistic competition). All model simulations are still driven by the same combination of TFP and monetary policy shocks.

As can be seen in Figure 7, nominal rigidities play a crucial role in the dynamics of the BGG model. Reducing the degree of Calvo price stickiness from $\alpha = 0.75$ (four quarters average duration) to $\alpha = 0.5$ (two quarters average duration) alone dramatically reduces the magnitude of the fluctuations in the external finance premium to levels that are comparable with those observed in the quarterly interest rate spread between the Baa corporate bond yield and the 20-year Treasury bill rate. It is interesting to note, however, that the simulations of the BGG model with the benchmark parameterization of price stickiness and with low price stickiness are nonetheless highly correlated. In turn, the FA model that abstracts entirely from price stickiness and monopolistic competition but maintains the financial friction generates a significantly different external finance premium which is weakly (and negatively) correlated with the data, but strongly and negatively correlated with the BGG simulations under standard or low price stickiness. This can also be seen for the benchmark parameterization of the BGG model in the results of Table 4B.

The difference between the two versions of the BGG model and the FA model plotted in Figure 7 is that the FA model preserves monetary neutrality (and therefore spreads are entirely driven by TFP shocks) while the BGG model variants do

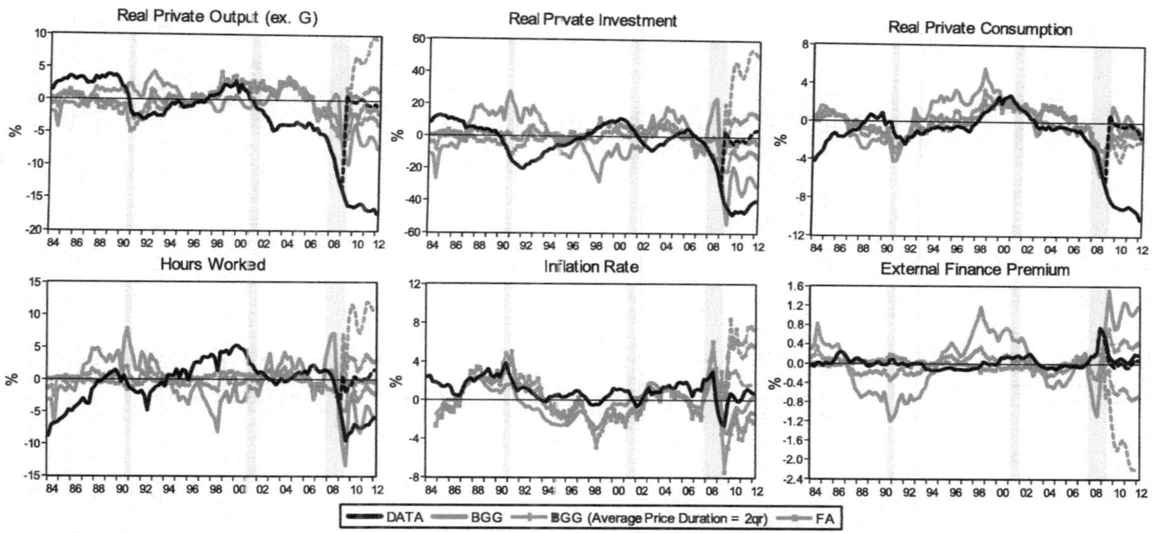

Fig. 7 Claim 2: the role of nominal rigidities.
This *graph plots in black* the detrended per capita real private output (ex. G) in logs, the detrended per capita real private investment in logs, the detrended per capita real private consumption in logs, the demeaned share of hours worked per capita in logs, the cyclical deviations of consumer (nondurables and services) price inflation in year-over-year terms relative to a 2 percent inflation target, and the demeaned Baa corporate quarterly spread over the 20-year Treasury Bill. I estimate the trends and means in the data for the Great Moderation period (1984:I–2007:IV), allowing for a level shift on 2009:II (except for the inflation variable). The *black solid line* represents the variables in deviations with respect to their Great Moderation trends, while the *black dashed line* indicates the variables in deviations accounding for a level shift in 2009:II. The *shaded areas* represent NBER recessions.
This *graph plots in solid red* the simulation of the BGG model with nominal rigidities (price stickiness implying an average price duration of 4 quarters and monopolistic competition) and financial frictions, in *solid gray with cross markers* the same model but assuming a lower degree of price stickiness that results in an average price duration of just 2 quarters, and in *solid gray with square filled markers* the FA model with flexible prices and perfect competition but with financial frictions. The *dashed line* represents the simulation of the corresponding model whenever the monetary shocks are derived under a Taylor rule which responds to a measure of output that has incorporated the adjustment for a level shift in economic activity occuring in 2009:II. For more details on the derivation of the realization of the monetary shock, see Appendix B 'U.S. Dataset'. I use Matlab R2012a (7.14.0.739) and Dynare v4.3.2 for the stochastic simulation

not. The evidence, therefore, reveals that the magnitude and cyclical patterns of the external finance premium depend nonlinearly on whether nominal rigidities are included in the model. One may be tempted to argue that the strong negative correlation between the spread simulated by the FA model and the spread derived from the two versions of the BGG model is attributable to monetary shocks—as those have real effects and influence the spread in the BGG case with price stickiness, but not in the FA case. However, as it was discussed in the previous sub-section (in regards to Figure 6), the contribution of monetary shocks alone cannot explain everything. In fact, the distortion of the demand allocation resulting from nominal rigidities produces a similar path for the spread even when TFP shocks are the sole drivers of the cycle.

Hence, what really matters the most is the size of the demand distortion that price stickiness produces under the prevailing policy rule. The evidence illustrated through Figure 7 suggests that a range of plausible values for the Calvo price stickiness parameter α that sets the average price duration between 2 and 4 quarters can nonetheless generate very significant differences in the simulated path of the macro variables of interest. The smaller the parameter α (the shorter the average price duration), the smaller the fluctuations of the external finance premium and the smaller the impact those have on investment. As a result, the effects of the financial mechanism become rapidly similar to those of the FA model without price stickiness for output, investment, consumption and even hours worked. More broadly one could argue that a parameterization of the BGG model that is consistent with the fluctuations of the spreads actually observed in the data is likely to assign a much more modest role to fluctuations of the external finance premium in explaining the cyclical patterns of the macro variables of interest than that found under the benchmark BGG model.

One explanation often postulated for the 2007 recession is that borrowing costs may have substantially increased since the onset of the 2007 recession due to constraints on loan supply that resulted from the concurrent banking crisis (the bank lending channel may indeed have been impaired) or due to the increased burden of financial regulation. I explicitly account for that possibility in the data by allowing for a level shift to have occurred in 2009:II. The BGG model, even with low price stickiness, seems unable to match the path of the observed spread better than the FA model—but both models have shortcomings in explaining output, investment and hours worked.

One could argue on the basis of these observations that a change in the specification of the financial friction that introduces exogenous shocks to the external finance premium may help overcome the BGG model's apparent inability to account for the 2007 recession. In order to understand the 2007 recession, however, something more than augmenting the specification of the model with shocks is needed—a more fundamental question must be addressed first. Are the increases in borrowing spreads documented in the data (see Figure 1) better thought of as endogenous responses, or can they be modelled as random exogenous shocks to the spreads—or to the supply of loans from deposits—that are unpredictable for the economic agents? If one goes through the route of endogenizing the bank lending channel, then an extension of the Bernanke et al. (1999) framework is clearly needed. That work is left for future research.

4.4 *Claim 3: The Role of the Monetary Policy Rule*

Another thought experiment that one could consider is whether the measurement of inflation and output in the Taylor (1993) monetary policy rule has any bearing on the economic impact of following that policy rule. That is the purpose of the simulations illustrated in Figure 8. There are many ways in which this general question could be addressed, but I decided to restrict myself to just two very specific issues of measurement in the specification: whether measuring inflation in terms of year-over-year growth rates or annualized quarter-over-quarter rates matters; and whether to use deviations of output from trend or deviations of output from its frictionless potential (i.e., the output gap) matters. Most of the theoretical literature, after all, describes the reaction function of policymakers to inflation in terms of quarter-over-quarter rates and to output fluctuations in terms of the output gap—while part of the empirical literature on Taylor rules, including Taylor (1993) himself, looks at responses in terms of year over year inflation rates and detrended output.

In all simulations I have plotted so far in Figures 5–7, the implicit assumption is that the U.S. monetary policy targets the year-over-year growth rate and detrended output as explicitly stated in Eq. (20). I report in Figure 8 the simulation of the BGG model (gray line) under that specification of the Taylor rule for reference. But, then, how do I evaluate the BGG model under alternative specifications of the policy rule that may differ on the measurement of inflation or the measurement of output fluctuations?

One way to address the importance of using annualized quarter-over-quarter growth rates instead of year-over-year growth rates is by re-estimating the Taylor rule residuals under this alternative inflation rate measure in the specification, assuming that all economic agents know and believe that the response to inflation is set in terms of quarter-over-quarter rates. Then I could simulate the model again, but feeding these Taylor rule deviations derived under the alternative measure of inflation into the corresponding policy functions that solve the linearized rational expectations equilibrium of the model. The disadvantage of following this route is that it obscures the exact contribution of the measure of inflation to the dynamics since the simulation of the endogenous variables would jointly reflect the change in the inflation rate used to set the monetary policy target under the rule as well as a different shock process (and realization) for the exogenous monetary policy deviations. And then, how does one disentangle the contribution of one from the other?

A alternative approach could be followed to compare a policy rule specification that responds to the output gap instead of detrended output. Instead of following that route, I undertake here a much more modest thought-experiment. I will simply take as given the monetary policy shock process and assume it is exactly the same one I

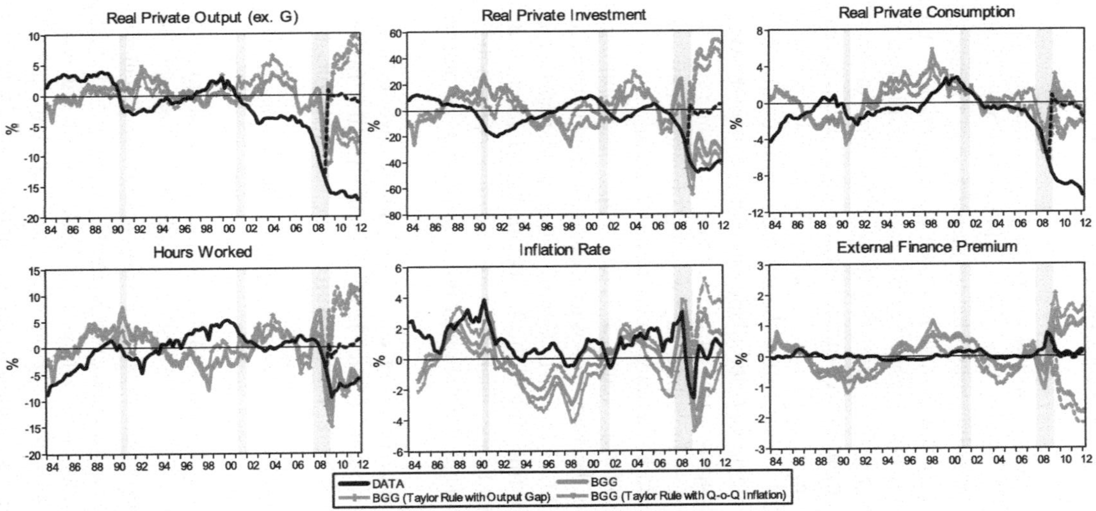

Fig. 8 Claim 3: the role of the monetary policy rule.
This *graph plots in black* the detrended per capita real private output (ex. G) in logs, the detrended per capita real private investment in logs, the detrended per capita real private consumption in logs, the demeaned share of hours worked per capita in logs, the cyclical deviations of consumer (nondurables and services) price inflation in year-over-year terms relative to a 2 percent inflation target, and the demeaned Baa corporate quarterly spread over the 20-year Treasury Bill. I estimate the trends and means in the data for the Great Moderation period (1984:I–2007:IV), allowing for a level shift on 2009:II (except for the inflation variable). The *black solid line* represents the variables in deviations with respect to their Great Moderation trends, while the *black dashed line* indicates the variables in deviations accounding for a level shift in 2009:II. The *shaded areas* represent NBER recessions.
This *graph plots in solid gray* the simulation of the BGG model with nominal rigidities (price stickiness and monopolistic competition) and financial frictions under the Taylor (1993) rule that reacts to year-over-year changes in inflation and the detrended output, in *solid gray with cross markers* the same model under an alternative specification of the Taylor rule that reacts to year-over-year inflation and the output gap, and in *solid gray with inverted triangle markers* the same model under the alternative Taylor rule that reacts to quarter-over-quarter annualized inflation and detrended output. The *dashed line* represents the simulation of the corresponding model whenever the monetary shocks are derived under a Taylor rule which responds to a measure of output that has incorporated the adjustment for a level shift in economic activity occuring in 2009:II. For more details on the derivation of the realization of the monetary shock, see Appendix B 'U.S. Dataset'. I use Matlab R2012a (7.14.0.739) and Dynare v4.3.2 for the stochastic simulation

have used thus far. Then I simulate the model under the assumption that monetary policy reacts to quarter-over-quarter annualized inflation rates and detrended output but with the same realization of the monetary shock (gray line with inverted triangle markers). Similarly, I take as given the monetary policy shock process and simulate the model under the assumption that monetary policy reacts to year-over-year inflation rates and the output gap with the same realization of the monetary shock (gray line with cross markers). This is a counterfactual exercise, but it gives me a sensible quantification of the impact that a change of this type in the monetary policy rule may have had.

Otherwise, the financial accelerator model that I simulate in all these variants corresponds exactly to the same parameterization that I describe in Table 1. One would conjecture that such a seemingly small change in the monetary policy rule specification cannot have major implications for the dynamics of the economy. The surprising thing is that just the opposite happens to be true, but not for all measurement changes considered. The plots in Figure 8 illustrate that the benchmark BGG model where monetary policy responds to year-over-year inflation, in fact, overlaps for the most part with the variant of the BGG model where monetary policy responds to quarter-over-quarter annualized inflation. Discrepancies between both simulations are only sizeable for the (year-over-year) inflation rate itself, but have only a marginal impact on all real variables.

In turn, the economic consequences of responding to the output gap instead of detrended output (as in the benchmark specification set in (20)) are clearly of first-order importance. I examine here the economy's response to two exogenous shocks—a monetary policy shock and a TFP shock—when the central bank follows the interest rate rule in (20) with output gap instead of detrended output on the right-hand side. Since output potential in a frictionless RBC environment under monetary neutrality is unaffected by the monetary policy shock, the response of output to monetary shocks matches that of the output gap. However, the same cannot be said for the productivity shocks. A positive TFP shock leads to a persistent decline in both inflation and the output gap, but increases output given the benchmark parameterization of the model (including the interest rate rule parameters).

Hence, whether the monetary policy rule responds to output or the output gap clearly matters when TFP shocks are one of the main drivers of business cycles because output and the output gap would tend to move in opposing directions. While the economic impact of both rule specifications is certainly important (as can be seen from Figure 8), the path of the macro variables of interest tends to be positively correlated under both specifications since the onset of the Great Moderation. As a result, it cannot be said that one specifications provides a clearly superior match for the observed data than the other one.

This counterfactual exercise is just one experiment on a broader set of questions about the role of monetary policy. While most of my previous experiments have been based on the interpretation of the discretionary component of monetary policy and how it is propagated in the presence of nominal rigidities, this counterfactual exercise comes to show that the systematic part of the policy rule can indeed have a major impact on the performance of the economic model. It also shows that issues like the proper measurement of output deviations used in the policy rule can—in turn—be fundamental for the outcome of the model.

5 Concluding Remarks

I investigate a synthesis of the Bernanke et al. (1999) model—the BGG model—with leveraged borrowers (entrepreneurs), financial frictions, and nominal rigidities. I also consider three economically-relevant variants of this framework: the Real Business Cycle (RBC) model without nominal rigidities or financial frictions, the Dynamic New Keynesian (DNK) model with nominal rigidities but no financial frictions, and the Financial Accelerator (FA) model with financial frictions but without nominal rigidities. I parameterize the model to be consistent with Bernanke et al. (1999) and with the available data for the U.S. during the Great Moderation period.

In mapping the model to the data, I linearly detrend the nonstationary variables—output, investment and consumption—and demean or express in deviations the others—share of hours worked, inflation and quarterly interest rate spread—based on their Great Moderation trends but allowing for the possibility of a level shift in the aftermath of the 2007 recession. I also derive a realization of the detrended TFP shock and the monetary shock from U.S. data that I subsequently use to simulate U.S. business cycles with the different variants of the Bernanke et al. (1999) model over the Great Moderation period (from 1984:I until 2007:IV) and since the 2007 recession (2008:I–2012:I). I evaluate the performance of each model by comparing the business cycle features it generates and its fit against the observed data.

On the basis of these simulations, I argue that the characterization of the reaction function of monetary policy has non-trivial implications for the performance of the model and that the interpretation of all monetary policy deviations as exogenous shocks is anything but trivial. I also find that the degree of price stickiness is crucial for the accentuation of the economic impact of financial frictions, but that the interaction between these two frictions works nonlinearly for plausible values implying an average duration of prices between 2 and 4 quarters. The evidence reported suggests very stark differences in the dynamics implied by the BGG model at both ends of that range.

However, I find otherwise limited support in favor of the financial accelerator model as a superior framework to account for the U.S. business cycle during the Great Moderation period and—especially—during the 2007 recession and its aftermath. In fact, in some dimensions it becomes clear that a plain-vanilla RBC model (or the FA variant that includes the financial friction but no nominal rigidities) gets closer to accounting for the endogenous variables observed in the data than the model of Bernanke et al. (1999) does. In a nutshell, the problem with the BGG model is that it generates fluctuations of the external finance premium that are too large relative to the data under the standard parameterization of price stickiness. These fluctuations of the external finance premium largely drive investment and hours worked.

It is ultimately the amplitude of the external finance premium movements over the business cycle that explains the wedge between the BGG model and the other nested alternatives—the RBC, the DNK and the FA models. The BGG model, however, also implies that the external finance premium ought to be strongly counter-cyclical. The large and counter-cyclical fluctuations in the premium generate, in turn, a strong incentive to substitute resources away from consumption and into investment in periods when the spread is "low" and the cost of funding investment is "cheaper". This tends to produce counter-cyclical consumption patterns on top of counter-cyclical spreads, both of which run contrary to the observed patterns in the U.S. data.

One can look at these broad results in two different ways. One can take the view that they cast the implications of the financial accelerator model of Bernanke et al. (1999) in a slightly less positive light and, therefore, that the BGG model is perhaps not yet ready for policy evaluation and analysis of the business cycles at the level we would like it to be. That is a reasonable reading of my results, but I would argue that it is still premature to claim on the basis of quantitative findings like the ones presented here that the financial accelerator mechanism is incompatible with the data or that it should be discarded altogether.

Another more sympathetic view would be that—indeed—there is a financial friction at play and it is important to account for it. The puzzle is, therefore, worse than it is conventionally thought because some source of randomness not accounted for or other features of the structural transmission mechanism that have not been explicitly modelled are still needed in order to bridge the gap between the model and the data. The problem could also be that monetary policy itself and monetary policy shocks in particular are not well-understood in this framework—e.g., if one estimates that the monetary policy regime may have shifted over time or doubts the extent to which policymakers have incorporated the possibility of a level shift in output in their calculations of the optimal target rate under Taylor (1993).

While the latter argument ends up creating more questions than it actually answers, the true fact is that more work needs to be done to better understand the role that financial frictions play on real economic activity and their interactions with monetary policy and with other frictions. My hope is that this paper will not be viewed as a closing chapter on the subject or on the model itself, but as an attempt to spur further interest towards a more quantitative evaluation of these questions and to encourage further development and integration of financial features in general equilibrium models.

Acknowledgements I thank Tor-Erik Bakke, Charles T. Carlstrom, Simona Cociuba, John V. Duca, Evan F. Koenig, María Teresa Martínez-García, Mark A. Wynne, Carlos Zarazaga, and many participants at the XXII Symposium Moneda y Crédito, 2010 Midwest Macro Meetings, 2010 WEAI meetings, 2010 Southern Economic Association meetings, 2010 SAEe meetings, 2011 EEA-ESEM Congress, Bowling Green State University, Kansas University, and Federal Reserve Banks of San Francisco, Cleveland and Dallas for their helpful suggestions. I also thank ZEW for their encouragement through the 2012 workshop on "Non-linear Economic Modeling: Theory and Applications". I gratefully acknowledge the research assistance provided by Valerie Grossman and Janet Koech, and the Federal Reserve Bank of Dallas's support. This research was initiated while part-time visiting the Department of Economics at the University of North Texas (UNT), whose hospitality is greatly appreciated. Ethan Cohen-Cole was involved in its inception, and his advice has been invaluable. All remaining errors are mine alone. The views expressed do not necessarily reflect those of the Federal Reserve Bank of Dallas, or the Federal Reserve System.

Appendix A: The Log-Linearized Model

As a notational convention, all variables identified with lower-case letters and a caret on top are expressed in logs and in deviations relative to the their steady state values.

A.1 The Financial Accelerator Model

Aggregate Demand Equations

$$\widehat{y}_t \approx (1-\gamma_x-\gamma_{c^e})\widehat{c}_t+\gamma_x\widehat{x}_t+\gamma_{c^e}\widehat{n}_{t+1},$$

$$\gamma_x \equiv \delta\left(\frac{\psi}{\mu(\upsilon(\gamma_n^{-1})\beta^{-1}-(1-\delta))}\right),$$

$$\widehat{c}_t \approx \mathbb{E}_t[\widehat{c}_{t+1}]-\sigma\widehat{r}_{t+1},$$

$$\mathbb{E}_t\left[\widehat{r}^k_{t+1}\right] \approx \widehat{r}_{t+1}+\vartheta(\widehat{q}_t+\widehat{k}_{t+1}-\widehat{n}_{t+1}), \qquad \vartheta \equiv \left(\frac{\upsilon'(\gamma_n^{-1})\gamma_n^{-1}}{\upsilon(\gamma_n^{-1})}\right),$$

$$\widehat{r}^k_t \approx (1-\epsilon)(\widehat{mc}_t+\widehat{y}_t-\widehat{k}_t)+\epsilon\widehat{q}_t-\widehat{q}_{t-1}, \qquad \epsilon \equiv \left(\frac{1-\delta}{\upsilon(\gamma_n^{-1})\beta^{-1}}\right),$$

$$\widehat{q}_t \approx \chi(\widehat{x}_t-\widehat{k}_t).$$

Aggregate Supply Equations

$$\widehat{y}_t \approx \widehat{a}_t+\psi\widehat{k}_t+(1-\psi-\varrho)\widehat{h}_t,$$

$$\widehat{mc}_t \approx \frac{1}{\varphi}\widehat{h}_t+\frac{1}{\sigma}\widehat{c}_t-(\widehat{y}_t-\widehat{h}_t), \qquad \varphi \equiv \eta\left(\frac{1-H}{H}\right),$$

$$\widehat{\pi}_t \approx \beta\mathbb{E}_t[\widehat{\pi}_{t+1}]+\left(\frac{(1-\alpha\beta)(1-\alpha)}{\alpha}\right)\widehat{mc}_t.$$

Law of Motion for State Variables

$$\widehat{k}_{t+1} \approx (1-\delta)\widehat{k}_t+\delta\widehat{x}_t,$$

$$\widehat{n}_{t+1} \approx \left(\zeta\beta^{-1}\gamma_n^{-1}\right)\left(\widehat{r}^k_t-\widehat{r}_t\right)+\widehat{r}_t+\widehat{n}_t+\cdots$$
$$\left(\upsilon(\gamma_n^{-1})-1\right)\gamma_n^{-1}\left(\widehat{r}^k_t+\widehat{q}_{t-1}+\widehat{k}_t\right)+\cdots$$
$$\frac{\varrho}{\psi}\left(\upsilon(\gamma_n^{-1})\beta^{-1}-(1-\delta)\right)\gamma_n^{-1}\widehat{y}_t+\widehat{mc}_t.$$

Monetary Policy Rule

$$
\begin{aligned}
\widehat{r}_{t+1} &\equiv \widehat{i}_{t+1} - \mathbb{E}_t[\widehat{\pi}_{t+1}],\\
4\widehat{i}_{t+1} &\approx \rho_i 4\widehat{i}_t + (1-\rho_i)[\phi_\pi \widetilde{\pi}_t + \phi_y \widetilde{y}_t] + \widehat{m}_t,\\
\widetilde{\pi}_t &\approx \begin{cases} (\widehat{p}_t - \widehat{p}_{t-4}) = \widehat{\pi}_t + \widehat{\pi}_{t-1} + \widehat{\pi}_{t-2} + \widehat{\pi}_{t-3},\\ 4\widehat{\pi}_t, \end{cases}\\
\widetilde{y}_t &\approx \begin{cases} \widehat{y}_t,\\ \widehat{x}_t \equiv \widehat{y}_t - \widehat{y}_t^F. \end{cases}
\end{aligned}
$$

A.2 The Frictionless Model

Aggregate Demand Equations

$$
\begin{aligned}
\widehat{y}_t^F &\approx (1-\gamma_x^F)\widehat{c}_t^F + \gamma_x^F \widehat{x}_t^F, \qquad \gamma_x^F \equiv \delta\left(\frac{\psi}{\beta^{-1}-(1-\delta)}\right),\\
\widehat{c}_t^F &\approx \mathbb{E}_t\left[\widehat{c}_{t+1}^F\right] - \sigma \widehat{r}_{t+1}^F,\\
\mathbb{E}_t\left[\widehat{r}_{t+1}^{kF}\right] &\approx \widehat{r}_{t+1}^F,\\
\widehat{r}_t^{kF} &\approx (1-\epsilon^F)\left(\widehat{mc}_t^F + \widehat{y}_t^F - \widehat{k}_t^F\right) + \epsilon^F \widehat{q}_t^F - \widehat{q}_{t-1}^F, \qquad \epsilon^F \equiv \left(\frac{1-\delta}{\beta^{-1}}\right),\\
\widehat{q}_t^F &\approx \chi\left(\widehat{x}_t^F - \widehat{k}_t^F\right).
\end{aligned}
$$

Aggregate Supply Equations

$$
\begin{aligned}
\widehat{y}_t^F &\approx \widehat{a}_t + \psi \widehat{k}_t^F + (1-\psi)\widehat{h}_t^F,\\
\widehat{mc}_t^F &\approx \frac{1}{\varphi}\widehat{h}_t^F + \frac{1}{\sigma}\widehat{c}_t^F - \left(\widehat{y}_t^F - \widehat{h}_t^F\right), \quad \varphi \equiv \eta\left(\frac{1-H}{H}\right),\\
\widehat{mc}_t^F &\approx 0.
\end{aligned}
$$

Law of Motion for State Variables

$$
\widehat{k}_{t+1}^F \approx (1-\delta)\widehat{k}_t^F + \delta \widehat{x}_t^F.
$$

Monetary Policy Rule

$$
\begin{aligned}
\widehat{r}_{t+1}^F &\equiv \widehat{i}_{t+1}^F - \mathbb{E}_t\left[\widehat{\pi}_{t+1}^F\right],\\
4\widehat{i}_{t+1}^F &= \rho_i 4\widehat{i}_t^F + (1-\rho_i)\left[\phi_\pi \widetilde{\pi}_t^F + \phi_y \widetilde{y}_t^F\right] + \widehat{m}_t,\\
\widetilde{\pi}_t^F &\approx \begin{cases} (\widehat{p}_t^F - \widehat{p}_{t-4}^F) = \widehat{\pi}_t^F + \widehat{\pi}_{t-1}^F + \widehat{\pi}_{t-2}^F + \widehat{\pi}_{t-3}^F,\\ 4\widehat{\pi}_t^F, \end{cases}\\
\widetilde{y}_t^F &\approx \begin{cases} \widehat{y}_t^F,\\ \widehat{x}_t^F \equiv \widehat{y}_t^F - \widehat{y}_t^F = 0. \end{cases}
\end{aligned}
$$

A.3 Shock Processes

$$\widehat{a}_t = \rho_a \widehat{a}_{t-1} + \varepsilon_t^a, \qquad \varepsilon_t^a \sim N\left(0, \sigma_a^2\right),$$

$$\widehat{m}_t = \rho_m \widehat{m}_{t-1} + \varepsilon_t^m, \qquad \varepsilon_t^m \sim N\left(0, \sigma_m^2\right).$$

Appendix B: U.S. Dataset

B.1 Macro Aggregates

I adapt the work of Cociuba et al. (2009) to construct the quarterly series on hours worked and that of Gomme and Rupert (2007) to derive the quarterly measures of the U.S. stock of physical capital, U.S. investment, U.S. Total Factor Productivity (TFP) and other macro aggregates. The calculations of U.S. TFP are based on standard technological constraints which I summarize as follows,

Production Function: $Y_t = A_t (K_t)^{\alpha} (\gamma^t H_t)^{1-\alpha}$,
Aggregate TFP: $A_t = A_{t-1}^{\rho} e^{\varepsilon_t^a}$, $|\rho| < 1$, A_0 given, $\varepsilon_t^a \sim N(0, \sigma_a^2)$,
Law of Motion for Capital in Structures: $K_{st+1} = (1 - \delta_{st}) K_{st} + X_{st}$, K_{s0} given,
Law of Motion for Capital in Equipment and Software: $K_{et+1} = (1 - \delta_{et}) K_{et} + X_{et}$, K_{e0} given,
Law of Motion for Capital in Housing: $K_{ht+1} = (1 - \delta_{ht}) K_{ht} + X_{ht}$, K_{h0} given,
Total Capital: $K_{t+1} = K_{st+1} + K_{et+1} + K_{ht+1}$,
Total Investment: $X_{t+1} = X_{st+1} + X_{et+1} + X_{ht+1}$,

where H_t denotes total hours worked, K_t is the total stock of physical capital, X_t is the total investment in physical capital, Y_t is real private output excluding government wages and salaries, and A_t is an aggregate TFP process that is thought to be stationary. I disaggregate capital and investment in three types: K_{st} is capital in market structures and X_{st} its corresponding real investment; K_{et} is capital in equipment and software and X_{et} real investment on equipment and software; K_{ht} is housing capital and X_{ht} is housing investment in real terms. I also consider labor-augmenting technological progress with a deterministic growth rate of γ.

The parameter δ_{it} denotes the time-varying depreciation rate of capital by type $i \in \{s, e, h\}$. Real investment X_{it} is computed by deflating the nominal aggregate investment series by the consumption (nondurables and services) deflator, i.e.,

$$X_{it} \equiv (\gamma_q)^t Q_{it} I_{it}, \quad \text{for all } i \in \{s, e, h\},$$

where the relative price of investment in units of consumption grows at the deterministic rate of γ_q and its stationary component is denoted Q_{it} for each capital type $i \in \{s, e, h\}$. While current investment uses I_{it} units of current real private output per type of capital i, it yields X_{it} units of capital for production. In this sense, the relative price of investment $(\gamma_q)^t Q_{it}$ reflects the current state of the technology for producing the different types of capital. I recognize that capital depreciation δ_{it} and the relative price of investment $(\gamma_q)^t Q_{it}$ can differ across capital types, so I construct the stock of capital K_t by adding up its components (K_{st}, K_{et}, K_{ht}). In this situation, the disaggregated capital types are treated as perfect substitutes in obtaining the total stock of physical capital.

Since the model abstracts from population changes, then output, capital, investment and total hours worked should be expressed in per capita terms for consistency. I denote the population size as L_t and define the variables of interest in per capita terms as,

$$y_t \equiv \frac{Y_t}{L_t}, \qquad k_t \equiv \frac{K_t}{L_t}, \qquad x_t \equiv \frac{X_t}{L_t}, \qquad h_t \equiv \frac{H_t}{L_t},$$

$$x_{st} \equiv \frac{X_{st}}{L_t}, \qquad x_{et} \equiv \frac{X_{et}}{L_t}, \qquad x_{ht} \equiv \frac{X_{ht}}{L_t},$$

$$k_{st} \equiv \frac{K_{st}}{L_t}, \qquad k_{et} \equiv \frac{K_{et}}{L_t}, \qquad k_{ht} \equiv \frac{K_{ht}}{L_t}.$$

An implication of the Cobb-Douglas production function is that the specification admits a per-capita representation,

$$y_t = A_t (k_t)^{\alpha} \left(\gamma^t h_t\right)^{1-\alpha},$$

$$A_t = A_{t-1}^{\rho} e^{\varepsilon_t^a}$$

so the TFP measure A_t is unaffected by the population adjustment. Other endogenous variables of the model such as consumption $c_t \equiv \frac{K_t}{L_t}$ are also expressed in per capita terms, while prices such as the interest rates or the consumer price index (CPI) do not admit a representation in per capita terms.

The strategy to recover the Solow residual (measured TFP) is to: (a) calculate total hours worked per capita, h_t; (b) reconstruct the stock of total physical capital in per capita terms, k_t, with the perpetual inventory method using the aggregate investment series (X_{st}, X_{et}, X_{ht}) given the vector of depreciation rates $(\delta_s, \delta_e, \delta_h)$ and the population series L_t; and (c) identify aggregate TFP from the production function as the part of real private output excluding government wages and salaries in per capita terms, y_t, that cannot be accounted for by the aggregate factors of production, h_t and k_t, given the capital income share α. In the process to calculate the Solow residual, I also derive the relevant macro aggregates for real private output per capita, consumption per capita, investment per capita, hours worked per capita, and inflation as discussed here.

B.1.1 Average Hours Worked and Working-Age Population

Working-Age Population Source: U.S. Department of Labor, Bureau of Labor Statistics (BLS). Employment and Earnings—Household Survey (Tables A-13 and A-22)

1. Download the following BLS series:
   Employment and Earnings—Household Survey, Selected Labor Statistics by Sex and Detailed Age Group (NSA, Monthly, Thous): Civilian Noninstitutional Population: 16 Years and Over; Civilian Noninstitutional Population: 65 Years and Over.
2. Compute quarterly data of Population 16 and over and Population 65 and over by averaging over the monthly data.
3. Obtain quarterly civilian noninstitutional population ages 16 to 64 by substracting one series from the other.
4. The series obtained is seasonally-adjusted using the Census X-12, multiplicative seasonal adjustment method.

Average Hours Worked per Capita Source: U.S. Department of Labor, Bureau of Labor Statistics (BLS). Current Population Survey (CPS)

1. Download the following BLS series:
   (Unadj.) Average Hours, Total At Work, All Industries
   (Unadj.) Number Employed, At Work
   Both data series are at monthly frequency since June 1976. I complete the series with the historical data from July 1947 to May 1976 collected by Cociuba et al. (2009).
2. Convert the monthly series into data on a quarterly basis (by averaging the monthly numbers).
3. Seasonally-adjust the quarterly series using the Census X-12, multiplicative seasonal adjustment method.
4. Total hours worked per quarter are given by the product of employed persons at work on a quarterly basis times the average hours worked per week on a quarterly basis times $\frac{52}{4}$.
5. Quarterly average hours worked per capita can be computed by dividing the total civilian hours worked by the civilian noninstitutional working-age population (16–64 years old). The quarterly hours worked per capita is divided by $\frac{5200}{4}$ ($\frac{52}{4}$ weeks per quarter times 100 productive hours per week) to express the per capita hours worked as a ratio.

B.1.2 Consumption (Nondurables and Services) Deflator and Inflation Rate

Source: U.S. Department of Commerce, Bureau of Economic Analysis (BEA), National Income and Wealth Division, National Income and Product Accounts, Domestic Product and Income (Table 1)

1. Download the following BEA series:
   Domestic Product and Income: Table 1.1.5, Gross Domestic Product (Bil. $, Annual): Personal Consumption Expenditures: Nondurable Goods; Personal Consumption Expenditures: Services.

Domestic Product and Income: Table 1.1.5, Gross Domestic Product (Bil. $, Quarterly, SAAR): Personal Consumption Expenditures: Nondurable Goods; Personal Consumption Expenditures: Services.

Domestic Product and Income: Table 1.1.6, Real Gross Domestic Product (Bil. Chn. 2005. $, Annual): Real Personal Consumption Expenditures: Nondurable Goods; Real Personal Consumption Expenditures: Services.

Domestic Product and Income: Table 1.1.6, Real Gross Domestic Product (Bil. Chn. 2005. $, Quarterly, SAAR): Real Personal Consumption Expenditures: Nondurable Goods; Real Personal Consumption Expenditures: Services.

2. Construct the annual consumption (nondurables and services) deflator. Divide ('personal consumption expenditures: nondurable goods' plus 'personal consumption expenditures: services') by ('real personal consumption expenditures: nondurable goods' plus 'real personal consumption expenditures: services'), using annual data.
3. Construct the quarterly consumption (nondurables and services) deflator. Divide ('personal consumption expenditures: nondurable goods' plus 'personal consumption expenditures: services') by ('real personal consumption expenditures: nondurable goods' plus 'real personal consumption expenditures: services'), using quarterly data.
4. Construct the year-over-year inflation rate for the consumption (nondurables and services) deflator in percentages,

$$\overline{\pi}_t \equiv \left(\frac{P_t - P_{t-4}}{P_{t-4}} \right) 100.$$

B.1.3 Private Output, Consumption and Investment

Source: U.S. Department of Commerce, Bureau of Economic Analysis (BEA), National Income and Wealth Division. National Income and Product Accounts, Domestic Product and Income (Table 1)

1. Download the following BEA series:

Domestic Product and Income: Table 1.1.5, Gross Domestic Product (Bil. $, Quarterly, SAAR): Gross Domestic Product; Personal Consumption Expenditures: Private Nonresidential Investment: Structures; Private Nonresidential Investment: Equipment and Software; Private Residential Investment.

Domestic Product and Income: Table 1.1.6, Real Gross Domestic Product (Bil. Chn. 2005. $, Quarterly, SAAR): Real Personal Consumption Expenditures: Nondurable Goods; Real Personal Consumption Expenditures: Services.

Domestic Product and Income: Table 1.12, National Income by Type of Income (Bil. $, Quarterly, SAAR): Government Wages and Salaries.

2. Construct the quarterly real output series for the U.S. Subtract 'government wages and salaries' from the 'gross domestic product'. Divide the resulting series by the quarterly consumption (nondurables and services) deflator computed before.

3. Construct the quarterly real consumption series for the U.S. Add 'real personal consumption expenditures: nondurable goods' plus 'real personal consumption expenditures: services'.
4. Construct the quarterly real investment series for structures, equipment and software, and housing. Use the quarterly (nominal) investment series for each type of capital ('personal consumption expenditures: private nonresidential investment: structures', 'private nonresidential investment: equipment and software', and 'private residential investment') and divide them by the quarterly consumption (nondurables and services) deflator computed before. The real investment sample starts in the first quarter of 1947. Calculate total real investment as the sum of the real investment for all three types of capital (structures, equipment and software, and housing).
5. Construct the series for real output, real consumption and real investment by capital type in per capita terms and at quarterly rates. Divide the quarterly real output, real consumption and real investment computed before by the civilian noninstitutional population between the ages of 16 and 64 (which was derived earlier). The civilian noninstitutional population is expressed in thousands and must be multiplied by 1000 to express it in number of individuals. The real output, consumption and investment series are expressed in billions of 2005 dollars and must be multiplied by 10^9 to express everything in units of 2005 dollars. Divide the resulting ratios by 4 to express the quarterly per capita real output, real consumption and real investment on a quarterly basis—rather than at an annualized rate.

B.1.4 Total Factor Productivity

Capital's Share of Income Source: U.S. Department of Commerce, Bureau of Economic Analysis (BEA), National Income and Wealth Division. National Income and Product Accounts, Domestic Product and Income (Table 1) and Supplemental Tables (Table 7)

1. Download the following BEA series:

   Domestic Product and Income: Table 1.12, National Income by Type of Income (Bil. $, Annual): Compensation of Employees, Paid; Government Wages and Salaries; Rental Income of Persons, with Capital Consumption Adjustments; Corporate Profits with Inventory Valuation and Capital Consumption Adjustments; Net Interest and Miscellaneous Payments on Assets.

   Domestic Product and Income: Table 1.3.5, Gross Value Added by Sector, (Bil. $, Annual): Gross Value Added: General Government.

   Domestic Product and Income: Table 1.7.5, Relation of Gross Domestic Product, Gross National Product, Net National Product, National Income, and Personal Income (Bil. $, Annual): Gross National Product; Net National Product.

   Domestic Product and Income: Table 1.9.5, Net Value Added by Sector (Bil. $, Annual): General Government: Net Domestic Product.

Supplemental Tables: Table 7.4.5, Housing Sector Output, Gross and Net Value Added (Bil. $, Annual): Gross Housing Value Added; Housing: Compensation of Employees; Housing: Rental Income of Persons with Capital Consumption Adjustments; Housing: Corporate Profits with Inventory Valuation and Capital Consumption Adjustments; Housing: Net Interest; Net Housing Value Added.

2. Exclude government labor income to be consistent with the concept of output defined in the model. NIPA includes an imputed capital income flow for owner occupied housing, but omits the corresponding labor income flows. This omission can introduce an upward bias in the derivation of the capital income share α. Instead of attempting to correct or adjust the data to account for the omission of labor income flows for owner occupied housing, exclude housing imputed rents from the capital income series for the purpose of computing the capital's share of income. Calculate nominal labor income, Y^{LP}, as 'compensation of employees, paid' minus 'housing: compensation of employees' minus 'government wages and salaries.' Calculate nominal capital income including depreciation, Y^{KP_d}, as 'rental income of persons, with capital consumption adjustments' plus 'corporate profits with inventory valuation and capital consumption adjustments' plus 'net interest and miscellaneous payments on assets' minus 'housing: rental income of persons with capital consumption adjustments' minus 'housing: corporate profits with inventory valuation and capital consumption adjustments' minus 'housing: net interest' plus depreciation. Compute depreciation as ('gross national product' minus 'gross value added: general government' minus 'gross housing value added') minus ('net national product' minus 'general government: net domestic product' minus 'net housing value added').
3. Under the standard assumptions of an aggregate Cobb-Douglas production function and perfect competition, the capital share in the production function α can be computed as the ratio of all capital income sources divided by private output. Compute the capital's share of income for each year as,

$$\alpha = \frac{Y^{KP_d}}{Y^{LP} + Y^{KP_d}}.$$

Calculate the average for the entire sample period after the Korean War (starting in 1954) in order to pin down the capital and labor shares in the production function (i.e., the constant parameter α).

Depreciation Rate Source: U.S. Department of Commerce, Bureau of Economic Analysis (BEA), National Income and Wealth Division. National Income and Product Accounts, Domestic Product and Income (Table 1) and Fixed Assets and Consumer Durable Goods (formerly called Fixed Reproducible Tangible Wealth in the U.S.), Capital Stock (Tables 4 and 5)

1. Download the following BEA series:

Domestic Product and Income: Table 1.1.5, Gross Domestic Product (Bil. $, Quarterly, SAAR): Personal Consumption Expenditures: Private Nonresidential

Investment: Structures; Private Nonresidential Investment: Equipment and Software; Private Residential Investment.

Fixed Assets and Consumer Durable Goods, Capital Stock: Tables 4.1, 4.2 and 4.3, Net Stock of Private Fixed Nonresidential Assets by Legal Form and Industry, Year-end Estimates at Current Cost (Bil. $, Annual): Net Stock: Private Fixed Nonresidential Structures; Net Stock: Private Fixed Nonresidential Equipment and Software.

Fixed Assets and Consumer Durable Goods, Capital Stock: 5.1, 5.2 and 5.3, Net Stock of Private Fixed Nonresidential Assets by Legal Form and Industry, Year-end Estimates at Current Cost (Bil. $, Annual): Net Stock: Private Residential Fixed Assets.

2. Construct the annual (year-end) stock of real capital on structures, equipment and software, and housing. Use the annual (nominal) stocks of capital at current cost for each category ('net stock: private fixed nonresidential structures', 'net stock: private fixed nonresidential equipment and software', and 'net stock: private residential fixed assets') and divide them by the annual consumption (nondurables and services) deflator computed before.
3. Construct the quarterly real investment series for structures, equipment and software, and housing. Use the quarterly (nominal) investment series for each category ('personal consumption expenditures: private nonresidential investment: structures', 'private nonresidential investment: equipment and software', and 'private residential investment') and divide them by the quarterly consumption (nondurables and services) deflator computed before. The real investment sample starts in the first quarter of 1947. The quarterly real investment series are expressed at quarterly—rather than annualized—rates. In other words, the series of quarterly annualized investment deflated by the quarterly consumption (nondurable and services) deflator must be divided by 4.
4. Recover the fixed quarterly depreciation rates by year. K_t is the stock of capital and K_{it} is the stock for each capital type $i \in \{s, e, h\}$ accumulated at the end of $t-1$ that becomes available for production during t. Consistently with this timing convention, assume that the annual year-end estimate of the stock of capital is equal to the capital available for production during the first quarter of the following year. Let $\{t, t+1, t+2, t+3\}$ be the quarters corresponding to a given year z, and $\{t+4, t+5, t+6, t+7\}$ those corresponding to year $z+1$. Then, for all $i \in \{s, e, h\}$ the capital available in each quarter subject to the fixed depreciation rate δ_{iz} can be expressed as,

$$t+1 \text{ (Year } z, \text{ Second Quarter): } \quad K_{it+1} = (1-\delta_{iz})K_{it} + X_{it},$$

$$t+2 \text{ (Year } z, \text{ Third Quarter): } \quad K_{it+2} = (1-\delta_{iz})K_{it+1} + X_{it+1},$$

$$t+3 \text{ (Year } z, \text{ Fourth Quarter): } \quad K_{it+3} = (1-\delta_{iz})K_{it+2} + X_{it+2},$$

$$t+4 \text{ (Year } z+1, \text{ First Quarter): } \quad K_{it+4} = (1-\delta_{iz})K_{it+3} + X_{it+3},$$

which can be written recursively in the form of a quartic equation,

$$K_{it+4} = K_{it}(1-\delta_{iz})^4 + X_{it}(1-\delta_{iz})^3 + X_{it+1}(1-\delta_{iz})^2 + X_{it+2}(1-\delta_{iz}) + X_{it+3}.$$

The depreciation rates for each one of the three categories—structures, software and equipment and housing—are computed using a quartic solver in Matlab. The solver returns eight numbers (in the complex plane) that satisfy the formula,

$$a_4\lambda^4 + a_3\lambda^3 + a_2\lambda^2 + a_1\lambda + a_0 = 0.$$

Take the deflated year-end stock of capital for the previous year—which is available for production during the first quarter of the current year—to be $a_4 (\equiv K_{it})$. Then, $a_3 (\equiv X_{it})$ represents the real investment in the first quarter of the year, $a_2 (\equiv X_{it+1})$ is the real investment in the second quarter, $a_1 (\equiv X_{it+2})$ is the real investment in the third quarter, and $a_0 \equiv (X_{it+3} - K_{it+4})$ is the difference between the real investment in the fourth quarter and the deflated year-end stock of capital available for production during the first quarter of the following year. The depreciation rate for a given year z is computed as one minus the largest real root on the quartic polynomial (i.e., $\delta_{iz} = 1 - \lambda$). While the depreciation rates are invariant within a year, they are allowed to vary from one year to the next in the calculations of the quarterly stock of capital. For more details on finding the solution to the quartic polynomial, see e.g. Gomme and Rupert (2007).

Stock of Capital

1. Construct the quarterly stock of real capital on structures, equipment and software, and housing. Start in the first quarter of 1947 with the available stock of capital for that quarter that corresponds to the deflated year-end stock of capital at current costs for each type of capital for the year 1946—which was calculated before. The first quarter of 1947 stock of capital net of depreciation (using the 1947 depreciation rates) plus the real investment in the first quarter of 1947 gives the capital available for production in the second quarter of 1947. Compute recursively the stock of capital available for production in a given quarter as the sum of the real investment in the previous quarter plus the stock of capital available in the previous quarter net of depreciation (at the corresponding depreciation rate of the year on which the previous quarter falls). The quarterly depreciation rates vary across years, but are constant within a given year and had been previously calculated. The quarterly real investment series by capital type have also been computed before.
2. Construct the quarterly stock of real capital on structures, equipment and software, and housing in per capita terms. Divide the quarterly real stock of capital for each type by the civilian noninstitutional population between the ages of 16 and 64 (which was derived before). The civilian noninstitutional population is expressed in thousands and must be multiplied by 1000 to express it in number of individuals. The capital stock and investment series are expressed in billions of 2005 dollars and must be multiplied by 10^9 to express them in units of 2005 dollars. Add the real stock of capital per capita disaggregated by type to obtain to total real stock of capital per capita available for production.

Solow Residual

1. Construct the quarterly Solow residual using the calculated series for per capita output, hours worked and total capital. Without loss of generality, transform all series into indexes where the first quarter of 1948 takes the value of 1 (i.e., 1948:I = 1). The average capital income share determines the parameter α. Then, calculate the Solow residual index $(\frac{S_t}{S_0})$ as,

$$\frac{S_t}{S_0} \equiv \left(\frac{(\frac{y_t}{y_0})}{(\frac{k_t}{k_0})^{\alpha}(\frac{h_t}{h_0})^{1-\alpha}} \right),$$

where $(\frac{y_t}{y_0})$ is the per capita real output index with the base year set at $t = 0$. Similarly, for the indexes of capital and hours worked in per capita terms, i.e. for $(\frac{k_t}{k_0})$ and $(\frac{h_t}{h_0})$ respectively. The level of the Solow residual does not convey any additional information beyond is contained in this index, so one can work directly with the index series.
2. Express the Solow residual in logs as,

$$\overline{s}_t \equiv \left(\ln\left(\frac{S_t}{S_0}\right)\right) 100,$$

and the levels of real per capita output (excluding government), real per capita consumption, real per capita total investment and total hours worked per capita in logs as,

$$\overline{y}_t \equiv \ln(y_t)100, \qquad \overline{c}_t \equiv \ln(c_t)100,$$
$$\overline{x}_t \equiv \ln(x_t)100, \qquad \overline{h}_t \equiv \ln(h_t)100.$$

All these variables are multiplied by 100 to be able to quote them in percentages.

B.2 *Financial and Monetary Variables*

B.2.1 Interest Rate Spread

Source: Treasury Department, and Board of Governors of the Federal Reserve System, "Selected Interest Rates", "Interest Rates Updated Before FRB Publication", H.15 (415).

1. Download the following Treasury Department and Federal Reserve Board series:

   Selected Interest Rates—FRB H.15 (NSA, Quarterly Average of Daily Data, Yields in Percent Per Annum): 20-Year Treasury Note Yield at Constant Maturity; Baa Corporate Bonds, Moody's Seasoned.

   Treasury Long-Term Composite (over 10 years) is no longer available on the FRB H.15 release, but the series continuous to be regularly updated by the Treasury Department. The composite is an unweighted average of all issues outstanding of bonds neither due nor callable in less than 10 years.

2. Complete the 20-year Treasury series. The nominal 20-year Treasury was discontinued between January 1, 1987 through September 30, 1993. Data for this period is calculated as an average of the 10- and 30-year constant maturity yields. Data prior to April 1953 corresponds to the Treasury Long-Term composite (over 10 years) yield. This Long-Term composite index is an unweighted average of all issues outstanding of bonds neither due nor callable in less than 10 years.
3. Compute the yields for the 20-Year Treasury, and Moody's Baa corporate yields on a quarterly basis. These nominal yields, R_t^j for each type $j = \{$20-year Treasury, Baa corporate$\}$, are quoted per annum, in percent. A typical transformation is to compute the quarterly compounded rate ρ_{jt} as follows,

$$\rho_{jt} = 0.25 \ln\left(1 + \frac{R_t^j}{100}\right).$$

4. Compute the spread between the Baa corporate yield and the 20-year Treasury yield. Moody's drops bonds if the remaining life falls below 20 years, if the bond is susceptible to redemption, or if the rating changes. Hence the spread with the 20-year Treasury indicates the risk of corporates at the margin (which are barely above investment grade), controlling for maturity. Compute the spread simply taking the difference between the quarterly compounded yields of two rates i, k as follows,

$$spread_t^{i,k} = 100 * (\rho_{it} - \rho_{kt}).$$

5. Compute the ratio $\frac{R^k}{R}$ with the spread of the Baa corporate bond rate and 20-year Treasury rate as,

$$\frac{R^k}{R} = 1 + \frac{spread^{\text{Baa Corporate, 20-year Treasury}}}{100}.$$

Compute the historical averages in order to calibrate the model.

B.2.2 Monetary Policy

Source: Board of Governors of the Federal Reserve System, "Selected Interest Rates", "Interest Rates Updated Before FRB Publication", H.15 (415).

1. Download the following Federal Reserve Board series:
   Selected Interest Rates—FRB H.15 (NSA, Quarterly Average of Daily Data, Yields in Percent Per Annum): Federal Funds (Effective).
2. Taylor (1993) proposes the rate of inflation over the previous four quarters as one of the key objectives for monetary policy. Construct the year-over-year inflation rate for the consumption (nondurables and services) deflator derived before in percentages,

$$\overline{\pi}_t \equiv \left(\frac{P_t - P_{t-4}}{P_{t-4}}\right) 100.$$

3. Taylor (1993) proposes detrended output as one of the key objectives for monetary policy. Express the quarterly real per capita output (excluding government compensation) in logs and multiplied by 100 as $\overline{y}_t \equiv \ln(y_t)100$. Then, estimate a linear time trend for the real per capita output index in logs as before (over the sample period: 1984:I–2007:IV),

$$\overline{y}_t = \widehat{\overline{y}}_t + u_t^y,$$
$$\widehat{\overline{y}}_t = \alpha_y + \beta_y t.$$

Detrended output as the percentage deviation of per capita real output (excluding government compensation) relative to its trend can be calculated as

$$\widehat{y}_t \equiv \overline{y}_t - \widehat{\overline{y}}_t.$$

4. Compute the Taylor rule rate following the reaction function proposed in Taylor (1993),

$$i_t^{TR} \equiv r + \overline{\pi}_t + \frac{1}{2}\widehat{y}_t + \frac{1}{2}(\overline{\pi}_t - \pi)$$
$$= (r + \pi) + \frac{1}{2}\widehat{y}_t + \frac{3}{2}(\overline{\pi}_t - \pi),$$

where i_t^{TR} is the target rate for monetary policy, $\overline{\pi}_t$ defines the year-over-year inflation rate of consumption (nondurables and services) for quarter t (in percentages), $(\pi_t - \pi)$ refers to the percentage deviation of inflation relative to its long-run target, and $\widehat{y}_t$ is the detrended output obtained as the percentage deviation of per capita real output (excluding government) relative to its log-linear trend. Implicit in this equation is the notion that the real (annualized) interest rate is $r \equiv 2$ % and the long-run inflation target is $\pi \equiv 2$ % (per annum), so the long-run nominal interest rate is equal to $(r + \pi) = 4$ %. This specification is otherwise isomorphic to an alternative policy that assumes a long-run inflation target of zero and a real interest rate of $r - \frac{1}{2}\pi = 2\ \% - \frac{1}{2}2\ \% = 1\ \%$.
5. Compute monetary policy shocks m_t relative to this policy rule as the difference between the federal funds rate (effective) and the Taylor-implied target rates in every period, i.e. $m_t \equiv i_t - i_t^{TR}$. Demean the monetary policy shock process, if the mean is statistically different from zero, in order to obtain $\widehat{m}_t$.
6. Estimate an $AR(1)$ process for the demeaned monetary policy shock series $\overline{m}_t$ without a constant term,

$$\widehat{m}_t = \rho_m \widehat{m}_{t-1} + \varepsilon_t^m,$$

in order to characterize the monetary policy shock process in the model.

B.3 Mapping the Non-stationary Data

The financial accelerator model of Bernanke et al. (1999) is stationary by construction, but most observed aggregate variables in the data—even in per capita terms—are not so. The non-stationary observed time series are mapped into the stationary series endogenously generated by the model by detrending the data first (to account for their upward trend). The way to solve the financial accelerator model which I use is to employ a log-linear approximation and express all variables in log deviations from their steady state. All stationary series in the data which are not subject to detrending, then, are demeaned to make them comparable with the endogenous series simulated by the model.

The implicit assumption about the long-run growth path is that the relative price of investment grows at the deterministic rate of γ_q for all $i \in \{s, e, h\}$ and that labor-augmenting technological progress attains a rate of long-run growth of γ. These assumptions are consistent with a deterministic balanced growth path where all variables grow at a constant—but not necessarily common—rate over the long-run (see, e.g., Gomme and Rupert 2007). Hence, when detrending the data these differences in long-run growth rates across macro variables must be taken into account.

Similar to Gomme and Rupert (2007), the constant growth rate of real output per capita g should be given by $g = (g_k)^{\alpha}(\gamma)^{1-\alpha}$ along a balanced growth path. For the growth rate of capital per capita g_k to be constant it must be the case that $g_k = g\gamma_q$. Hence, these balanced growth path relations imply that,

$$g = (\gamma_q)^{\frac{\alpha}{1-\alpha}}\gamma.$$

The long-run growth of the relative price of investment, γ_q, and the long-run rate of labor-augmenting technological progress, γ, both affect the balanced growth path of per capita real output y_t.

All real investment series—x_{st}, x_{et}, x_{ht} and x_t—as well as real consumption per capita c_t must grow at the same common growth rate as real output, i.e. must grow at g. The stock of capital for each type (k_{st}, k_{et}, k_{ht}) and total capital k_t grow in turn at a different fixed rate $g_k = (\gamma_q)^{\frac{1}{1-\alpha}}\gamma$ (where $g_k \neq g$ if $\gamma_q \neq 1$). The share of hours worked per capita h_t is stationary, and bounded within the unit interval. The Solow residual S_t trends upwards at the rate of technological progress,

$$S_t \equiv \frac{y_t}{(k_t)^{\alpha}(h_t)^{1-\alpha}} = A_t\left(\gamma^{1-\alpha}\right)^t,$$

even when working with variables in per capita terms (i.e., y_t, k_t and h_t). However, the rate of long-run growth on real output per capita given by $g = (\gamma_q)^{\frac{\alpha}{1-\alpha}}\gamma$ and the rate of growth of S_t given by $\gamma^{1-\alpha}$ do not generally coincide.

I do not impose this long-run relationships directly on the data for detrending, but I infer from the logic of the argument that: hours worked do not require detrending, only need to be demeaned; that real output per capita, y_t, real investment per capita, x_t, and real consumption per capita, c_t, all should be detrended using a common trend growth; and that the Solow residual measured as S_t is non-stationary as well but detrending it requires a different trend growth than that of output, investment and consumption.

I focus my investigation on the period of the Great Moderation since 1984:I. This period is characterized by stable trends in the data until the 2007 recession. I model the possible presence of nonlinearities in the data since 2007:IV by allowing for a break in the level, but not in the growth rate, of the trend component. For variables that do not trend upwards, I allowed for the possibility that the historical mean of the Great Moderation may have shifted as well.

Detrending

1. Fit a linear time trend to the index series for the Solow residual, $\overline{s}_t$, in logs as follows,

$$\overline{s}_t \equiv \left(\ln\left(\frac{S_t}{S_0}\right)\right)100,$$
$$\overline{s}_t = \alpha_s + \beta_s t + \widehat{a}_t.$$

   Estimate this linear trend for the sample period 1984:I–2007:IV, abstracting from the break in the data during the 1970s. While growth resumed during the Great Moderation period, it was at a slower pace than in the 1950s and 1960s. I do not account for that break or model explicitly the transition dynamics implied by it. Estimate an $AR(1)$ process for the detrended Solow residual series without a constant term,

$$\widehat{a}_t = \rho_a \widehat{a}_{t-1} + \varepsilon_t^a.$$

   This characterizes the aggregate TFP shock process in the model.
2. Fit a common linear time trend to the series for the real output (excluding government) per capita in logs, $\overline{y}_t$, for the real consumption per capita in logs, $\overline{c}_t$, and for the real investment per capita in logs, $\overline{x}_t$, as follows,

$$\overline{y}_t \equiv \ln(y_t)100, \qquad \overline{c}_t \equiv \ln(c_t)100, \qquad \overline{x}_t \equiv \ln(x_t)100,$$
$$\begin{pmatrix} 100\frac{c_t}{y_t} \\ 100\frac{x_t}{y_t} \end{pmatrix} = \begin{pmatrix} \gamma_c \\ \gamma_x \end{pmatrix} + \begin{pmatrix} u_t^{c\_share} \\ u_t^{x\_share} \end{pmatrix},$$
$$\begin{pmatrix} \overline{y}_t \\ \overline{c}_t \\ \overline{x}_t \end{pmatrix} = \begin{pmatrix} 100\ln(\alpha_y) \\ 100\ln(\alpha_y \frac{\gamma_c}{100}) \\ 100\ln(\alpha_y \frac{\gamma_x}{100}) \end{pmatrix} + \beta_y t + \begin{pmatrix} u_t^y \\ u_t^c \\ u_t^x \end{pmatrix},$$

   adjusting the intercept to be consistent with the average consumption share, γ_c, and investment share, γ_x. Estimate this trend specification for the sample period 1984:I–2007:IV. Re-estimate the linear trend including all data available until now, but allowing for the possibility of a level shift in the intercept occurring during the 2007 recession both resulting from a permanent change in the output level, α_y, or a permanent shift in the consumption and investment shares, γ_c and γ_x.

Demeaning

1. Transform the series on hours worked into an index where the first quarter of 1948 takes the value of 1 (i.e., 1948:I = 1). Demean the series on hours worked in logs and multiplied by 100 by estimating the following relationship,

$$\overline{h}_t \equiv \ln(h_t)100,$$
$$\overline{h}_t = \alpha_h + u_t^h,$$

over the sample period 1984:I–2007:IV. Allow for the possibility of a level shift in the intercept occurring during the 2007 recession.

2. Compute the percentage deviation of inflation relative to the long-run inflation target inflation by substracting π from the series π_t,

$$\widehat{\overline{\pi}}_t \equiv (\overline{\pi}_t - \pi) = \left(\frac{P_t - P_{t-4}}{P_{t-4}}\right)100 - \pi,$$

where the standard practice is to set the long-run inflation target during the Great Moderation period (1984:I–2007:IV) at $\pi \equiv 2\,\%$ (per annum).

Sample Period Notice that 1971:III signifies also the advent of a distinctly different monetary policy regime in the U.S. On February 1, 1970, Arthur F. Burns became chairman of the Fed replacing the long-serving William McChesney Martin. Then, on August 15, 1971, the U.S. unilaterally terminated convertibility of the U.S. dollar to gold. The dollar becoming a fiat currency in 1971:III ended the Bretton Woods international monetary system that had been in place since the end of World War II. The onset of the Great Moderation period of low macro volatility and low inflation is often traced back to the appointment of Paul Volcker in August 6, 1979, who decidedly brought the inflation of the 1970s under control. The start of the Great Moderation is generally dated to 1984:I, so it coincided for the most part with the tenure of Alan Greenspan as chairman of the Fed which began in August 11, 1987. And, as Taylor (1993) famously noted, the monetary policy during the period of the Great Moderation is fairly well-described by the simple policy rule that I have adopted in this paper.

1. Detrend/demean all series further back to 1971:III with the estimates of the linear trend/level obtained for the Great Moderation period (1984:I–2007:IV).
2. Set 1971:III as the initial period for the simulation of the model because actual output is closest to its Great Moderation trend at that point than at any other quarter prior to 1984:I and because it occurs after the trend break in productivity of the 1970s.

References

Basu, S. (1996). Procyclical productivity: increasing returns or cyclical utilization? *The Quarterly Journal of Economics*, *111*(3), 719–751.

Bernanke, B. S., & Woodford, M. (1997). Inflation forecasts and monetary policy. *Journal of Money, Credit, and Banking, Part 2*, *29*(4), 653–684.

Bernanke, B. S., Gertler, M. L., & Gilchrist, S. (1999). The financial accelerator in a quantitative business cycle framework. In J. B. Taylor & M. Woodford (Eds.), *Handbook of macroeconomics* (Vol. 1, pp. 1341–1393). Amsterdam: Elsevier. Chap. 21.

Calvo, G. A. (1983). Staggered prices in a utility-maximizing framework. *Journal of Monetary Economics*, *12*(3), 383–398.

Carlstrom, C. T., & Fuerst, T. S. (1997). Agency costs, net worth, and business fluctuations: a computable general equilibrium analysis. *The American Economic Review*, *87*(5), 893–910.

Carlstrom, C. T., & Fuerst, T. S. (2001). Monetary shocks, agency costs, and business cycles. *Carnegie-Rochester Conference Series on Public Policy*, *54*, 1–27.

Christiano, L. J., Motto, R., & Rostagno, M. (2003). The great depression and the Friedman-Schwartz hypothesis. *Journal of Money, Credit, and Banking*, *35*, 1119–1198.

Cociuba, S. E., Prescott, E. C., & Ueberfeldt, A. (2009). *U.S. hours and productivity behavior using CPS hours worked data: 1947-III to 2009-III Mimeo*. Federal Reserve Bank of Dallas.

Faia, E., & Monacelli, T. (2007). Optimal interest rate rules, asset prices, and credit frictions. *Journal of Economic Dynamics & Control*, *31*(10), 3228–3254.

Galí, J. (2008). *Monetary policy, inflation, and the business cycle: an introduction to the new Keynesian framework*. Princeton: Princeton University Press.

Gomes, J. F., Yaron, A., & Zhang, L. (2003). Asset prices and business cycles with costly external finance. *Review of Economic Dynamics*, *6*(4), 767–788.

Gomme, P., & Rupert, P. (2007). Theory, measurement and calibration of macroeconomic models. *Journal of Monetary Economics*, *54*(2), 460–497.

Kiyotaki, N., & Moore, J. (1997). Credit cycles. *Journal of Political Economy*, *105*(2), 211–248.

McConnell, M. M., & Pérez-Quirós, G. (2000). Output fluctuations in the United States: what has changed since the early 1980's? *The American Economic Review*, *90*(5), 1464–1476.

Meier, A., & Müller, G. J. (2006). Fleshing out the monetary transmission mechanism: output composition and the role of financial frictions. *Journal of Money, Credit, and Banking*, *38*(8), 2099–2133.

Modigliani, F., & Miller, M. H. (1958). The cost of capital, corporation finance and the theory of investment. *The American Economic Review*, *48*(3), 261–297.

Taylor, J. B. (1993). Discretion versus policy rules in practice. *Carnegie-Rochester Conference Series on Public Policy*, *39*, 195–214.

Walentin, K. (2005). *Asset pricing implications of two financial accelerator models*. Unpublished paper, New York University.

Woodford, M. (2003). *Interest and prices. Foundations of a theory of monetary policy*. Princeton: Princeton University Press.

Early Warning Signals of Financial Stress: A "Wavelet-Based" Composite Indicators Approach

Marco Gallegati

Abstract In this paper we exploit the usefulness of wavelet multi resolution analysis in providing early warning signals of financial stress (conditions). The proposed "wavelet-based" approach gives rise to a composite indicator obtained by aggregating several "scale-based" sub-indexes whose individual components are selected on the basis of their cross-correlations properties at different frequency bands. The performance of the "wavelet-based" composite indicator is assessed by evaluating its predictive power relative to the individual financial variables taken in isolation through an out-of-sample forecasting exercise for the US financial stress index. The findings indicate that the wavelet-based composite indicator largely outperforms any individual financial variable taken in isolation in early detecting financial stress at every horizon and that the gain tends to increase as the time horizon increases.

1 Introduction

Composite indicators are summary measures widely used in social science research. They are obtained by combining a set of individual indicators with the aim to provide a measure of complex unobservable phenomena, like the current economic situation or the overall level of financial conditions. Examples of the former are provided by the indicators approach to business cycle analysis firstly introduced by Burns and Mitchell (1946), which is still a major component of the NBER business cycle program, and the OECD systems of composite leading indicators (Nilsson 1987). As regards financial stress conditions, several government and international institutions have recently developed several measures of the degree of stress prevailing in the overall financial system following the recurrent crisis episodes in financial markets, especially the US financial crisis of 2007–2008 (see Kliesen et al. 2012, for a recent survey of financial stress indexes for the US).

M. Gallegati (✉)
Department of Economics and Social Sciences, Faculty of Economics "G. Fua", Polytechnic University of Marche, Piazzale Martelli 8, 60121 Ancona, Italy
e-mail: marco.gallegati@univpm.it

F. Schleer-van Gellecom (ed.), *Advances in Non-linear Economic Modeling*, Dynamic Modeling and Econometrics in Economics and Finance 17,
DOI 10.1007/978-3-642-42039-9_3, © Springer-Verlag Berlin Heidelberg 2014

The timing classification and selection of these "reliable" individual indicators is exclusively based on the time domain information content provided by time series aggregate data.[1] Although this reliability is regularly tested by checking whether the timing pattern of each individual indicator changes or weakens,[2] the stability across frequencies of the relationship between each indicator and the reference series is completely omitted in the construction of composite indexes. The implicit assumption underlying the selection of these "reliable" individual indicators is that the remaining part of information, obtainable by analyzing the data in the frequency domain, is not relevant.

Many economic and financial time series are nonstationary and, moreover, exhibit changing frequencies over time. Much of the usefulness of wavelet analysis has to do with its ability to handle a variety of nonstationary signals. Indeed, as wavelets are constructed over finite intervals of time and are not necessarily homogeneous over time, they can reveal the localized time and frequency information without requiring the time series to be stationary. Thus, two interesting features of wavelet time scale decomposition for economic variables are that (i) the data need not be detrended or differenced, and (ii) the nonparametric nature of wavelets takes care of potential nonlinear relationships without losing detail (Schleicher 2002).

A useful property of wavelet analysis is that, by decomposing a signal into its time scale components, it allows to exploit the different informative content of individual indicators at different time scales. Separating these scales can reveal patterns and features in the data beyond those evidenced from the usual aggregate perspective and may be useful in analyzing situations in which the degree of association between two time series is likely to change across frequency bands.[3] Indeed, the time scale decomposition property of wavelets allows identification of different relationships between variables on a scale-by-scale basis so that, as evidenced by Gençay et al. (2005), a seemingly nonlinear relationship between variables may be a scale specific phenomenon.

In this paper we propose a "wavelet-based" approach to the construction of composite indicators that fully exploits all available information stemming from individual indicators over time and across scales. The overall composite index is obtained

[1]See Moore and Shiskin (1967) for a comprehensive illustration of the selection criteria.

[2]The composition of composite indexes is periodically revised due to the evolutive nature of economic systems.

[3]That economic and financial relationships can differ across frequencies has been firstly documented by Engle (1974, 1978) using band spectrum regressions, and more recently by Ramsey and Lampart (1998a, 1998b) using wavelet analysis. Similar findings using time scale regression analysis are also reported in the several recent works of Ramsey and his co-authors for the wage Phillips curve (Gallegati et al. 2011), the link between stock and bond market prices for investments (Gallegati and Ramsey 2013), and the relationship between interest rate spreads and output (Gallegati et al. 2013).

by simply aggregating several "scale-based" sub-indexes whose components are selected on the basis of their statistical performance at each frequency band. The result is a composite indicator that, by applying a frequency domain based variable selection technique, can potentially identify new and unexplored sources of information and, by retaining parsimony, also reduce the number of false signals.[4]

The proposed "wavelet-based" methodology is applied to the construction of an early warning composite index of financial stress. For that purpose, we first construct a financial stress index for the US that is able to capture well known historical episodes of severe distress in the US financial markets and display a high degree of similarity with several well known alternative US financial stress indexes. Then, we show how the same aggregate financial variables used in the construction of the financial stress indexes can be used to construct a leading composite indicator of financial stress by simply aggregating several "scale-based" sub-indexes whose components have been selected according to their the scale-by-scale leading relationship with the overall level of financial stress.

The leading indicator properties for financial stress of the "wavelet-based" early warning composite indicator are examined by evaluating its predictive power (relative to the individual financial variables taken in isolation) through a pseudo out-of-sample forecasting exercise. Specifically, we compare its out-of-sample performance fit with that of a simple autoregressive function for a number of US financial stress indexes, that is KCFSI, STLFSI and CFSI. The findings indicate that the "wavelet-based" composite indicator largely outperforms any individual financial variable taken in isolation in predicting financial stress at every horizon and that the gain tends to increase as the horizon increases.

The paper is organized as follows. Section 2 explains the motivation for using wavelets in economic applications and the methodology used to construct the wavelet-based composite index. In Section 3, after describing our FSI for the US and comparing it with alternative financial stress indexes, we present an application of the proposed methodology by constructing an early-warning composite indicator of financial stress for the US and evaluating its performance in forecasting financial stress conditions. Section 4 concludes.

2 "Wavelet-Based" Composite Indexes: Motivation and Methodology

Analysis of the data in the time domain yields only part of the information embedded in the data. In order to obtain the remaining part of the information, the data need to be analyzed in the frequency domain. In such a context, joint time-frequency

[4] This latter property relies on the ability of the wavelet transform to decompose the observed time series into a combination of structural and random error components (Ramsey et al. 2010).

analysis methods combining information from both time-domain and frequency-domain simultaneously, like wavelets, can be very appealing.

Wavelets are mathematical functions that allows the decomposition of a signal into its different frequency components and to study the dynamics of each of these components separately.[5] Although widely used in many other disciplines like geophysics, climatology, etc., wavelets remain relatively unknown to practitioners in the fields of economics and finance notwithstanding the results obtained using wavelet analysis in empirical analyses clearly show that the separation by time scale decomposition analysis can be of great benefit for a proper understanding of economic relationships that are actually recognized as puzzling. Indeed, the analysis in the time-frequency domain is able to produce more accurate information on the presence of (highly localized) patterns, possibly only at certain scales, and dominant scales of variation in the data, like characteristic scales (Keim and Percival 2010), as well as on the relationships between variables.[6]

In the non-parametric indicators approach[7] to the construction of composite indicators the attention of the researcher is focused on finding those indicators providing the best statistical performance with respect to the phenomenon to be explained. This search process relies on application of conventional time domain methods and tools. However, if the variable realizations depend on different frequency components rather than just one, as it is implicitly stated in the standard time domain approach, any information regarding the different frequency components contained into the original data series will be lost (Masset 2008).[8] The most important gain from using the multiresolution decomposition properties of wavelet transform for the indicators approach is that it allows to efficiently process all available information by exploiting the different informative content of individual indicators at different time scales. In particular, wavelet analysis can reveal structure to a problem and/or relationships between variables that standard time series analysis is not able to capture.

The methodology proposed in this paper differs from the traditional "indicators approach" as to the construction and variable selection procedures in that the overall composite index is obtained by aggregating several sub-indices, each corresponding

[5]A brief introduction to wavelets in general, and the maximal overlap discrete transform in particular, is provided in the Appendix.

[6]See Ramsey and Lampart (1998a, 1998b), Kim and In (2003), Gençay et al. (2005), Aguiar-Conraria and Soares (2011).

[7]The parametric alternative is represented by the multivariate logit or probit regression approach.

[8]The economic intuition supporting the application of time-frequency domain techniques in macroeconomics and finance is that many economic and financial processes are the results of decisions of agents with different, sometimes very different, time horizons information. For example, in financial markets the presence of heterogeneous agents with different trading horizons may generate very complex patterns in the time-series of economic and financial variables, e.g. in Muller et al. (1993) and Lynch and Zumbach (2003).

to a different frequency band.[9] Specifically, each sub-index is obtained by combining those indicators displaying the best statistical cross-correlations performance on a scale-by-scale basis. The result is a composite indicator that, by combining the best-performing individual components at each scale, is able to process all available information efficiently and parsimoniously. Moreover, by allowing decomposition of the observed time series into a sum of structural and random error components, a "wavelet-based" approach can also deal with the problem of false signals and lessen the incidence of erratic movements.[10]

The methodology to construct a "wavelet-based" composite indicator requires going through several steps: first, each individual variable need to be decomposed into its time scale components by applying the Discrete Wavelet Transform (DWT). Next, a set of variables is chosen at each scale level according to some statistical significance criteria previously selected. Later, at each scale level the selected variables are aggregated into a composite "scale-based" sub-index using a weighting methodology; and finally, the various composite sub-indexes are aggregated by simple sum into a "wavelet-based" composite index.

Hence, let $I_1, I_2, \ldots, I_N$ be the set of N "reliable" indicators available for the construction of the composite index.[11] By applying a J-level multi resolution decomposition analysis we can provide a complete decomposition of each individual indicator I_i into a smoothed version of the original signal and a set of detail information at different scales:

$$I_i \approx S_J[I_i] + D_J[I_i] + \cdots + D_j[I_i] + \cdots + D_2[I_i] + D_1[I_i] \tag{1}$$

where $S_J[I_i]$ contains the "smooth component" of the signal, and $D_j[I_i]$, with $j = 1, 2, \ldots, J$, the detail signal components at ever-increasing level of detail. The scale-by-scale analysis of the relationships between each variable and the reference series allows to identify the "best-performing" variables on a scale-by-scale basis. These "best-performing" individual variables at each scale level can then be used to construct a "scale-based" composite sub-index CI_{D_j}, one for each level of decomposition $j = 1, 2, \ldots, J$, by means of a weighted aggregation. Thus, with $k \leq n$ statistically significant "reliable" indicators at scale level j the composite sub index

[9]Previous examples of wavelet-based methods to construct composite indicators are provided in Picchetti (2008) and Rua and Nunes (2005). Both approaches differ from that proposed in the present contribution.

[10]By separating those time scales at which the relationships are significant from those scales at which it is not, a scale selection process is allowed because scales corresponding to finest details contain mostly noise.

[11]Individual indicators, in order to be included in a composite index, are required to satisfy a number of economic and statistical criteria such as economic rationale (reasoning), consistently timing behavior, statistical adequacy (in terms of quality and reliability of the data), importance of irregular and non-cyclical factors (smoothness) and timeliness (see Moore and Shiskin 1967).

CI_{D_j} will be given by

$$\mathrm{CI}_{D_j} = \omega_{1,j} D_j[I_1] + \omega_{2,j} D_j[I_2] + \omega_{3,j} D_j[I_3] + \cdots + \omega_{k,j} D_j[I_k] \qquad (2)$$

where ω_{ij} is the weight of each indicator i at scale j. Finally, by aggregating the j "scale-based" composite sub-indexes CI_{D_j} we can obtain the "wavelet-based" composite index CI^W, that is

$$\mathrm{CI}^W = \mathrm{CI}_{S_J} + \mathrm{CI}_{D_J} + \cdots + \mathrm{CI}_{D_j} + \cdots + \mathrm{CI}_{D_1} \qquad (3)$$

Two main features are likely to characterize a "wavelet-based" composite index CI^W: first, each "scale based" composite sub-index can include just a subset, sometimes a very small subset, of all "reliable" indicators, since it is highly unlikely that all indicators be statistically significant at each time scale. In this sense it can be considered a parsimonious index. Second, since this subset of statistically significant indicators is likely to be smaller (at limit also zero) the lower the scale level, the sub-indexes corresponding to the lowest scales could be excluded from the overall composite index without any significant lost of information. In such a case, since the noise component is generally restricted to the scales corresponding to the highest frequencies, e.g. Ramsey et al. (2010) and Gallegati and Ramsey (2012), the "wavelet-based" composite index will be characterized by a higher degree of smoothness than the corresponding traditional composite index.

3 Constructing a "Wavelet-Based" Early Warning Composite Indicator of Financial Stress for the US

Measuring deterioration in financial conditions has come to the forefront of policy discussions after the recent global financial crisis.[12] A number of alternative indexes of financial stress have been proposed by several US and international institutions with the aim to identify stressing periods and determine whether financial stress is high enough to be a serious concern for financial stability.[13] Such composite indexes are generally constructed by combining a number of indicators representative of different financial markets, i.e. interbank, credit, equity, foreign exchange, etc., in order to capture various aspects of systemic financial stress such as information asymmetry, flight to quality, flight to liquidity, increased uncertainty, etc. (see Hakkio

[12] As recently stated by Bernanke (2011, pp. 13–14): "The crisis has forcefully reminded us that the responsibility of central banks to protect financial stability is at least as important as the responsibility to use monetary policy effectively in the pursuit of macroeconomic objectives".

[13] Prominent examples are the Financial Stress Indexes of the Federal Reserve Bank of Kansas City (KCFSI), St. Louis (STLFSI) and Cleveland (CFSI), the Federal Reserve Bank of Chicago National Financial Conditions Index (NFCI) and International Monetary Fund Advanced Economies Financial Stress Index (IMFFSI).

Table 1 Alternative FRB financial stress indexes for the US

Name (source)	Number of components	Period	Frequency	Weighting methodology
KCFSI (Hakkio and Keeton 2009)	11	1990:2–	Monthly	Principal components
STLFSI (Kliesen and Smith 2010)	18	1993:12–	Weekly	Principal components
CFSI (Oet et al. 2011)	12	1994:1–	Weekly	US credit weights

and Keeton 2009).[14] They differ as to the number of components, time span, frequency of observations and weighting methodology since different trade-offs have to be faced by researchers when deciding how to construct a financial stress measure.[15] Table 1 shows several alternative US financial stress indexes along with their main features. For example, the Kansas City Fed's FSI (KCFSI), starting in 1990:2, is constructed using 11 monthly data series weighted by the principal component method where coefficients are chosen so that the index explains the maximum possible amount of the total variation in the 11 variables over the sample period. The implicit assumption underlying this methodology is that financial stress is assumed to be the (common) factor responsible for the co-movement of the group of variables.

Although the literature on early warning systems for currency and banking crises has been extensive,[16] developed economies have generally received limited attention mainly because of the relatively rare occurrence of financial crises in these countries. And much less attention has been paid to early warning composite indicators for detection of early signs of periods of financial distress. A rare example is the Kaminsky's (1999) leading indicator approach to early warning system where the warning signals have to be transformed into a binary variable. Another exception is represented by Brave and Butters' (2012) NFCI's nonfinancial leverage subindex, an early warning index of financial instability that is made up of two nonfinancial leverage measures included in the Chicago Fed's NFCI.[17]

The wavelet-based methodology described in the previous section can be used to construct an early warning composite indicator of financial stress by simply identifying at each time scale those variables that, according to purely statistical grounds, display the best information content as leading indicators of financial stress. There-

[14] Survey articles about financial stress indexes are provided in a number of recent articles by Hatzius et al. (2010), Oet et al. (2011), Hollo et al. (2012), and Kliesen et al. (2012).

[15] As an example, the inclusion of relatively new measures can limit the time span covered by the index, but can better represent the evolving pattern of financial markets.

[16] Early warning signal models can be based on composition of leading indicators as in Kaminsky et al. (1998), and Kaminsky and Reinhart (1999), or on probit/logit models, as in Berg and Pattillo (1999).

[17] See also Berti et al. (2012) for an early-warning composite index developed for the early detection of fiscal stress.

fore, in what follows we assess the benefits of the methodology presented in the previous section by constructing a "wavelet-based" early warning composite index of financial stress for the US and testing its ability to provide early warning signals of financial distress. In order to emphasize the ability of wavelet multi resolution analysis to selectively extract the relevant information, we first create an index of financial stress for the US and then construct a "wavelet-based" early-warning composite index from the same variables used in the construction of the financial stress index.

3.1 An Index of Financial Stress for the US

The construction of a comprehensive and synthetic measure of financial stress requires a number of criteria to be met while selecting among the long list of potential individual indicators (see Kliesen et al. 2012).[18] Following these criteria we construct a financial stress index that include, in addition to stock market returns, several indicators in the form of interest rate spreads since they represent quality, term and liquidity premium indicators that are likely to reflect stress situations on the bond, equity and interbank markets. Specifically, the variables included in the composite index of financial stress represent bank related stress variables, like the difference between the 3-month interbank interest rate and the same maturity government bill rate (TED spread) and the difference between the 3-month and 10-year Treasury bills rate (the inverted term spread), as well as securities related stress variables, like the difference between Aaa corporate bond yield and long-term 10-year government bond yield (corporate bond spread), the Baa-Aaa spread (default risk spreads) and stock market returns (measured as declines).[19]

Figure 1 shows the composite index obtained by aggregating the monthly values of the five variables through the equal variance weights method[20] over the period 1990:2–2012:08. Values above (below) 0 indicate stress (normal) periods, with crisis periods identified with values of the index exceeding the threshold of one standard deviation. Following the approach generally used in the current literature, the financial stress index presented in Figure 1 is evaluated on the basis of its ability

[18]These criteria generally refer to availability of data for a sufficiently long period so as to include several periods of financial distress, and coverage of the different markets of the financial system, i.e. money, bond, stock and foreign exchange markets.

[19]Exchange rate stress variables are excluded because of the limited relevance of the exchange rate market as a factor of stress for the US.

[20]Weights are computed by dividing each component by its standard deviation. Then the coefficients are scaled so that the standard deviation of the index is unity. This is the method used for example by IMF in constructing its financial stress index for different countries (Cardarelli et al. 2009). A similar method is also used for the construction of Conference Boards' composite indicators (2001).

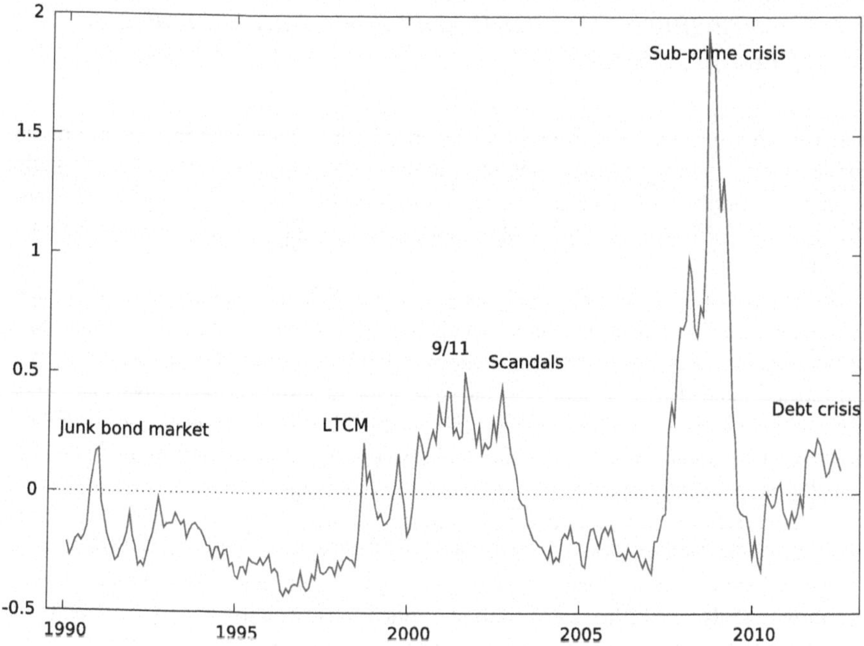

Fig. 1 The composite index and historical periods of financial stress for the US

to identify past known episodes of financial stress both in absolute and in relative terms. Visual inspection of Figure 1 suggests that well known periods of significant financial stress in the US financial history, such as the US saving and loans crisis in early '90s, the Long-Term Capital Management collapse in late 1998, the 9/11 terrorist attacks, the corporate accounting scandals of the early 2000s (WorldCom, Enron, and Arthur Anderson), the recent financial crisis from mid-2007 through mid-2009 and the 2010 European sovereign debt crisis, are all clearly identified by the historical record of the index.

The visual evidence emerging from Figure 1 is confirmed by the simple correlation analysis between our index and several alternative measures of US financial stress, that is the CFSI, the STLFSI and the KCFSI. The degree of correlation is generally high, very high with the STLFSI and the KCFSI, with contemporaneous coefficient values ranging from 0.74 for the CFSI to about 0.90 for the STLFSI and the KCFSI. These values provide indication of a coincident timing behavior between our index and the measures of US financial stress taken for comparison. Thus, we can conclude that the composite index obtained by aggregating the Ted spread, the inverted term spread, the corporate bond spread, the Baa-Aaa spread and stock market returns declines provides a good approximation of the degree of stress prevailing in the US overall financial system when compared with other well-known financial stress indexes.

3.2 Constructing a "Wavelet-Based" Early Warning Indicator of Financial Stress

The first step in the construction of a "wavelet-based" composite indicator consists in identifying the information content of individual components on a scale-by-scale basis. To comply with such a task we preliminarily need to decompose each individual indicator into its time scale components by applying the discrete wavelet transform.

Before performing wavelet analysis a number of decisions must be made: what type of wavelet transform to apply, which family of wavelet filters to use, and how boundary conditions at the end of the series are to be handled. Because of the practical limitations of DWT we perform the time scale decomposition analysis by using the maximal overlap discrete wavelet transform (MODWT). Indeed, the MODWT is a non-orthogonal variant of the classical DWT that, unlike the orthogonal DWT, is translation invariant, as shifts in the signal do not change the pattern of coefficients and is not restricted to a dyadic sample length. The wavelet filter used in the decomposition is the Daubechies least asymmetric (LA) wavelet filter of length $L = 8$, or LA(8) wavelet filter, based on eight non-zero coefficients (Daubechies 1992), the most widely used filter in economic applications. Finally, in order to calculate wavelet coefficient values near the end of the series boundary conditions are to be assumed. We apply the reflection boundary condition.[21] A 4-level decomposition produces four sets of N wavelet coefficients d_4, d_3, d_2, d_1 and a set of N scaling coefficients s_4 which provide information on the short, medium and long term features of the signal.[22]

After performing the time scale decomposition for any individual component we are able to investigate the timing of the comovements between each individual component and the reference series at different frequency bands. Just as the usual unconditional cross-correlation coefficients provides a measure of association between variables in the time domain, wavelet cross-correlation coefficients provide information on the lead/lag relationships between two processes on a scale-by-scale basis. In this way it is possible to separate those time scales at which the relationship is statistically significant from those scales at which it is not. In particular, the results from wavelet cross-correlation analysis are expected to provide very useful insights into the timing behavior of different variables at different frequency bands, especially as to their leading behavior. Indeed, given that our aim is to construct a

---

[21] In this method an extension is made by pasting a reflected (time-reversed) version of the original signal of length T at its end and then applying a periodic boundary condition on the first T elements of the reconstructed signal (and omitting the remaining T elements).

[22] Given that the level of the transform defines the effective scale λ_j of the corresponding wavelet coefficients, for all families of Daubechies compactly supported wavelets the level j wavelet coefficients are associated with changes at scale 2^{j-1}. Since scale 2^{j-1} corresponds to frequencies in the interval $f \in [1/2^{j+1}, 1/2^j]$, using monthly data scale 1 wavelet coefficients are associated to 2–4 month periods, while scales 2 to 4 are associated to 4–8, 8–16 and 16–32 month periods, respectively. Scaling coefficients correspond to periods greater than 32 months.

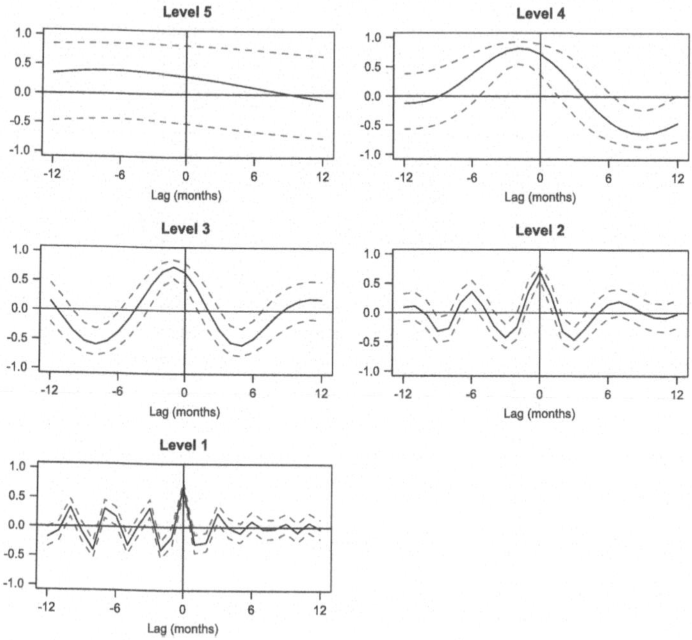

Fig. 2 Wavelet cross correlation between TED spread and KCFSI

composite indicator providing early warning signals of financial stress we are interested in finding those variables that at each frequency band are leading the reference series.

In Figures 2, 3, 4, 5 and 6 we report the MODWT-based wavelet cross-correlation coefficients (solid lines), with the corresponding approximate confidence intervals (dotted lines), between each financial variable and the KCFSI against time leads and lags up to 12 months for all scales.[23] The results from the analysis of wavelet cross-correlation coefficients may be summarized as follows:

- the magnitude of the association with KCFSI tend to differ widely between variables and across scales;
- the leading relationships are restricted to the coarse scales, i.e. scale levels 3, 4 and 5, and the leading period tends to increase as the time scale increases, e.g. Gallegati (2008);
- at the shortest scales, that at scale levels 1 and 2, no leading relationship is in evidence.

These results suggest two main features for the creation of a leading composite indicator: first, the time scale components corresponding to the lowest scales can be omitted since the magnitude of the association is generally close to zero at all

[23]The indicator of financial stress developed by Hakkio and Keeton (2009), called KCFSI, is taken as reference series for the US financial stress conditions.

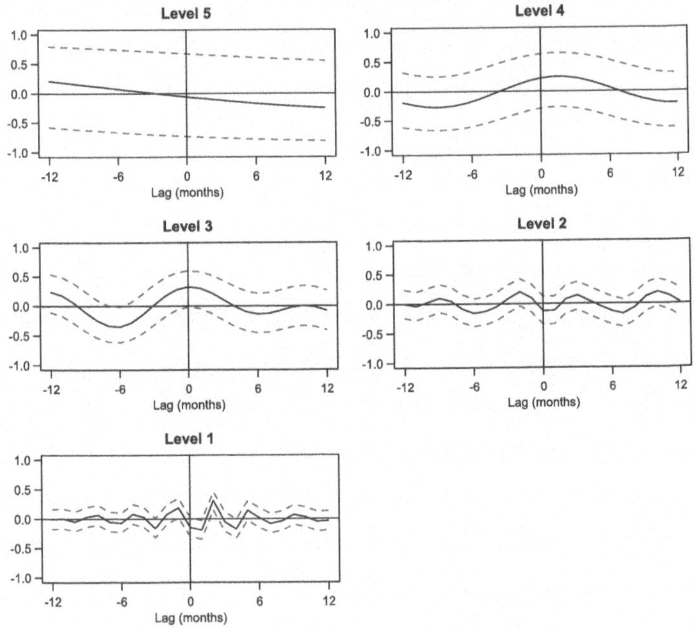

Fig. 3 Wavelet cross correlation between inverted term spread and KCFSI

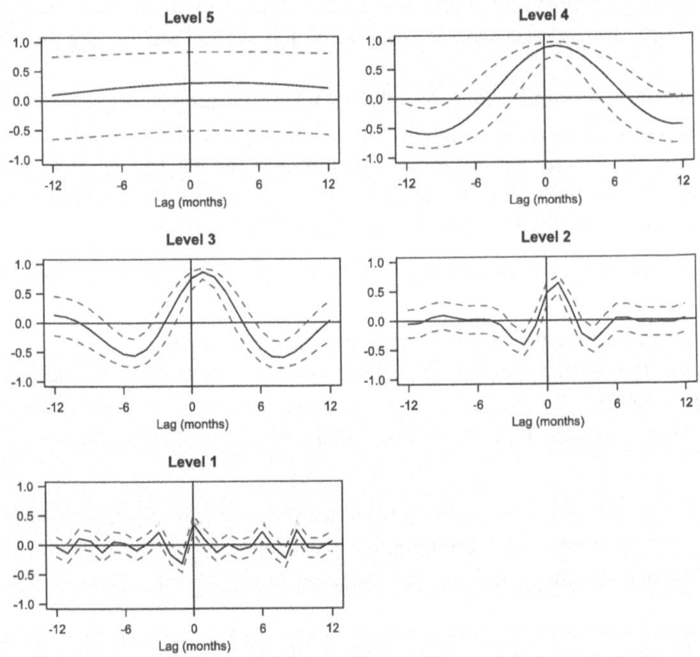

Fig. 4 Wavelet cross correlation between default risk spread and KCFSI

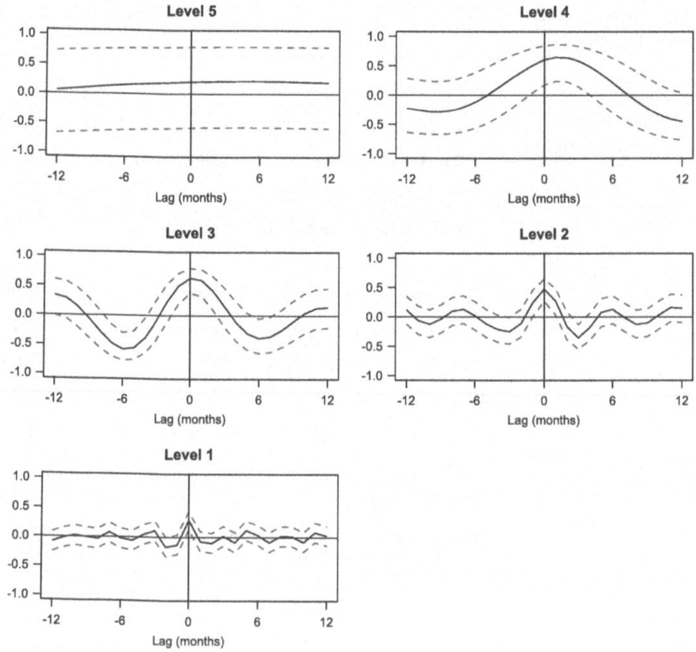

Fig. 5 Wavelet cross correlation between corporate bond spread and KCFSI

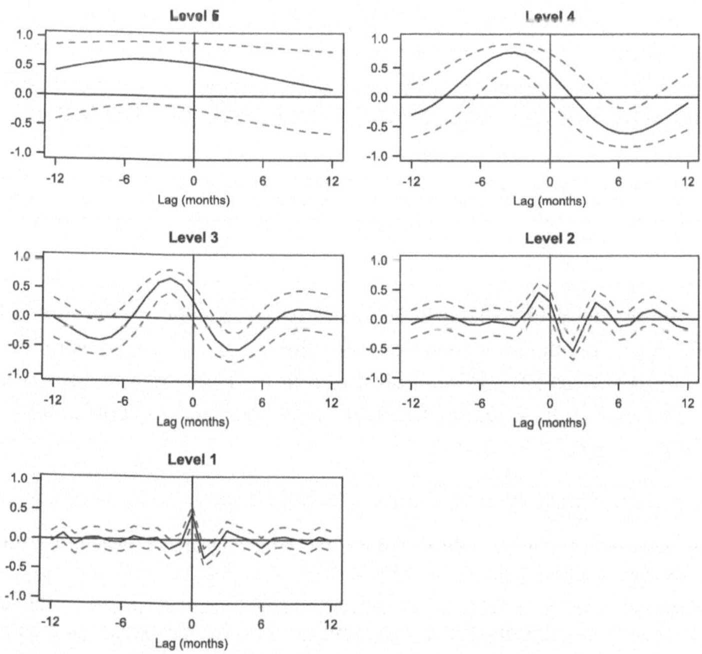

Fig. 6 Wavelet cross correlation between stock market returns (decline) and KCFSI

Table 2 Components of the sub-indexes EWI_{D_3}, EWI_{D_4} and EWI_{S_4} of the wavelet-based composite leading indicator with their scale-by-scale standardization factors

	Stdz. factors
EWI_{D_3} components	
3-month Libor-3-month T-bill spread (inverted)	0.2568
Aaa-10-year Treasury spread (inverted)	0.3386
Baa-Aaa spread (inverted)	0.3798
Stock market returns (declines)	0.0248
EWI_{D_4} components	
3-month Libor-3-month T-bill spread	0.5264
Aaa-10-year Treasury spread (inverted)	0.4142
Stock market returns (declines)	0.0593
EWI_{S_4} components	
3-month Libor-3-month T-bill spread	0.3658
Aaa-10-year Treasury spread	0.2484
Baa-Aaa spread	0.2814
Stock market returns (declines)	0.1044

leads. Second, the different behavior, in terms of leading time and statistical significance, displayed by the individual components across different frequency bands can be usefully exploited to construct a parsimonious and "informationally efficient" composite indicator.[24] In particular, it is possible to construct a "wavelet-based" composite indicator by combining the sub-indexes corresponding to scale levels 3, 4 and 5, where each sub-index is composed by a reduced number of individual components selected using wavelet cross-correlation results. The individual components of the "wavelet-based" composite leading indicator selected from wavelet cross-correlation analysis and used to construct the sub-indexes EWI_{S_4}, EWI_{D_4} and EWI_{D_3} are listed in Table 2 along with their scale-by-scale standardization factors.[25]

The leading indicator properties of the "wavelet-based" early warning composite indicator can be easily visualized in Figure 7 where cross-correlation results with several alternative financial stress indexes for the US are shown. A very clear leading relationship between EWI^W and all different measures of financial stress is in evidence. The leading time ranges from 4 to 6 months, and the correlation value at the peak is between 0.70 to 0.90.

[24]Parsimony and efficiency are related to the ability of reducing redundant information in the construction of the composite index.

[25]The standardization factor in Table 2 indicates the weight given to each component according to the equal variance weight method so that adjustments of the monthly values of each individual components are based on the inverse of the standard deviation of the component contribution and are also normalized to sum to one.

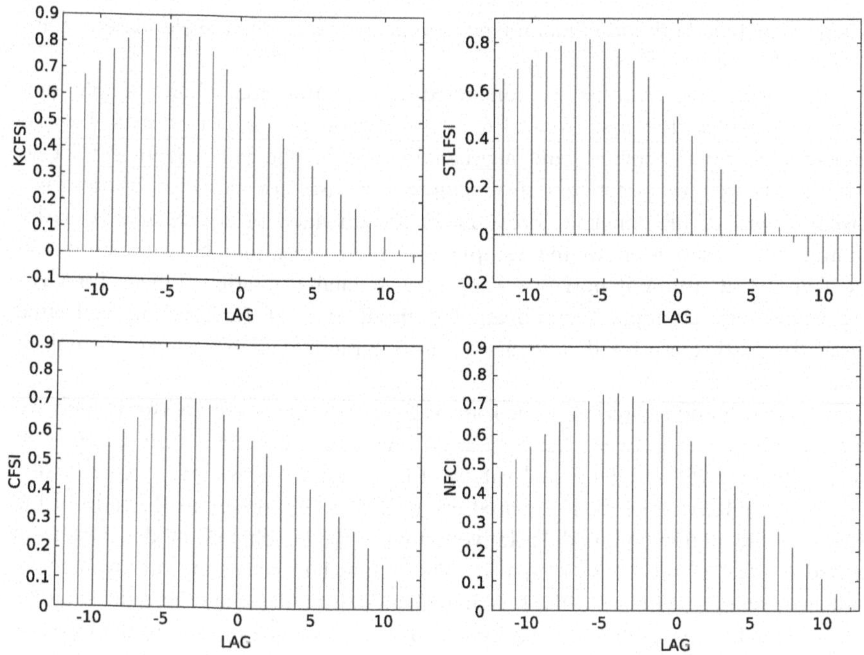

Fig. 7 Cross correlation of EWI^W vs. KCFSI (*top left*), STLFSI (*top right*), CFSI (*bottom left*) and NFCI (*bottom right*)

3.3 Forecasting Financial Stress

Financial stress indexes, as measures of current conditions of financial stress, can also enter early warning models as predicted variables and thus used to evaluate the ability of potential indicators to provide early warning signals of financial instability. In order to evaluate the predictive ability of the "wavelet-based" composite indicator EWI^W we follow the forecasting framework of Misina and Tkacz (2009) and Hatzius et al. (2010) by performing a pseudo out-of-sample forecasting exercise.[26] Hence, we compare a simple autoregression model where the current FSI is a function of its k-quarter lagged value

$$\mathrm{FSI}_t = \alpha + \beta \mathrm{FSI}_{t-k} + \epsilon_{1,t} \tag{4}$$

against an augmented model in which several additional explanatory variables are individually added to the baseline autoregressive specification

$$\mathrm{FSI}_t = \alpha + \beta \mathrm{FSI}_{t-k} + \gamma X_{t-k} + \epsilon_{2,t} \tag{5}$$

[26]For the forecasting experiment we use quarterly data obtained by converting monthly data through averaging.

where X include both individual financial indicators and the "wavelet-based" composite indicator EWI^W.

The forecasting exercise is conducted by splitting the whole sample into two sub-samples: the first, from 1990:2 to 2006:4, is used to estimate the parameters of the benchmark and augmented models, the latter, from 2007:1 to 2012:2, is used to generate out-of-sample forecasts for the FSI. These forecasted values of the financial stress index are obtained on a recursive basis by adding one observation to the sample and then repeating the estimation and the forecast at different horizons k of 1, 2, 4, and 6 quarter(s). This forecasting framework attempts to replicate the production of forecasts in real time since forecasts are provided using all information available at each point in time.[27]

For an evaluation criterion of the forecasting performances of the two models we use the differences between the forecasted financial stress index originating from models (4) and (5) and the actual values of the financial stress index. Specifically, we use the relative root mean squared error (RMSE) expressed as the ratio of the RMSE of the augmented model relative to that of the benchmark model. This ratio provides a test of model fit, so that a value below 1 indicates an improvement in forecast accuracy relative to the benchmark model, while when the ratio is above 1 this indicates a worsening of the forecasting performance relative to the baseline autoregressive model.

Table 3 presents the relative RMSE's ratios for this forecasting exercise The forecasted FSI is listed at the top of each panel (KCFSI, STLFSI and CFSI), while column headings denote the variable as explanatory variables in the augmented model, that is EWI^W, TED spread, inverted term spread (INVTS), corporate bond spread (CBS), default risk spread (SPR) and stock returns decline (STOCK). The rows in each panel denote the forecast horizon (k) measured in quarters beyond the end of the sample. The main finding is that the "wavelet-based" composite indicator EWI^W largely outperforms any individual financial variable taken in isolation in predicting (early detecting) financial stress at every horizon. Moreover, we find that the gain tends to increase as the time horizon increases. In contrast, at individual variable level, the differences between the benchmark model and the augmented model are generally small, with the exceptions of TED spread and stock market returns for the KCFSI and STLFSI at longer horizons, and of stock market returns for the CFSI at any horizon.

4 Conclusion

In this paper we propose a new methodology for the construction of composite indicators that take advantage of the ability of wavelet analysis to fully ex-

[27] Moreover, the issue of the impact of data revisions does not apply in this case since the observations of our financial variables are not revised.

Table 3 Out-of-sample forecasting properties of EWI^W

k	EWI^W	TED	INVTS	CBS	SPR	STOCK
Index = KCFSI						
1	0.6703	0.9510	0.9976	1.0138	0.9752	0.9834
2	0.5208	0.8646	0.9964	1.0758	1.0241	0.9543
4	0.3450	0.7188	0.9797	1.2634	1.0759	0.8740
6	0.2817	0.5901	0.9614	1.5621	1.0761	0.8135
Index = STLFSI						
1	0.6290	1.0416	1.0249	1.1869	1.0386	0.9808
2	0.4272	0.9839	1.0717	1.4640	1.0429	0.9356
4	0.2269	0.8847	1.1763	2.6212	0.9323	0.8157
6	0.1574	0.7763	1.3339	5.1519	0.7600	0.7184
Index = CFSI						
1	0.7858	0.9565	0.9981	1.0783	1.0080	0.8747
2	0.6370	0.9380	1.0077	1.0252	1.0155	0.8380
4	0.5019	0.9572	1.0033	0.9567	1.0525	0.7919
6	0.4687	1.0219	1.0000	0.9030	1.0843	0.7826

Notes: The table displays the relative ratio of the root mean squared error (RMSE) of model (5) relative to that of model (4). The forecasted FSI is listed at the top of each panel, while column headings denote the variable added to the baseline model. The rows in each panel denote the forecast horizon (k) measured in quarters beyond the end of the sample

ploit the whole information present in the data. The proposed methodology differs from the traditional one as to the construction and data selection procedure. Indeed, the "wavelet-based" composite indicator is obtained by aggregating several "scale-based" sub-indexes whose components are selected on the basis of their cross-correlations leading properties as to the overall level of financial stress at each frequency band.

We apply this "wavelet based" procedure to the construction of a measure for early detection of periods of financial distress with US data. In order to emphasize the ability of wavelet multi resolution analysis to provide an efficient data reduction technique we use for the construction of the "wavelet-based" leading composite index the same individual financial indicators that at aggregate level are able to provide information on the current level of financial stress. The out-of-sample forecasting exercise shows that the composite leading indicator of financial stress, obtained by aggregating several "scale-based" subindexes, provide early warning signals of future financial stress episodes. In particular, the findings indicate that the "wavelet-based" composite indicator largely outperforms at every horizon any individual financial variable taken in isolation in predicting financial stress, and that the gain tends to increase as the time horizon increases.

We interpret this findings as a confirm that wavelets can be a powerful tool for empirical macroeconomics because of their ability to unveil relationships between variables that would be otherwise hard to detect using standard time domain methods.

Acknowledgements I would like to thank the participants at SEEK Workshop on "Non-Linear Economic Modeling", ZEW Mannheim 11–12 December 2012, and at 4th Conference on "Recent Development in Macroeconomics", ZEW Mannheim 18–19 July 2013, for providing very useful comments. This research project has received funding from the European Union's Seventh Framework Programme (SSH.2012.1.3-1) under grant agreement n°320278, the RASTANEWS project.

Appendix: Basic Notions of Wavelets

Given a stochastic process $\{X\}$, if we denote with $H = (h_0, \ldots, h_{L-1})$ and $G = (g_0, \ldots, g_{L-1})$ the impulse response sequence[28] of the wavelet and scaling filters h_l, and g_l, respectively, of a Daubechies compactly supported wavelet (with L the width of the filters), when $N = L^2$ we may apply the orthonormal discrete wavelet transform (DWT) and obtain the wavelet and scaling coefficients at the jth level defined as[29]

$$w_{j,t} = \sum_{l=0}^{L-1} h_{j,l} X_{t-l}$$

$$v_{J,t} = \sum_{l=0}^{L-1} g_{j,l} X_{t-l}$$

where $h_{j,l}$ and $g_{j,l}$ are the level j wavelet and scaling filters and, due to downsampling by 2^J, we have $\frac{N}{2^J}$ scaling and wavelet coefficients.[30]

The DWT is implemented via a filter cascade where the wavelet filter h_l is used with the associated scaling filter g_l[31] in a pyramid algorithm (Mallat 1989) consisting in an iterative scheme in which, at each iteration, the wavelet and scaling coefficients are computed from the scaling coefficients of the previous iteration.[32]

[28]The impulse response sequence is the set of all filter coefficients. The filter coefficients must satisfy three properties: zero mean ($\sum_{l=0}^{L-1} h_l = 0$), unit energy ($\sum_{l=0}^{L-1} h_l^2 = 1$) and orthogonal to its even shifts ($\sum_{l=0}^{L-1} h_l h_{l+2k} = 0$).

[29]The expressions used for DWT (and MODWT) wavelet and scaling coefficients refer to functions defined over the entire real axis, that is $t \in \Re$ as in this case $X_t = X_{t \bmod N}$ when $t < 0$.

[30]At the jth level the inputs to the wavelet and scaling filters are the scaling coefficients from the previous level $(j-1)$ and the output are the jth level wavelet and scaling coefficients.

[31]The wavelet and scaling filter coefficients are related to each other through a quadrature mirror filter relationship, that is $h_l = (-1)^l g_{L-1-l}$ for $l = 0, \ldots, L-1$.

[32]The only exception is at the unit level $(j = 1)$ in which wavelet and scaling filters are applied to original data.

However the orthonormal discrete wavelet transform (DWT), even if widely applied to time series analysis in many disciplines, has two main drawbacks: the dyadic length requirement (i.e. a sample size divisible by 2^J),[33] and the fact that the wavelet and scaling coefficients are not shift invariant due to their sensitivity to circular shifts because of the decimation operation. An alternative to DWT is represented by a non-orthogonal variant of DWT: the maximal overlap DWT (MODWT).[34]

In the orthonormal Discrete Wavelet Transform (DWT) the wavelet coefficients are related to nonoverlapping differences of weighted averages from the original observations that are concentrated in space. More information on the variability of the signal could be obtained considering all possible differences at each scale, that is considering overlapping differences, and this is precisely what the maximal overlap algorithm does.[35] Thus, the maximal overlap DWT coefficients may be considered the result of a simple modification in the pyramid algorithm used in computing DWT coefficients through not downsampling the output at each scale and inserting zeros between coefficients in the wavelet and scaling filters.[36] In particular, the MODWT wavelet and scaling coefficients $\tilde{w}_{j,t}$ and $\tilde{w}_{j,t}$ are given by

$$\tilde{w}_{j,t} = \frac{1}{2^{j/2}} \sum_{l=0}^{L-1} \tilde{h}_{j,l} X_{t-l}$$

$$\tilde{v}_{J,t} = \frac{1}{2^{j/2}} \sum_{l=0}^{L-1} \tilde{g}_{J,l} X_{t-l}$$

where the MODWT wavelet and scaling filters $\tilde{h}_{j,l}$ and $\tilde{g}_{j,l}$ are obtained by rescaling the DWT filters as follows:[37]

$$\tilde{h}_{j,l} = \frac{h_{j,l}}{2^{j/2}}$$

[33]This condition is not strictly required if a partial DWT is performed.

[34]The MODWT goes under several names in the wavelet literature, such as the "non-decimated DWT", "stationary DWT" (Nason and Silverman 1995), "translation-invariant DWT" (Coifman and Donoho 1995) and "time-invariant DWT".

[35]Indeed, the term maximal overlap refers to the fact that all possible shifted time intervals are computed. As a consequence, the orthogonality of the transform is lost, but the number of wavelet and scaling coefficients at every scale is the same as the number of observations.

[36]The DWT coefficients may be considered a subset of the MODWT coefficients. Indeed, for a sample size power of two the MODWT may be rescaled and subsampled to obtain an orthonormal DWT.

[37]Whereas DWT filters have unit energy, MODWT filters have half energy, that is $\sum_{l=0}^{L-1} \tilde{h}_{j,l}^2 = \sum_{l=0}^{L-1} \tilde{g}_{j,l}^2 = \frac{1}{2^j}$.

$$\tilde{g}_{j,l} = \frac{g_{j,l}}{2^{j/2}}$$

The MODWT wavelet coefficients $\tilde{w}_{j,t}$ are associated with generalized changes of the data on a scale $\lambda_j = 2^{j-1}$. With regard to the spectral interpretation of MODWT wavelet coefficients, as the MODWT wavelet filter $h_{j,l}$ at each scale j approximates an ideal high-pass with passband $f \in [1/2^{j+1}, 1/2^j]$,[38] the λ_j scale wavelet coefficients are associated to periods $[2^j, 2^{j+1}]$.

MODWT provides the usual functions of the DWT, such as multiresolution decomposition analysis and variance analysis based on wavelet transform coefficients, but unlike the classical DWT it

- can handle any sample size;
- is translation invariant, as a shift in the signal does not change the pattern of wavelet transform coefficients;
- provides increased resolution at coarser scales.[39]

In addition, MODWT provides a larger sample size in the wavelet variance and correlation analyses and produces a more asymptotically efficient wavelet variance estimator than the DWT.[40]

In addition to the features stated above wavelet transform is able to analyze the variance of a stochastic process and decompose it into components that are associated to different time scales. In particular, given a stationary stochastic process $\{X\}$ with variance σ_X^2 and defined the level j wavelet variance $\sigma_X^2(\lambda_j)$, the following relationship holds

$$\sum_{j=1}^{\infty} \sigma_X^2(\lambda_j) = \sigma_X^2$$

where $\sigma_X^2(\lambda_j)$ represent the contribution to the total variability of the process due to changes at scale λ_j. This relationship says that wavelet variance decomposes the variance of a series into variances associated to different time scales.[41] By definition,

[38]On the other hand at scale λ_J the scaling filter $g_{J,l}$ approximates an ideal low-pass filter with passband $f \in [0, 1/2^{j+1}]$.

[39]Unlike the classical DWT which has fewer coefficients at coarser scales, it has a number of coefficients equal to the sample size at each scale, and thus is over-sampled at coarse scales.

[40]For a definition of wavelet variance see Percival (1995).

[41]The wavelet variance decomposes the variance of certain stochastic processes with respect to the scale $\lambda_j = 2^{j-1}$ just as the spectral density decompose the variance of the original series with respect to frequency f, that is

$$\sum_{j=1}^{\infty} \sigma_X^2(\lambda_j) = \text{var}\, X = \int_{-1/2}^{1/2} S_X(f)df$$

where $\sigma_X^2(\lambda_j)$ is wavelet variance at scale λ_j and $S_{(\cdot)}$ is the spectral density function.

the (time independent) wavelet variance for scale λ_j, $\sigma_X^2(\lambda_j)$, is defined to be the variance of the j-level wavelet coefficients

$$\sigma_X^2(\lambda_j) = \text{var}\{\tilde{w}_{j,t}^2\}$$

As shown in Percival (1995), provided that $N - L_j \geq 0$, an unbiased estimator of the wavelet variance based on the MODWT may be obtained, after removing all coefficients affected by the periodic boundary conditions,[42] using

$$\tilde{\sigma}_X^2(\lambda_j) = \frac{1}{\tilde{N}_j} \sum_{t=L_J}^{N} \tilde{w}_{j,t}^2$$

where $\tilde{N}_j = N - L_j + 1$ is the number of maximal overlap coefficients at scale j and $L_j = (2_j - 1)(L - 1) + 1$ is the length of the wavelet filter for level j.[43] Thus, the jth scale level j wavelet variance is simply the variance of the nonboundary or interior wavelet coefficients at that level (Percival 1995 and Serroukh et al. 2000). Both DWT and MODWT can decompose the sample variance of a time series on a scale-by-scale basis via its squared wavelet coefficients, but the MODWT-based estimator has been shown to be superior to the DWT-based estimator (Percival 1995).

Starting from the spectrum $Sw_{X,j}$ of the scale j wavelet coefficients it is possible to determine the asymptotic variance V_j of the MODWT-based estimator of the wavelet variance (covariance) and construct a random interval which forms a $100(1 - 2p)$ % confidence interval.[44]

The formulas for an approximate $100(1 - 2p)$ % confidence intervals MODWT estimator robust to non-Gaussianity for $\tilde{\sigma}_{X,j}^2$ are provided in Gençay et al. (2002).[45]

---

[42] As MODWT employs circular convolution, the coefficients generated by both beginning and ending data could be spurious. Thus, if the length of the filter is L, there are $(2^j - 1)(L - 1)$ coefficients affected for 2^{j-1}-scale wavelet and scaling coefficients, while $(2^j - 1)(L - 1) - 1$ beginning and $(2^j - 1)(L - 1)$ ending components in 2^{j-1}-scale details and smooths would be affected (Percival and Walden 2000).

[43] The quantity estimated in Eq. (2) is time-independent even in case of nonstationary processes but with stationary dth-order differences, provided that the length L of the wavelet filter is large enough to make the wavelet coefficients $\tilde{w}_{j,t}$ a sample of stationary wavelet coefficients (Serroukh et al. 2000). This is because Daubechies wavelet filters may be interpreted as generalized differences of adjacent averages and are related with difference operator (Whitcher et al. 2000).

[44] For a detailed explanation of how to construct the confidence intervals of wavelet variance, see Gençay et al. (2002).

[45] The empirical evidence from the wavelet variance suggest that $N_j = 128$ is a large enough number of wavelet coefficients for the large sample theory to be a good approximation (Whitcher et al. 2000).

References

Aguiar-Conraria, L., & Soares, M. J. (2011). Oil and the macroeconomy: using wavelets to analyze old issues. *Empirical Economics*, *40*, 645–655.

Berg, A., & Pattillo, C. (1999). Predicting currency crises: the indicators approach and alternative. *Journal of International Money and Finance*, *18*, 561–586.

Bernanke, B. S. (2011). The effects of the great recession on Central Bank doctrine and practice. In *Speech delivered at the FRB of Boston 56th economic conference*, Boston, 18 October 2011.

Berti, K., Salto, M., & Lequien, M. (2012). *An early-detection index of fiscal stress for EU countries* (Working Paper Series No. 475). European Commission.

Brave, S. A., & Butters, R. A. (2012). *Detecting early signs of financial instability* (Chicago Fed Letter No. 305). Federal Reserve Bank of Chicago.

Burns, A. F., & Mitchell, W. C. (1946). *Measuring business cycles*. New York: National Bureau of Economic Research.

Cardarelli, R., Elekdag, S., & Lall, S. (2009). *Financial stress, downturns, and recoveries* (Working Paper 09/100). International Monetary Fund.

Coifman, R. R., & Donoho, D. L. (1995). Time-invariant wavelet de-noising. In A. Antoniadis & G. Oppenheim (Eds.), *Wavelets and statistics* (pp. 125–150). New York: Springer.

Conference-Board (2001). Business cycles indicators handbook.

Daubechies, I. (1992). Ten lectures on wavelets. In *Series in applied mathematics: Vol. 61. CBSM-NSF regional conference*. Philadelphia: SIAM.

Engle, R. F. (1974). Band spectrum regression. *International Econonic Review*, *15*, 1–11.

Engle, R. F. (1978). Testing price equations for stability across spectral frequency bands. *Econometrica*, *46*, 869–881.

Gallegati, M. (2008). Wavelet analysis of stock returns and aggregate economic activity. *Computational Statistics & Data Analysis*, *52*, 3061–3074.

Gallegati, M., & Ramsey, J. B. (2012). Errors-in-variables and the wavelet multiresolution approximation approach: a Monte Carlo study. In B. H. Baltagi, R. Carter Hill, W. K. Newey, & H. L. White (Eds.), *Advances in econometrics: essays in honor of Jerry Hausman* (Vol. 29, pp. 149–171). Bingley: Emerald.

Gallegati, M., & Ramsey, J. B. (2013). Bond vs stock market's Q: testing for stability across frequencies and over time. *Journal of Empirical Finance*, *24*, 138–150.

Gallegati, M., Gallegati, M., Ramsey, J. B., & Semmler, W. (2011). The US wage Phillips curve across frequencies and over time. *Oxford Bulletin of Economics and Statistics*, *73*, 489–508.

Gallegati, M., Ramsey, J. B., & Semmler, W. (2013). Exploratory analysis by time scale using wavelets: interest rate spreads and output. *Axioms*, *2*, 182–207.

Gençay, R., Selçuk, F., & Whitcher, B. (2002). *An introduction to wavelets and other filtering methods in finance and economics*. San Diego: Academic Press.

Gençay, R., Selçuk, F., & Whitcher, B. (2005). Multiscale systemic risk. *Journal of International Money and Finance*, *24*, 55–70.

Hakkio, C. S., & Keeton, W. R. (2009). Financial stress: what is it, how can it be measured, and why does it matter? *Economic Review*, *94*, 5–50.

Hatzius, J., Hooper, P., Mishkin, F. S., Schoenholtz, K. L., & Watson, M. W. (2010). *Financial conditions indexes: a fresh look after the financial crisis* (Working Paper No. 16150). National Bureau of Economic Research.

Hollo, D., Kremer, M., & Duca, M. L. (2012). *CISS: a composite indicator of systemic stress in the financial system* (Working Paper No. 1426). European Central Bank.

Kaminsky, G. L. (1999). *Currency and banking crises: the early warning of distress* (Working Paper 99/1778). International Monetary Fund.

Kaminsky, G. L., & Reinhart, C. M. (1999). The twin crises: the causes of banking and balance-of-payments problems. *The American Economic Review*, *89*, 473–500.

Kaminsky, G. L., Lizondo, S., & Reinhart, C. M. (1998). IMF Staff Papers No. 45, International Monetary Fund.

Keim, M. J., & Percival, D. B. (2010). Assessing characteristic scales using wavelets. arXiv: 1007.4169.

Kim, S., & In, F. H. (2003). The relationship between financial variables and real economic activity: evidence from spectral and wavelet analyses. *Studies in Nonlinear Dynamics and Econometrics*, *7*, 1–18.

Kliesen, K. L., & Smith, D. C. (2010). *Measuring financial market stress* (Economic Synopses 2). Federal Reserve Bank of St. Louis.

Kliesen, K. L., Owyang, M. T., & Vermann, E. K. (2012). *Disentangling diverse measures: a survey of financial stress indexes* (Review 94). Federal Reserve Bank of St. Louis.

Lynch, P., & Zumbach, G. (2003). Market heterogeneities and the causal structure of volatility. *Quantitative Finance*, *3*, 320–331.

Mallat, S. (1989). A theory for multiresolution signal decomposition: the wavelet representation. *IEEE Transactions on Pattern Analysis*, *11*, 674–693.

Masset, P. (2008). Analysis of financial time-series using Fourier and wavelet methods.

Misina, M., & Tkacz, G. (2009). Credit, asset prices, and financial stress. *International Journal of Central Banking*, *4*, 95–122.

Moore, G. H., & Shiskin, J. (1967). *Indicators of business expansions and contractions*. New York: National Bureau of Economic Research.

Muller, U. A., Dacorogna, M. M., Dave, R. D., Pictet, O. V., Olsen, R. B., & Ward, J. R. (1993). *Fractals and intrinsic time—a challenge to econometricians*. Paper presented at the 39th international conference of the applied econometric association, Luxembourg, 22–26 November 1993.

Nason, G. P., & Silverman, B. W. (1995). The stationary wavelet transform and some statistical applications. In A. Antoniadis & G. Oppenheim (Eds.), *Wavelets and statistics* (pp. 281–300). New York: Springer.

Nilsson, R. (1987). *OECD leading indicators* (OECD Economic Studies No. 9). Organization for Economic Co-operation and Development.

Oet, M. V., Eiben, R., Bianco, T., Gramlich, D., & Ong, S. J. (2011). *The financial stress index: identification of system risk conditions* (Working Paper No. 1130). Federal Reserve Bank of Cleveland.

Percival, D. B. (1995). On estimation of the wavelet variance. *Biometrika*, *82*, 619–631.

Percival, D. B., & Walden, A. T. (2000). *Wavelet methods for time-series analysis*. Cambridge: Cambridge University Press.

Picchetti, P. (2008). *Wavelet based leading indicators of industrial activity in Brazil*. Paper presented at the 29th CIRET conference on business tendency surveys and policy formation, Santiago, 8–11 October 2008.

Ramsey, J. B., & Lampart, C. (1998a). The decomposition of economic relationships by time scale using wavelets: money and income. *Macroeconomic Dynamics*, *2*, 49–71.

Ramsey, J. B., & Lampart, C. (1998b). The decomposition of economic relationship by time scale using wavelets: expenditure and income. *Studies in Nonlinear Dynamics and Econometrics*, *3*, 23–42.

Ramsey, J. B., Gallegati, M., Gallegati, M., & Semmler, W. (2010). Instrumental variables and wavelet decomposition. *Economic Modeling*, *27*, 1498–1513.

Rua, A., & Nunes, L. C. (2005). Coincident and leading indicators for the euro area: a frequency band approach. *International Journal of Forecasting*, *21*, 503–523.

Schleicher, C. (2002). *An introduction to wavelets for economists* (Working Paper Series 2002-3). Bank of Canada.

Serroukh, A., Walden, A. T., & Percival, D. B. (2000). Statistical properties and uses of the wavelet variance estimator for the scale analysis of time series. *Journal of the American Statistical Association*, *95*, 184–196.
Whitcher, B., Guttorp, P., & Percival, D. B. (2000). Wavelet analysis of covariance with application to atmospheric series. *Journal of Geophysical Research. Atmospheres*, *105*, 14941–14962.

Part II
Non-linearities in Other Fields of Research

Least Absolute Deviation Based Unit Root Tests in Smooth Transition Type of Models

Rickard Sandberg

Abstract Building on work by Phillips (Econ. Theory 7:450–463, 1991), we derive LAD based unit root tests in a first-order ESTAR model with strong mixing innovations. Further theoretical results are derived and LAD based unit root tests in general nonlinear first-order dynamic models admitting a Taylor-series approximation are thereby easily obtained.

Finite sample properties of the tests are explored using Monte Carlo experiments. The results show that the size properties of the tests are satisfactory, and the power against stationary ESTAR alternatives with innovational outliers is significantly higher than the power of the LS based unit root tests by Kapetanios et al. (J. Econ. 112:359–379, 2003) and Rothe and Sibbertsen (Allg. Stat. Arch. 90:439–456, 2006). In contrast, the LS based tests are more powerful than our tests in the case of stationary ESTAR models with Gaussian errors (no outliers).

In an empirical application to eight real effective exchange rates (for major economies), evidence of the PPP hypothesis is supported for six of the countries using our tests. If LS based tests are instead used, the PPP hypothesis is supported for three countries only (countries for which the PPP hypothesis is also supported by our tests).

1 Introduction

Nowadays, there is a rich literature on unit root testing in nonlinear models, and the tests involved have played a critical role in disentangling the resilience of the unit root hypothesis in economic time series; see Enders and Granger (1998), Bec et al. (2004), Kapetanios et al. (2003), Kapetanios and Shin (2006), Rothe and Sibbertsen (2006) and Kiliç (2011) among others. However, the unit root tests in the aforementioned work are, in one way or another, based on least squares (LS). In the context of aberrant observations, this might be an inappropriate estimation method. For instance, it is well known that classical LS based unit root tests in linear models

R. Sandberg (✉)
Department of Economic Statistics, Department of Economics, Stockholm School of Economics, 113 83 Stockholm, Sweden
e-mail: rickard.sandberg@hhs.se

F. Schleer-van Gellecom (ed.), *Advances in Non-linear Economic Modeling*, Dynamic Modeling and Econometrics in Economics and Finance 17,
DOI 10.1007/978-3-642-42039-9_4,

(Dickey-Fuller type of tests) are sensitive to the effect of outliers in finite samples. That is, outliers may not only give rise to (seriously) size distorted tests, the power properties of the tests may also be adversely affected (see, e.g., Lucas 1995a, 1995b; Herce 1996; Franses and Haldrup 1994; Shin et al. 1996; Vogelsang 1999 and Haldrup et al. 2011). Naturally, one may expect that similar small sample problems transfer to LS based unit root tests in nonlinear models, and other more robust estimation methods may be called for.

A step towards alleviating the problem of outliers in linear unit root regressions was taken by Herce (1996) who derived tests based on the least absolute deviation (LAD) estimator. This approach is novel in that an LAD estimator is more robust in the case of non Gaussian errors and yields the maximum likelihood estimator assuming double exponentially distributed errors. As such, the LAD estimator is expected to be robust against errors with heavy tails or so-called innovational outliers (IOs), but it is not necessarily robust against outliers in the factor space or so-called additive outliers (AOs) (see, e.g., Lucas 1995a; also discussed in our section of simulations).

Our goal with this work is to consider LAD based unit root tests in smooth transition regression (STR) models in the presence of IOs. Both economic theory and (overwhelming) empirical findings support nonlinearities in economic and financial time series, and STR models are found to be successful at rendering such behavior, see van Dijk et al. (2002), Lundbergh et al. (2003), and Teräsvirta et al. (2010) and the references therein. For the same reasons, in the last decade or so, the STR models have also become a powerful alternative to linear models when testing the unit root hypothesis (see the aforementioned references on unit root testing in nonlinear models). Furthermore, economic and financial time series are often observed during periods of turbulence and uncertainty and the presence of IOs, which often represents the onset of an external cause (Tsay 1988) like a financial crisis (say), is quite likely.[1] Towards this end, even though nonlinearities and IOs are two features that are likely to be present in financial and economic data, IO robust unit root tests (and outlier robust tests in general) in STR models are, to the best of our knowledge, lacking in the literature.[2]

In our framework, IOs have a permanent effect under the null hypothesis and yield a unit root process with level shift(s). Under the alternative hypothesis, they have a temporary effect due to the (presumed) ergodicity properties of the STR models, but the regime switching behavior is affected. For instance, IOs can offset the pulling forces from one regime toward another regime, and it may be the case that a recession (one extreme regime, say) becomes even deeper instead of turning into a boom (another extreme regime, say). In addition, observations that are close to one regime can be pushed away to another regime, thus causing "extra" regime shifts. Such phenomena are exemplified in our application.

[1] Whereas a typical example of an AO is the recording error (Tsay 1988).

[2] van Dijk et al. (1999) consider outlier robust linearity tests in various STR models, but they assume a stationary autoregressive process under the null hypothesis of linearity.

Having noticed that IOs may add to the complexity of the nonlinear time series model, using Monte Carlo experiments, we will examine small sample properties of LS and LAD based unit root tests in STR models with IOs. In these simulations, it is demonstrated that IOs under the null hypothesis, irrespective of a non-robust or a robust approach, constitute a less severe problem (i.e. IOs that are mistaken for nonlinearities and causing spurious rejections of the null hypothesis is not a severe problem). However, it is shown that IOs under the alternative hypothesis can make the power of the LS based tests substantially inferior to the power of the LAD based tests (i.e. using an LAD based unit root test yields that nonlinear processes contaminated with IOs are less frequently mistaken for unit root processes). These simulation based findings are expected to be important to practitioners and the analysis of economic variables (say) where economic theory supports a nonlinear behavior (e.g. nonlinear adjustments around a long-run level) but the time series is not overly long (e.g., quarterly based post Bretton Woods variables only having about 160 observations) yet spanning over, e.g., the 2008–2009 global financial crises. Finally, in our simulation studies, we also show that the LAD based tests are relatively efficient in the case of no outliers (NOs).

Our analysis of LAD estimators is facilitated by a first-order Taylor-series approximation of STR models in the context of linearity testing as described in Luukkonen et al. (1988) and Granger and Teräsvirta (1993), as well as "shortcut" asymptotics of Phillips (1991) which entail a generalized first-order approximation of the objective function (which is not differentiable in the common sense). We will use the exponential smooth transition autoregressive (ESTAR) model (see Tong 1990) as a benchmark model throughout this work. The main reasons for this benchmark are implied tractable results and that it is widely used among practitioners. Nonetheless, general asymptotic theory is presented in the Appendix such that LAD based unit root tests in a wide class of first-order dynamic STR models are easily derived.

In an application, one of our LAD tests as well as other common unit root tests are applied to real effective exchange rates (REER) for eight (major) economies. The outcomes of the LAD test are advantageous in the sense that they give the strongest support among all tests for the purchasing power parity (PPP) hypothesis. Furthermore, the Australian REER series (one of the eight series) is analyzed in more detail, and a number of outliers (some of them are related to macroeconomic events) are identified. The impact of these outliers on the regime switching behavior of an estimated ESTAR model is also discussed.

The rest of the paper is organized as follows: Section 2 defines the ESTAR model and the LAD estimator. The tests and theoretical results are established in Section 3. Monte Carlo experiments are conducted in Section 4. An application is considered in Section 5. Concluding remarks are presented in Section 6. Mathematical proofs are given in the Appendix.

A few words on the notation in this work: $\xrightarrow{p}$ abbreviates convergence in probability, $x \triangleq y$ indicates that x is defined by y, $\mathcal{D}\lfloor[0, 1]$ denotes the space of cadlag functions on $[0, 1]$, $\Rightarrow$ signifies weak convergence with respect to the Skorohod metric (as defined in Billingsley 1968), $\|X\|_p = (\sum_i \mathrm{E}|X_i|^p)^{1/p}$ denotes the L_p-norm for vectors, A^q abbreviates the qth-order Hadamard product ($\odot$) of a vector or a matrix A. Integrals of the type $\int_0^1 B(s)ds$ and $\int_0^1 B(s)dB'(s)$ are denoted $\int B$ and $\int BdB'$ for short. Finally, $[x]$ signifies the integer part of x.

2 ESTAR Model

Consider a stochastic process $\{X_t\}$ generated by a first-order ESTAR model

$$X_t = \alpha X_{t-1} + \gamma X_{t-1} G(X_{t-1};\theta) + u_t, \quad t = 1, \ldots, n, \tag{1}$$

where the starting value X_0 is assumed to be known (either fixed or stochastic), (α, γ) are unknown parameters, and u_t is an error term, the properties of which will be discussed in detail below. The continuous function $\psi(X_{t-1}, \theta) : \mathbb{R} \times \mathbb{R}_+ \to [0, 1]$ is defined by

$$G(X_{t-1};\theta) \triangleq 1 - \exp\{-\theta X_{t-1}^2\}, \tag{2}$$

and is symmetrically u-shaped in X_{t-1} with limiting properties

$$\lim_{|X_{t-1}|\to\infty} G(X_{t-1};\theta) = 1 \quad (\theta \in \mathbb{R}_+) \quad \text{and} \quad \lim_{|X_{t-1}|\to 0} G(X_{t-1};\theta) = 0.$$

As such, the model in (1) with (2) defines a regime-switching model with two outer regimes and one corridor regime characterized by the dynamic roots $\alpha + \gamma$ and α, respectively. Parameter θ is a smoothness parameter and controls for the velocity of transition between regimes. It is now straightforward to show that $|\alpha + \gamma| < 1$ is a sufficient condition for the process $\{X_t\}$ to be geometrically ergodic (see e.g. Kapetanios et al. 2003, p. 362). Furthermore, as is common in practice, the restriction $\alpha = 1$ is also imposed and implies that the process $\{X_t\}$ accommodates a unit root in the corridor regime but remains geometrically ergodic provided that $-2 < \gamma < 0$. These restrictions can be justified by economic theory, see Balke and Fomby (1997), Michael et al. (1997), and Kapetanios et al. (2003), and are further discussed in our application. The model of interest can now be written as

$$\Delta X_t = \gamma X_{t-1} G(X_{t-1};\theta) + u_t. \tag{3}$$

It is evident that the unit root hypothesis in (3) can either be tested by $H_0 : \gamma = 0$ or $H_0 : \theta = 0$, under which (3) reduces to the random walk $\Delta X_t = u_t$. Under the alternative hypothesis, it is assumed that $-2 < \gamma < 0$ and $\theta > 0$. To circumvent the problem of unidentified parameters under the null hypothesis of a random walk and facilitate the testing situation, a Taylor-series approximation of the model in (3) is used. Other possible solutions are the simulation based methods by e.g. Andrews and Ploberger (1994) and Hansen (1996). However, numerical recipes are not tractable in our case. Substitute for a first-order Taylor-series approximation of G around $\theta = 0$ into (3) to obtain

$$\Delta X_t = \pi X_{t-1}^3 + u_t^*, \tag{4}$$

where $\pi = \theta\gamma$, $u_t^* = u_t + R(\theta)$, and $R(\theta)$ is a remainder term satisfying $R(0) = 0$. Thus, $u_t^* = u_t$ whenever $\theta = 0$. It also follows that the originally stated null hypothesis of a unit root is transformed to $\tilde{H}_0 : \pi = 0$. Noteworthy, as pointed out in Luukkonen et al. (1988), the auxiliary regression equation in (4) should not be seen as any sensible time series model; it only serves as an auxiliary regression equation for our testing situation. The LAD estimator of π in (4) is now defined by

$$\hat{\pi}_n \triangleq \arg\min_{\pi \in \mathbb{R}} \left(n^{-1} \sum_{t=1}^{n} \left| \Delta X_t - \pi X_{t-1}^3 \right| \right). \tag{5}$$

A main goal of this work is to derive the limiting distribution for $\hat{\pi}_n$ under a unit root assumption. It should be mentioned that Kapetanios et al. (2003) and Rothe and Sibbertsen (2006) have already derived the limiting distribution for the LS estimator of π in (4) when the underlying process is a random walk.

In order to derive the limiting distribution for $\hat{\pi}_n$, the techniques of Phillips (1991) or Herce (1996) may be used. For convenience, the approach by the former author is followed closely, and we will therefore rely upon shortcut asymptotics for LAD estimators. In this sequel, it is noted that a first-order condition for $\hat{\pi}_n$ in (5) is given by

$$n^{-1} \sum_{t=1}^{n} \operatorname{sgn}(\Delta X_t - \hat{\pi}_n X_{t-1}^3) X_{t-1}^3 = 0, \tag{6}$$

where $\operatorname{sgn}(x) = -1$ if $x < 0$ and $\operatorname{sgn}(x) = 1$ if $x \geq 0$ (the sign function). Proceed by a generalized first-order Taylor-series expansion (see Phillips 1991, pp. 452–453) of (6) around zero to obtain

$$0 = n^{-1} \sum_{t=1}^{n} \operatorname{sgn}(u_t^*) X_{t-1}^3 - 2n^{-1} \left(\sum_{t=1}^{n} \delta(u_t^*) X_{t-1}^6 \right) \hat{\pi}_n + \xi, \tag{7}$$

where $d/dx(\operatorname{sgn}(x)) = 2\delta(x)$ is used and $\delta(x)$ is the (generalized) delta function (see Gelfand and Shilov 1964), and $\xi = 2 \sum_{k=2}^{\infty} (-1)^k (k!)^{-1} n^{-1} \sum_{t=1}^{n} \delta^{(k-1)}(u_t^*) \times X_{t-1}^3 (X_{t-1}^3 \hat{\pi}_n)^k$ is a remainder term where $\delta^{(j)}(u_t)$ denotes successive derivatives of the delta function. The expression (7) is a key result on which the asymptotic theory for $\hat{\pi}_n$ is built.

3 Tests

To accommodate various forms of temporal dependence and heteroscedasticity, the following assumption is imposed on the error term in the ESTAR model.

Assumption (i) For some $p > \beta > 2$, $\{u_t\}$ is a zero-mean, strong mixing sequence with mixing coefficients α_m of size $-p\beta/(p-\beta)$ and $\sup_{t\geq 1}\|u_t\|_p = \kappa < \infty$. (ii) The errors $\{u_t\}$ have the median zero and probability density f that is positive and analytic at zero.

Assumption (i) is frequently utilized in a unit root context (see, e.g., Phillips 1987 and Phillips and Perron 1988) and corresponds to Assumption 1(i) in Herce (1996) except the sequence $\{u_t\}$ not being strictly stationary. Moreover, Assumption (ii) is similar to Assumption 1(ii) in Herce (1996) except that f is assumed to be analytic instead of continuous at zero. This is a somewhat stronger assumption but, on the other hand, yields that we may adopt the theory of generalized Taylor-series expansions as in Phillips (1991).

Next, define $w_t^* \triangleq \text{sgn}(u_t^*)$, where it is noted that $w_t = \text{sgn}(u_t)$ holds under $\tilde{H}_0$. Moreover, define the vector processes $v_t \triangleq (u_t, w_t)'$ and $V_t \triangleq \sum_{i=1}^{t} v_i = (U_t, W_t)'$, where $\{v_t\}$ defines a strong vector mixing sequence of the same order as $\{u_t\}$ (see Theorem 14.1 in Davidson 1994) and also assuming that

$$\Omega = \begin{bmatrix} \sigma_1^2 & \sigma_{12} \\ \sigma_{12} & \sigma_2^2 \end{bmatrix} = \lim_{n\to\infty} (1/n)\mathrm{E}V_n(1)V_n(1)' < \infty,$$

a multivariate invariance principle yields (see, e.g., Theorem 2.1 in Phillips and Durlauf 1986)

$$V_n(s) = n^{-0.5}V_{[ns]} \quad \Rightarrow \quad B(s) = \begin{bmatrix} B_1(s) & B_2(s) \end{bmatrix}',$$

where $s \in [0,1]$ and $B(s)$ is a bivariate Brownian motion with the covariance matrix Ω. In the following discussion, it also turns out to be convenient to define the matrix containing the serial correlation, viz

$$\Lambda = \begin{bmatrix} \Lambda_{11} & \Lambda_{12} \\ \Lambda_{21} & \Lambda_{11} \end{bmatrix} \triangleq \lim_{n\to\infty} (1/n) \sum_{i=1}^{n} \sum_{j=i+1}^{\infty} \mathrm{E}v_i v_j'.$$

We are now ready to define LAD estimator based test statistics for testing the unit root hypothesis in (3). Thus, a coefficient and a t-type of test statistic are defined by $C_\pi \triangleq n^2\hat{\pi}_n$ and $t_\pi \triangleq \hat{\pi}_n/\sqrt{V_n}$, respectively, where $V_n = \hat{\sigma}_{1,1}^2(\sum_t y_{t-1}^6)^{-1}$ and $\hat{\sigma}_{1,1}^2$ is a consistent estimator of the error variance $\sigma_{1,1}^2 = \lim_{n\to\infty} n^{-1}\sum_t \mathrm{E}u_t^2$. Obviously, the choice of variance estimator is not unique, but it is noted that V_n has the same variance estimator expression as that employed by Kapetanios et al. (2003) and Rothe and Sibbertsen (2006) to construct an LS based t-type of unit root tests in the ESTAR model. Thereby, a transparent comparison of LS and LAD based unit root tests becomes possible since any differences in the performance of the tests can (more or less) be attributed to differences in the performance of LS and LAD estimators.

We now have the following large sample results for our tests.

Theorem 1 *Let* $\{u_t\}$ *satisfy assumption with* $\beta = 6$. *Consider the model* (4) *when* $\tilde{H}_0$ *holds. Then,*

$$\text{(i)} \quad C_\pi \Rightarrow \left(2f(0)\int B_1^6\right)^{-1}\left(\int B_1^3 dB_2 + 3\Lambda_{12}\int B_1^2\right),$$

$$\text{(ii)} \quad t_\pi \Rightarrow (2f(0))^{-1}\left(\sigma_{1,1}^2\int B_1^6\right)^{-1/2}\left(\int B_1^3 dB_2 + 3\Lambda_{12}\int B_1^2\right).$$

Proof See the Appendix. □

The moment condition $\beta = 6$ in Theorem 1 is a sufficient condition to yield weak convergence results. The same condition was imposed by Kapetanios et al. (2003) and Rothe and Sibbertsen (2006) to ensure weak convergence results for LS based unit root tests in the ESTAR model. Furthermore, it is noted that the limiting distributions in (i) and (ii) depend on a number of nuisance parameters as well as the density function f. This makes the test virtually useless in practice, and even though the term Λ_{12} equals zero in the case of a martingale difference sequence or independent and identically distributed (iid) errors, the parameter measuring the covariance between the marginal processes B_1 and B_2 (i.e. $\sigma_{12} = \mathrm{E}(u_t w_t) = \mathrm{E}|u_t|$) is non-zero. To remedy the problem of nuisance parameters in the results of Theorem 1, we define the modified test statistics

$$\tilde{C}_\pi \triangleq 2\hat{f}(0)\hat{\sigma}_1^3\hat{\sigma}_2^{-1} \times C_\pi - 3n^2\hat{\Lambda}_{12}\hat{\sigma}_1^3\hat{\sigma}_2^{-1}\sum_t X_{t-1}^2\left(\sum_t X_{t-1}^6\right)^{-1}, \tag{8}$$

and

$$\tilde{t}_\pi \triangleq 2\hat{f}(0)\hat{\sigma}_{1,1}\hat{\sigma}_2^{-1} \times t_\pi - 3\hat{\Lambda}_{12}\hat{\sigma}_2^{-1}\sum_t X_{t-1}^2\left(\sum_t X_{t-1}^6\right)^{-1/2}, \tag{9}$$

where $\hat{f}(0)$, $\hat{\sigma}_1^2$, $\hat{\sigma}_2^2$, $\hat{\sigma}_{1,1}^2$, and $\hat{\Lambda}_{12}$ signify some consistent estimators (discussed in more detail below) of $f(0)$, σ_1^2, σ_2^2, $\sigma_{1,1}^2$, and Λ_{12}, respectively. The following results are central.

Corollary 2 *Under the conditions of Theorem* 1,

$$\text{(i)} \quad \tilde{C}_\pi \Rightarrow \left(\int b_1^6\right)^{-1}\left(\int b_1^3 db_2\right),$$

$$\text{(ii)} \quad \tilde{t}_\pi \Rightarrow \left(\int b_1^6\right)^{-1/2}\left(\int b_1^3 db_2\right),$$

where $b_1 = B_1/\sigma_1$ *and* $b_2 = B_2/\sigma_2$.

Proof See the Appendix. □

It is seen that the limiting distributions in Corollary 2 are similar to the limiting distributions in Kapetanios et al. (2003) and Rothe and Sibbertsen (2006) except that the measure of integration in our case equals b_2 (and not b_1). In fact, this implies that our large sample results still depend on the correlation between the marginal processes b_1 and b_2, say ρ, which varies with the distributional choice of u_t. This is somewhat unfortunate, but in the simulation studies below, it is demonstrated that the limiting distributions in Corollary 2 assuming Gaussian errors yield fairly close approximations to limiting distributions assuming non Gaussian errors (provided that the correlation in the case of Gaussian errors is reasonably close to the correlation in the case of non Gaussian errors). It should finally be noted that the problem of the nuisance parameter ρ can be circumvented in a similar way as outlined in Herce (1996). More precisely, also in our case it is possible to construct test statistics that are nuisance parameter free and have either conditional or unconditional normal limiting distributions.[3] Albeit such large sample results being tractable, it can be shown that the resultant tests are not consistent against ESTAR models with Gaussian errors (cf. the discussion in Herce 1996 on inconsistent LAD based unit root tests in linear autoregressive models with Gaussian errors). In addition, it can be shown that the power against ESTAR alternatives accommodating outliers of such tests is inferior (for a wide range of parameter values) to the power obtained by LS based tests; and the power of the LAD based tests increases much more slowly with the sample size than the power for LS based tests.[4]

Using the techniques of Park and Phillips (1988), the results in Corollary 2 can be generalized so that they also apply to time series with trends. As such, let $\{X_t\}$ be a random walk without drift, then the rth-order detrended series, X_t^r, reads $X_t^0 = X_t$ for $r = 0$ and $X_t^r = X_t - \hat{\gamma}_0 - \hat{\gamma}_1 t - \cdots - \hat{\gamma}_{r-1} t^{r-1}$ for $r = 1, 2, \ldots$, where $(\hat{\gamma}_0, \hat{\gamma}_1, \ldots, \hat{\gamma}_{r-1})'$ are the LS estimators obtained by regressing X_t on $(1, t, \ldots, t^{r-1})$. The continuous time-analogue is now given by an rth-order detrended Brownian motion $B_{1,r}(s) = B_1(s) - \gamma_0 - \gamma_1 s - \cdots - \gamma_{r-1} s^{r-1}$, and under the conditions of Theorem 1, it follows that

$$\tilde{C}_{\pi,r} \Rightarrow \left(\int b_{1,r}^6 \right)^{-1} \left(\int b_{1,r}^3 db_2 \right), \tag{10}$$

and

$$\tilde{t}_{\pi,r} \Rightarrow \left(\int b_{1,r}^6 \right)^{-1/2} \left(\int b_{1,r}^3 db_2 \right), \tag{11}$$

where $\tilde{C}_{\pi,r}$ and $\tilde{t}_{\pi,r}$ are computed as in (8) and (9), respectively, but X_t^r is used instead of X_t, and $b_{1,r}$ signifies an rth-order detrended standard Brownian motion.

---

[3] Such theoretical results are available upon request.

[4] Such simulation results are available upon request.

Table 1 Critical values for the $\tilde{C}_{\pi,r}$ and $\tilde{t}_{\pi,r}$ statistics in the case of Gaussian errors

Statistic	n	Probability of a smaller value			Statistic	n	Probability of a smaller value		
		0.01	0.05	0.10			0.01	0.05	0.10
$\tilde{C}_{\pi,0}$	100	−100.71	−41.22	−23.45	$\tilde{t}_{\pi,0}$	100	−2.99	−2.25	−1.88
	200	−110.19	−43.21	−24.05		200	−2.92	−2.22	−1.87
	500	−112.80	−43.40	−24.10		500	−2.85	−2.18	−1.85
	∞	−115.02	−45.52	−25.25		∞	−2.54	−1.90	−1.58
$\tilde{C}_{\pi,1}$	100	−178.52	−89.81	−59.67	$\tilde{t}_{\pi,1}$	100	−3.58	−2.84	−2.48
	200	−204.03	−99.15	−64.43		200	−3.53	−2.83	−2.47
	500	−216.28	−102.32	−65.58		500	−3.44	−2.79	−2.46
	∞	−221.43	−107.99	−68.29		∞	−3.13	−2.53	−2.24
$\tilde{C}_{\pi,2}$	100	−315.90	−184.22	−134.51	$\tilde{t}_{\pi,2}$	100	−4.03	−3.25	−2.86
	200	−377.69	−212.96	−153.49		200	−3.98	−3.24	−2.86
	500	−416.42	−229.38	−162.14		500	−3.88	−3.21	−2.85
	∞	−418.47	−237.73	−189.87		∞	−3.59	−3.01	−2.70

Note: The results are based on 1,000,000 replications

4 Simulation Experiments

Empirical distribution for the $\tilde{C}_{\pi,r}$ and $\tilde{t}_{\pi,r}$ test statistics and the cases $r = 0, 1, 2$, are obtained by simulations using the DGP

$$X_t = X_{t-1} + u_t, \quad t = 1, \ldots, n, \qquad X_0 = 0$$

where $u_t \overset{\text{iid}}{\sim} N(0, 1)$. Conventional critical values for these empirical distributions and the sample sizes $n \in \{100, 200, 500\}$ are reported in Table 1. In addition, asymptotic critical values are also reported in Table 1, and they are obtained using sums with $n = 100{,}000$ as discrete approximations of standard Brownian motions. Furthermore, in order to operationalize $\tilde{C}_{\pi,r}$ and $\tilde{t}_{\pi,r}$, the LAD estimator $\hat{\pi}_n$ is computed by an iterative weighted least square algorithm (see, e.g., Huber 1981). The covariance matrix Ω is then consistently estimated by the Newey-West estimator (see Newey and West 1987) where the lag-truncation parameter is set at $[4(n/100)^{2/9}]$.[5] As a consistent kernel estimator of the density $f(0)$, we used the Gaussian kernel with a bandwidth equal to $0.9T^{-1/5}\min\{\hat{\sigma}_1, \text{IQR}/1.34\}$ (see Silverman 1986) where IQR denotes the interquartile range.

In the second simulation experiment, the empirical size and power properties of the $\tilde{t}_{\pi,1}$ (demeaned data) statistic are examined. The finite sample properties of the

[5] Using the decomposition $\Omega = \Sigma_0 + \Lambda + \Lambda'$ where $\Sigma_0 = \lim_{n\to\infty}(1/n)\sum_{i=1}^{n} \mathrm{E}v_i v_i'$, it follows that the Newey-West estimator also yields consistent estimators of $\hat{\sigma}^2_{1,1}$ (the (1,1)-element of $\hat{\Sigma}_0$) and $\hat{\Lambda}_{12}$ (the (1,2)-element of $\hat{\Lambda}$).

other tests are similar and therefore not reported. But, it should be mentioned that the $\tilde{C}_{\pi,r}$ statistic is generally more powerful than the $\tilde{t}_{\pi,r}$ statistic but with worse size properties, and the power of the $\tilde{C}_{\pi,r}$ and $\tilde{t}_{\pi,1}$ statistics decreases with the order of r.

The DGP in these simulations is the ESTAR model in (3) with parameter values $\theta = 0$ (size); $\theta \in \{0.01, 0.05, 0.10, 1\}$ and $\gamma \in \{-0.10, -0.50\}$ (power) which amount to a sub-set of parameters used in Kapetanios et al. (2003), except for the exclusion of the (less challenging) cases $\gamma = -1.0$ and $\gamma = -1.5$.[6] The parameter values under the alternative hypothesis comprise ESTAR models with a unit root in the middle regime and an AR(1) process with dynamic roots 0.90 (when $\gamma = -0.10$) or 0.50 (when $\gamma = -0.50$) in the two outer regimes, and the velocity of transitions between regimes is varied from relatively low ($\theta = 0.01$) to relatively high ($\theta = 1$). Moreover, for the error term in (3), we will consider four distributions: (i) $u_t \overset{\text{iid}}{\sim} N(0,1)$, (ii) $u_t = \epsilon_t + 5\omega_t$ where $\epsilon_t \overset{\text{iid}}{\sim} N(0,1)$ and $P(\omega_t = -1) = P(\omega_t = 1) = 0.025$ and $P(\omega_t = 0) = 0.95$ and ϵ_t and ω_t are independent, (iii) $u_t \overset{\text{iid}}{\sim} t(6)$, and (iv) $u_t \overset{\text{iid}}{\sim} t(2.5)$. The first case is a benchmark case and yields a model with NOs, and the latter cases all give rise to models with IOs. More specifically, the design in case (ii) gives rise to negative and positive IOs proportional to a factor 5 occurring with probability 0.025. If the error term follows this distribution we write $u_t \sim \omega$. Cases (iii) and (iv) imply that large (absolute) values of the innovations are more frequently observed and yield heavy tail models (yet with finite variances). In fact, case (iv) yields a situation where the moment condition in Theorem 1 is violated, but is interesting from the point of view that evidence of heavy tails is frequently found in economic and financial time series. Finally, as compared to the $\tilde{t}_{\pi,1}$ test, the power results for the LS based unit root t-tests by Rothe and Sibbertsen (2006, p. 445), abbreviated $Z_{NL}(t)$, are also reported.[7] It is noted that the $Z_{NL}(t)$ test is designed to direct the power against ESTAR models in a less outlier robust way. The experiments are conducted for sample sizes $n \in \{100, 200\}$, and 5 % asymptotic critical values from Table 1 are used. The outcomes are reported in Tables 2A and 2B.

Consider first the size-properties of the tests shown in Tables 2A and 2B and the column with a unit root heading. Here, it is seen that the empirical size of the $t_{\pi,1}$ test is close to the nominal size for all distributional cases. A main reason for such results is that the correlation between b_1 and b_2 in cases (ii)–(iv) is reasonably close to the correlation in the case of Gaussian errors (case (i); used in the simulations of the critical values).[8] That is, even though the limiting distributions in Corollary 2 depend on ρ, it appears that the critical values in Table 1 can be used for distributions

[6] The power results for these cases are available upon request.

[7] The nuisance parameters appearing in the definition of the $Z_{NL}(t)$ statistic are estimated by the Newey-West estimator with a lag-truncation parameter set at the same value as that for the $t_{\pi,1}$ statistic.

[8] The correlation between b_1 and b_2 for the four cases is: $\rho_{(\text{i})} = 0.7982$, $\rho_{(\text{ii})} = 0.6704$, $\rho_{(\text{iii})} = 0.7499$, and $\rho_{(\text{iv})} = 0.5465$.

Table 2A Rejection frequencies of LS and LAD based unit root tests in the ESTAR model

$n=100$ Dist. u_t	Test	Unit root $\theta=0$	ESTAR $\gamma=-0.50$ $\theta=0.01$	$\theta=0.05$	$\theta=0.10$	$\theta=1$	$\gamma=-0.10$ $\theta=0.01$	$\theta=0.05$	$\theta=0.10$	$\theta=1$
$N(0,1)$	$Z_{NL}(t)$	0.055	0.232 (0.219)	0.712 (0.702)	0.880 (0.871)	0.945 (0.936)	0.092 (0.101)	0.148 (0.135)	0.182 (0.171)	0.231 (0.165)
	$\tilde{t}_{\pi,1}$	0.065	0.192 (0.158)	0.406 (0.360)	0.524 (0.480)	0.591 (0.559)	0.086 (0.081)	0.139 (0.109)	0.145 (0.112)	0.220 (0.137)
ω	$Z_{NL}(t)$	0.080	0.510 (0.358)	0.925 (0.862)	0.970 (0.933)	0.969 (0.937)	0.118 (0.072)	0.160 (0.095)	0.171 (0.104)	0.177 (0.106)
	$\tilde{t}_{\pi,1}$	0.059	0.531 (0.498)	0.831 (0.808)	0.877 (0.860)	0.872 (0.855)	0.165 (0.145)	0.242 (0.212)	0.274 (0.241)	0.259 (0.231)
$t(6)$	$Z_{NL}(t)$	0.079	0.575 (0.428)	0.930 (0.865)	0.966 (0.921)	0.952 (0.901)	0.118 (0.071)	0.157 (0.087)	0.169 (0.097)	0.170 (0.100)
	$\tilde{t}_{\pi,1}$	0.058	0.587 (0.548)	0.823 (0.794)	0.868 (0.846)	0.845 (0.819)	0.213 (0.181)	0.288 (0.245)	0.292 (0.253)	0.288 (0.251)
$t(2.5)$	$Z_{NL}(t)$	0.100	0.790 (0.597)	0.969 (0.898)	0.979 (0.915)	0.970 (0.901)	0.116 (0.046)	0.122 (0.053)	0.126 (0.052)	0.126 (0.053)
	$\tilde{t}_{\pi,1}$	0.057	0.821 (0.800)	0.953 (0.944)	0.955 (0.946)	0.941 (0.929)	0.394 (0.357)	0.468 (0.424)	0.460 (0.420)	0.446 (0.410)

Notes: The results are based on 10,000 replications, and 5 % asymptotic critical values are used. Size-adjusted power is reported in parentheses

Table 2B Rejection frequencies of LS and LAD based unit root tests in the ESTAR model

$n=200$ Dist. u_t	Test	Unit root $\theta=0$	ESTAR $\gamma=-0.50$ $\theta=0.01$	$\theta=0.05$	$\theta=0.10$	$\theta=1$	$\gamma=-0.10$ $\theta=0.01$	$\theta=0.05$	$\theta=0.10$	$\theta=1$
$N(0,1)$	$Z_{NL}(t)$	0.054	0.675 (0.669)	0.988 (0.986)	0.998 (0.997)	0.998 (0.996)	0.168 (0.164)	0.400 (0.395)	0.509 (0.507)	0.537 (0.531)
	$\tilde{t}_{\pi,1}$	0.064	0.381 (0.341)	0.737 (0.704)	0.841 (0.816)	0.852 (0.835)	0.155 (0.127)	0.255 (0.215)	0.292 (0.252)	0.293 (0.257)
ω	$Z_{NL}(t)$	0.065	0.949 (0.921)	0.999 (0.998)	1.00 (1.00)	1.00 (1.00)	0.267 (0.196)	0.477 (0.372)	0.491 (0.393)	0.462 (0.367)
	$\tilde{t}_{\pi,1}$	0.045	0.861 (0.870)	0.989 (0.990)	0.996 (0.997)	0.993 (0.994)	0.367 (0.385)	0.523 (0.544)	0.554 (0.573)	0.534 (0.555)
$t(6)$	$Z_{NL}(t)$	0.077	0.964 (0.912)	0.999 (0.995)	1.00 (0.999)	1.00 (0.997)	0.268 (0.155)	0.419 (0.271)	0.438 (0.290)	0.401 (0.269)
	$\tilde{t}_{\pi,1}$	0.048	0.893 (0.897)	0.989 (0.991)	0.996 (0.998)	0.988 (0.992)	0.428 (0.438)	0.565 (0.578)	0.573 (0.586)	0.561 (0.573)
$t(2.5)$	$Z_{NL}(t)$	0.098	0.994 (0.971)	1.00 (0.997)	1.00 (0.999)	1.00 (0.999)	0.282 (0.103)	0.336 (0.128)	0.328 (0.126)	0.305 (0.116)
	$\tilde{t}_{\pi,1}$	0.041	0.985 (0.988)	1.00 (1.00)	1.00 (1.00)	0.998 (1.00)	0.720 (0.761)	0.796 (0.832)	0.805 (0.841)	0.785 (0.829)

Notes: The results are based on 10,000 replications, and 5 % asymptotic critical values are used. Size-adjusted power is reported in parentheses other than the Gaussian. Finally, it is seen that the size of the $Z_{NL}(t)$ test is (about) satisfactory in the case of Gaussian errors, and somewhat too liberal in the case of IOs.

Next, consider the power of the tests shown in Table 2A ($n = 100$) and the columns with an ESTAR heading. Since the $Z_{NL}(t)$ is somewhat over-sized in the case of IOs, we shall focus on size-adjusted power (shown in parentheses). Starting with the results for the benchmark case of NOs, it is evident that the $Z_{NL}(t)$ test is the preferable test for all parameter values considered. This is to be expected, and even though substantial power gains (e.g., 0.391 units in the case $\gamma = -0.50$ and $\theta = 0.10$) by the $Z_{NL}(t)$ test over the $\tilde{t}_{\pi,1}$ test are seen, the power of the $\tilde{t}_{\pi,1}$ is non trivial.[9]

Next, turn to the power results in the cases of IOs in Table 2A. For case (ii) and all parameter values considered, it is interesting to note a marked increase in power for the $\tilde{t}_{\pi,1}$ test as compared to the benchmark. For the $Z_{NL}(t)$ test, an increase in power as compared to the benchmark case is also seen if less persistent outer regimes ($\gamma = -0.50$) are considered, whereas a decrease in power is demonstrated in the presence of more persistent regimes ($\gamma = -0.10$). Taken together, this implies that the $Z_{NL}(t)$ test is, in general, more powerful than the $\tilde{t}_{\pi,1}$ test when $\gamma = -0.50$, and vice versa when $\gamma = -0.10$. For case (iii), the findings for the tests are similar to those for case (ii). Finally, even though the results for case (iv) should be interpreted with caution, it is seen that the performance of the $\tilde{t}_{\pi,1}$ test is advantageous. In fact, when $\gamma = -0.10$, the power of the $Z_{NL}(t)$ test is trivial or close to trivial and substantial power gains (e.g., 0.371 units in the case $\gamma = -0.10$ and $\theta = 0.05$) can be obtained using the $\tilde{t}_{\pi,1}$ test.

Increasing the sample size, it is seen in Table 2B ($n = 200$) that the ranking of the tests found in Table 2A, more or less, also holds here. Furthermore, when $\gamma = -0.50$ the power is satisfactory for both tests in most of the cases (i)–(iv) (an exception is for the $\tilde{t}_{\pi,1}$ test in a vicinity of the null hypothesis and NOs). It is also noted that when $\gamma = -0.10$, the power of the $Z_{NL}(t)$ test in the case of NOs has increased faster with the sample size than the power of the $\tilde{t}_{\pi,1}$ test, and vice versa in the cases of IOs. In particular, for case (iv) the power of the $Z_{NL}(t)$ is still noticeably low and evident power gains (e.g., 0.715 units in the case $\gamma = -0.10$ and $\theta = 0.10$) from the $t_{\pi,1}$ test are found.

In further simulation studies, which are not reported here, the $\tilde{t}_{\pi,1}$ and $Z_{NL}(t)$ test statistics were found to be too conservative (approximately of the same magnitude) in the case of autoregressive errors and to yield far too high (but similar) rejection frequencies in the case of moving average errors; properties that are characteristic for semi-parametric LS and LAD based unit root tests (e.g., see Schwert 1989 and Herce 1996). This also implies that LAD and LS tests are expected to be grossly over-sized in the cases of AOs because a moving average type autocorrelation component is then introduced (see Franses and Haldrup 1994).

[9]The findings on non trivial power for our test stand in contrast to the findings by the LAD based unit root tests of Herce (1996) in linear regressions with Gaussian errors. The reason for this is that we use another form of an LAD based test statistic which can be shown to be consistent against alternatives with Gaussian errors.

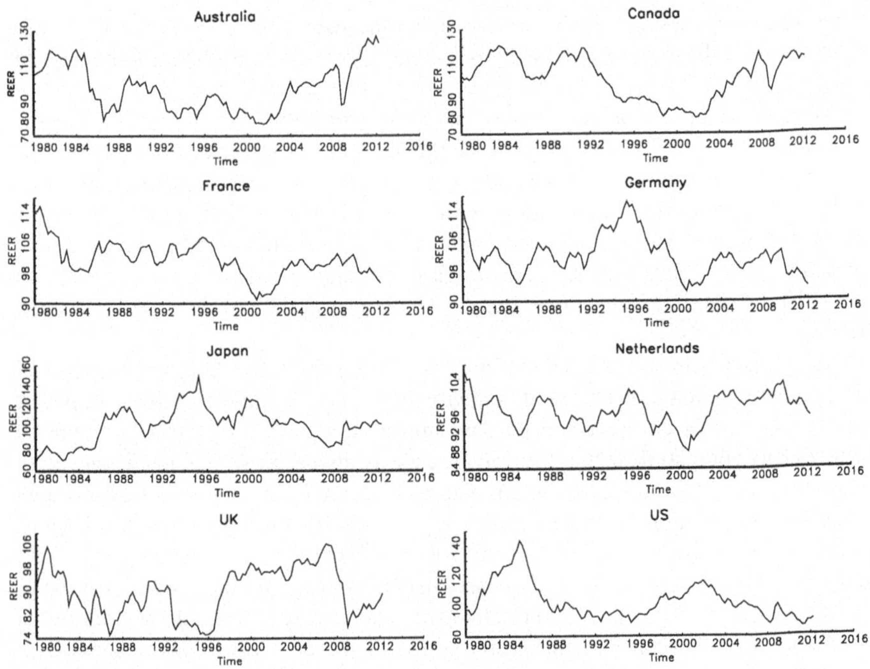

Fig. 1 Real effective exchange rates

5 Application: Real Effective Exchange Rates

In this empirical study, REER index series (CPI based and index 100 at 2005) are analyzed for the following countries: Australia (AU), Canada (CA), France (FR), Germany (GE), Japan (JA), Netherlands (NE), United Kingdom (UK), and the United States of America (US). The series are obtained from IMF's *International Financial Statistics* and they range from 1980Q1 to 2012Q2 (quarterly based and yield 130 observations). The series are displayed in Figure 1.

The notion of nonlinearities in exchange rate adjustments around a long-run level (the PPP) is not controversial, and it can be justified by economic theory. For instance, Taylor et al. (2001) and Kapetanios et al. (2003) argue that there exists a band for the equilibrium real exchange rate where there is no tendency of the real rate to revert to its equilibrium value, but outside this band commodity arbitrage becomes profitable, forcing the real exchange rate towards the band. An obvious candidate to resemble such nonlinear adjustments is the geometrically ergodic ESTAR model described above with a unit root in the corridor regime and a stationary autoregressive process in the outer regimes. Moreover, a pure unit root hypothesis in exchange rate series also plays a key role and simply means that PPP does not hold. Not surprisingly, testing the unit root hypothesis in exchange rate series has attracted a considerable amount of research. But, despite the theoretical foundations of nonlinear adjustment processes and allowing for nonlinear alternatives (rather

Table 3 Testing the unit root hypothesis in real effective exchange rate series

Country	PP	H	$Z_{NL}(t)$	$\tilde{t}_{\pi,1}$
AU	−1.207	−0.054	−1.407	−3.008**
CA	−0.739	−0.849	−1.981	−1.285
FR	−2.651*	−0.822	−2.040	−2.861**
GE	−2.247	−0.317	−2.290	−2.489*
JA	−2.265	−0.595	−2.842*	−2.612*
NE	−2.939**	−0.841	−3.22**	−3.268**
UK	−1.965	−0.034	−2.685*	−3.098*
US	−1.208	−0.541	−1.405	−1.872

**Significant at the 5 percent level. *Significant at the 10 percent level

than linear alternatives) when the PPP hypothesis is tested, the resilience of the unit root hypothesis is still noteworthy in many exchange rate series. Therefore, it is interesting to examine whether more outlier robust methods help shed some light on the PPP puzzle.

The persistence of the eight REER series is examined by the $\tilde{t}_\pi$ and $Z_{NL}(t)$ tests, the unit root test in Phillips and Perron (1988) (hereafter the *PP* test), and the LAD based unit root test in linear models by Herce (1996) (signified as the H test).[10] By visual inspection of the graphs in Figure 1, it appears that the series have a non-zero mean and are non-trending. As such, the $\tilde{t}_\pi$ and $Z_{NL}(t)$ tests are based on demeaned data ($r = 1$), and the *PP* and H tests are based on first-order autoregressive models with an intercept. The results for these unit root tests are reported in Table 3.

In Table 3, it is seen that the unit root hypothesis is rejected for six of the countries (AU, FR, GE, NE, and UK) using our robust test. Instead using the $Z_{NL}(t)$ test, support for the PPP hypothesis is only found for three countries (JA, NE, and UK). The *PP* test also entails weak support for the PPP hypothesis, and only two rejections are encountered (FR and NE). The H test does not seem very useful, and sometimes even the "wrong" sign of the test is obtained. Finally, we conclude that a case of a rejection by any of the other tests is also a case of a rejection by our test.

Inspired by the strong rejection of our test and the non-rejection of the other tests, we proceed by maximum likelihood estimation (MLE) of the ESTAR model in (3) for the AU series with $Y_t \triangleq X_t - a_0$, where X_t is the observed series and a_0 is an equilibrium parameter corresponding to the level of the PPP. In order to allow for IOs, we assume that the error term follows the modified t-distribution: $u_t \overset{\text{iid}}{\sim} \sigma\sqrt{v/(v-2)}t(v)$, where σ is a scale parameter capturing "excess" variation in the data.[11] The estimation is carried out in GAUSS 12 using the BHHH algorithm

[10] All these tests are similar in the respect that they entail a non-parametric estimation of serial correlation and heteroscedasticity (and the Newey-West estimator is applied for all tests in this work).

[11] Using one of the specification tests given below, we also conclude that Gaussian errors are not appropriate for the AU series.

in the CML library. We obtain the estimates:

$$\hat{a}_0 = \underset{\substack{(5.228)\\ [7.326]}}{102.687},$$

$$\Delta Y_t = \underset{\substack{(0.034)\\ [0.055]}}{-0.0689}\, Y_{t-1}\big(1-\exp\{\underset{\substack{(0.021)\\ [0.029]}}{-0.0786}\, Y_{t-1}^2\}\big) + \hat{u}_t, \tag{12}$$

$$\hat{\sigma} = \underset{\substack{(0.681)\\ [0.762]}}{4.077}, \qquad \hat{\upsilon} = \underset{\substack{(1.248)\\ [1.591]}}{3.567},$$

where ordinary and heteroscedasticity-robust standard errors are given below the parameter estimates in parentheses and brackets, respectively.[12] In (12), the estimated equilibrium of the series equals 102.687, and the magnitude of adjustments towards this equilibrium is modest because the estimated dynamic coefficient in the outer regimes is highly persistent ($\hat{\gamma} = -0.0689$) and comprises half-lives of PPP deviations of 2.43 years. Moreover, the estimated speed of transition between regimes ($\hat{\theta} = 0.0786$) yields a smooth u-shaped function in Y_{t-1} (illustrated in Figure 2).[13] The estimated model also supports heavy tails ($\hat{\sigma} = 4.077$, $\hat{\upsilon} = 3.567$) (yet with finite variances). To this end, because a relatively small sample is considered, and nonlinearities with highly persistent outer regimes and heavy tails are entertained (all parameters enter significantly into the model), it is not surprising that it is only our test that rejected the null hypothesis of a unit root.[14]

To assess the model adequacy in (12), we also computed (figures in parentheses after the test statistics denote p-values): AIC = 1.219 and BIC = 1.220 which can be compared to AIC = 1.244 and BIC = 1.246 for a unit root process; the Ljung-Box (LB) test of no remaining 4th and 8th-order serial correlation LB(4) = 6.216(0.185) and LB(8) = 8.681(0.369), respectively; the Lomnicki, Jarque, and Bera (LJB) normality test for the residual series LJB = 36.821(0.000); the LM test of no autoregressive conditional heteroscedasticity of first and fourth-order ARCH(1) = 0.002(0.964) and ARCH(4) = 2.388(0.648), respectively; the LM-type tests of Eitrheim and Teräsvirta (1996, pp. 61–69) for the hypothesis of no remaining nonlinearity (LM_1) and that of parameter constancy (LM_2) to obtain $LM_1 = 1.665(0.178)$ and $LM_2 = 1.481(0.223)$.[15] Taking these diagnostic checking results together, it appears that the estimated ESTAR model with a unit root corridor

[12] Asymptotic normality of the MLE's follows since the estimated ESTAR model satisfies the ergodicity condition discussed in Section 2.

[13] As is common in practice, the estimate of the scale-free version of the speed of transition parameter is reported in (12), i.e. the estimate of θ/s_{Y_t}, where s_{Y_t} denotes the standard deviation of the Y_t series.

[14] The PP and H tests are designed to direct power against linear models using non-robust and robust estimation methods, respectively. Moreover, the $Z_{NL}(t)$ test is powerful against ESTAR alternatives, but as demonstrated in our simulations, it is not necessarily powerful against ESTAR alternatives with persistent outer regimes and error terms with heavy tails.

[15] We used the F-test versions of the LM tests.

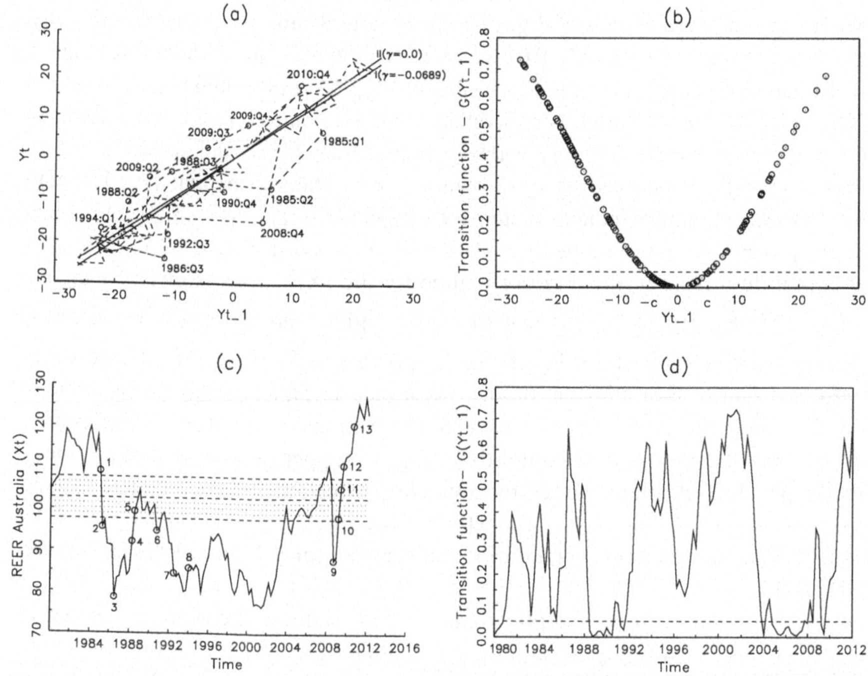

Fig. 2 (**a**) Scatter plot, estimated skeleton, and outliers, (**b**) speed of transition and the outer and the corridor regimes, (**c**) Australian REER series, band of equilibrium, and outliers, (**d**) regime switching behavior

regime and heavy tails in (12) approximates the properties of the AU REER series fairly well.

To facilitate the identification of possible outliers in the AU REER series and provide further insights of the dynamic behavior of the estimated ESTAR model in (12), we will use the graphs in Figure 2.

Panel (a) in Figure 2 shows a scatter plot (dashed-lines) and the estimated skeleton (solid lines) of the ESTAR model corresponding to the two extreme regimes ($\gamma = 0$ and $\gamma = -0.0689$ comprise the unit root regime and the stationary regime, respectively). By visual inspection, a number of outlying observations not belonging to the bulk of the data are discerned. Proceeding in a more formal way, we sequentially searched for IOs using the test of Tsay (1988) (at a 5 % significance level) with a linear first-order autoregressive filter.[16] Using his test, we found 13 IOs (all marked by a circle and a date). In panel (b), the estimated transition function $G(Y_{t-1})$ is shown (and each circle represents an observation) and, as already mentioned, the speed of transition between the extreme regimes is smooth, and the outer extreme regimes are, in fact, never reached. Furthermore, in a subsequent anal-

[16] Obviously, a linear filter is not appropriate in our case since we found rather strong evidence in favor of an ESTAR model. As such, the outcomes of the Tsay test should only be seen as indicative.

ysis, it turns out to also be useful to identify an equilibrium band (and thereby also the outer regimes) for the AU REER series. This band can be determined using the estimated $G(Y_{t-1})$ function. More specifically, the equilibrium band is a region where $\Delta Y_t \approx u_t$ holds which is equivalent to assume that $G(Y_{t-1}) \approx 0$, and as an arbitrary choice we let $G(Y_{t-1}) = 0.05$ (dashed line) (≈ 0) and solve for a (absolute) value Y_{t-1}, denoted by c, to obtain $c \approx 5$. This means that $\{Y_t\}$ is a unit root process (or approximately a unit root process) if $Y_{t-1} \in [-5, 5]$ and a stationary process (or approximately a stationary process) if $Y_{t-1} \notin [-5, 5]$. Hence, the equilibrium band or the corridor regime for the $\{X_t\}$ series is given by $\hat{a}_0 \pm c$ or $II \triangleq [97.687, 107.687]$, and that the outer regimes are given by the intervals $I^- \triangleq (-\infty, 97.687)$ and $I^+ \triangleq (107.687, \infty)$. In panel (c) the AU REER series is graphed (solid line) with the outliers identified in panel (a) (given by a circle and chronological numbering) as well as the estimated equilibrium ($\hat{a}_0$; central dashed line) and the band of equilibrium ($\hat{a}_0 \pm c$; upper and lower dashed lines). Finally, panel (d) displays the regime switching behavior of the estimated ESTAR model (solid line). Using $G(Y_{t-1}) = 0.05$ (dashed line), a unit root behavior for the AU REER series is roughly seen during the periods: early 1980's, 1989–1992, and 2004–2007.

As a last exercise in this application, we will examine the outliers (shown in panels (a) and (b) in Figure 2) and their potential effects on the regime switching behavior. This is done by linking each of the two coordinates (X_{t-1}, X_t) of an outlier to one of the regimes I^+, I^-, and II, and thereafter check: (i) if a regime switch has occurred or not from time $t-1$ to time t, and (ii) characterize the impact of the outlier in terms of adjustments toward (or possibly away from) the PPP. Further comments and a summary of this exercise are given in Table 4. Naturally, the analysis adherent to Table 4 is "naive" in the sense that it depends on our (arbitrary) procedure to determine c defining the equilibrium band and the different regimes, our outlier detection procedure, and that some of the events described below can also occur in the absence of outliers.

The contents of Table 4 are best analyzed using some examples. Starting with outlier one, we see that it has coordinates $(117.58, 109.14) \in (I^+, I^+)$, and thereby it is concluded that no regime shift has occurred. Moreover, since ΔX_t is negative, an adjustment towards the equilibrium takes place. Outlier two has the coordinates $(109.14, 95.67) \in (I^+, I^-)$, and no regime shift has occurred. But, ΔX_t is in this case so largely negative that the target zone is missed (under-shooting). As a final example, the third outlier causes no regime shift since $(91.01, 78.63) \in (I^-, I^-)$, but since ΔX_t is negative in this case, an adjustment away from the target zone takes place and the depreciation of the Australian dollar is instead magnified. Towards this end, some of the outliers that we found can be linked to macroeconomic events. For instance, the first three outliers are related to a sizeable decline in terms-of-trade (around 15 %) between March 1985 and March 1987 (which in fact was concentrated to two large movements in February 1985 and July 1986 which correspond to outliers 2 and 3, respectively). The last four outliers can be attributed to the global financial crises during 2008–2009 and its aftermath.

Table 4 Outliers and regime switching behavior

Outlier number	Date (t)	Coordinates	$X_{t-1} \to X_t$	Regime shift	Characterization	Adjustment (ΔX_t)
1	1985:Q1	(117.58, 109.14)	$I^+ \to I^+$	no	adjustment towards equilibrium	−8.44
2	1985:Q2	(109.14, 95.67)	$I^+ \to I^-$	no	under-shooting	−13.47
3	1986:Q3	(91.01, 78.63)	$I^- \to I^-$	no	adjustment away from equilibrium[a]	−12.38
4	1988:Q2	(84.83, 92.11)	$I^- \to I^-$	no	adjustment towards equilibrium	7.28
5	1988:Q3	(92.11, 99.43)	$I^- \to II$	yes	adjustment towards equilibrium	7.32
6	1990:Q4	(101.01, 94.75)	$II \to I^-$	yes	adjustment away from equilibrium	−6.26
7	1992:Q3	(91.53, 84.50)	$I^- \to I^-$	no	adjustment away from equilibrium[a]	−7.03
8	1994:Q1	(80.35, 85.75)	$I^- \to I^-$	no	adjustment towards equilibrium	5.40
9	2008:Q4	(107.72, 87.70)	$I^+ \to I^-$	no	under-shooting	−20.02
10	2009:Q2	(88.42, 98.11)	$I^- \to II$	yes	adjustment towards equilibrium	9.69
11	2009:Q3	(98.11, 105.20)	$II \to II$	no	adjustment around equilibrium	7.09
12	2009:Q4	(105.20, 110.81)	$II \to I^+$	yes	adjustment away from equilibrium	5.61
13	2010:Q4	(113.96, 120.41)	$I^+ \to I^+$	no	adjustment away from equilibrium[b]	6.45

Notes: [a]signifies depreciation increased, [b]signifies appreciation increased

6 Concluding Remarks

In this work, we have established the asymptotic distributions for LAD based unit root tests in a first-order ESTAR model with mixing innovations. Inspired by the shortcut asymptotics for LAD estimators of Phillips (1991) in linear models, the derivation of our results is considerably simplified by using a generalized first-order Taylor-series expansion of the objective function defining the LAD estimator. Further theoretical results are derived (given in the Appendix), and LAD based unit root tests in general nonlinear first-order dynamic models (and not only the ESTAR model) admitting a Taylor-series approximation are thereby easily obtained.

By means of simulations, we obtain finite sample and asymptotic critical values for the tests. Even though these critical values depend upon the distributional choice of the error term, it is demonstrated that the critical values obtained by assuming Gaussian errors can be adopted also in non-Gaussian cases without altering the size-properties of the tests too much. In further simulation studies, we assess the finite sample performance of one of our tests and the LS based test in ESTAR models of Kapetanios et al. (2003) and Rothe and Sibbertsen (2006). It is shown that the empirical size of our test in the case of IOs is reasonably close to the nominal size, whereas the LS based test is somewhat over-sized. In terms of power, our test is found to be advantageous if the DGP is an ESTAR model with IOs and persistent outer regimes. In the case of an ESTAR model with NOs or IOs and less persistent outer regimes, the LS based test is preferable.

In an application to real effective exchange rate series for eight (major) economies, we found that our LAD based unit root test rejected the unit root hypothesis for six series, whereas the corresponding LS based test only comprised a rejection for three series (series for which also our test entailed rejections). For the Australian exchange rate series, we also proceeded with model estimation rendering a first-order ESTAR process with persistent outer regimes and heavy tails. In addition, 13 outliers (some of them could be linked to macroeconomic events) were identified, and the potential impact of these outlying observations on the regime switching behavior of the estimated ESTAR was discussed.

Acknowledgements An earlier version of this work was presented at Aarhus University and the Center for Research in Econometric Analysis of Time Series (CREATES). Comments by the CREATES seminar participants are gratefully acknowledged. I am also thankful for financial support from Jan Wallander's and Tom Hedelius' Research Foundations, Grant No. P2012-0085:1.

Appendix

The below Lemma facilitates the proof of Theorem 1.

Lemma *Define*

$$\int_0^s V_n^q dV_n' \triangleq n^{-(q+1)/2} \sum_{i=1}^{[ns]} V_i^q v_{i+1}', \tag{13}$$

where v_i and V_i are defined in Section 3, and $q \in \mathbb{Z}_+$.[17] *Then, given assumption with $\beta = \max(3, 2q)$ and $\mathrm{E}V_n^{2q-4} = O(n^{q-2})$ for integers $q > 3$, the weak convergence result*

$$\int_0^s V_n^q dV_n' \Rightarrow \int_0^s B^q dB' + q \int_0^s \left[B^{q-1} \quad B^{q-1} \right] dr \odot \Lambda, \tag{14}$$

holds as $n \to \infty$.

Remark Because the result in the Lemma holds for $q \in \mathbb{Z}_+$, it is straightforward to derive LAD based unit root tests in models other than a first-order ESTAR model (as long as the models admit a Taylor-series expansion) and the order of the approximation may be set arbitrarily.

Proof Set $\mathcal{F}_t = \sigma(v_i : i \leq t)$ to be the smallest sigma-algebra containing the past history of $\{v_t\}$. Denote $\mathrm{E}(X|\mathcal{F}_i)$ by $\mathrm{E}_i X$, and define $\epsilon_i \triangleq \sum_{k=0}^{\infty} (\mathrm{E}_i v_{i+k} - \mathrm{E}_{i-1} v_{i+k})$ and $z_i \triangleq \sum_{k=1}^{\infty} \mathrm{E}_i v_{i+k}$. By Hansen (1992), it follows that we can write

$$v_i = \epsilon_i + z_{i-1} - z_i, \tag{15}$$

where ϵ_i is an $\mathcal{F}_i$-measurable sequence with $\mathrm{E}_{i-1}\epsilon_i = 0$, and $\{\epsilon_i, \mathcal{F}_i\}$ is a martingale difference sequence. Using the decomposition in (15) and letting $Y_n(s) = n^{-1/2} \sum_{i=1}^{[ns]} \epsilon_i$, the stochastic integral in (13) can be expressed as

$$\int_0^s V_n^q dV_n' = \int_0^s V_n^q dY_n' + \Lambda_n^*(s), \tag{16}$$

where $\tilde{\Lambda}_n^*(s)$ is a bias term, and it is not hard to show that

$$\tilde{\Lambda}_n^*(s) = n^{-(q+1)/2} \sum_{i=1}^{[ns]} (V_i^q - V_{i-1}^q) z_i' + o_p(1).$$

The first term on the rhs in (16) obeys, by Theorem 3.1 of Hansen (1992), $\int_0^s V_n^q dY_n' \Rightarrow \int_0^1 B^q dB'$ since $V_n^q(s) \Rightarrow B^q(s)$ by a continuous mapping theorem (CMT). The bias term on the rhs in (16) is analyzed substituting for the Binomial expansion of $V_i^q = (V_{i-1} + v_i)^q$ to obtain (ignoring the $o_p(1)$-term)

$$\begin{aligned} \tilde{\Lambda}_n^*(s) &= n^{-(q+1)/2} \sum_{i=1}^{[ns]} (V_i^q - V_{i-1}^q) z_i' \\ &= n^{-(q+1)/2} \sum_{i=1}^{[ns]} \left(\sum_{m=1}^{q} c_m^q V_{i-1}^{q-m} \odot v_i^m \right) z_i', \end{aligned} \tag{17}$$

[17] It is assumed that the zero element is contained in $\mathbb{Z}_+$.

where c_m^q is a Binomial coefficient. Moreover, define $e_i \triangleq v_i z_i' - \Lambda$, then the term corresponding to $m = 1$ in (17) yields (where $c_1^q = q$)

$$\begin{aligned}
\frac{q}{n^{(q+1)/2}} \sum_{i=1}^{[ns]} (V_{i-1}^{q-1} \odot v_i) z_i' &= \frac{q}{n^{(q+1)/2}} \sum_{i=1}^{[ns]} [\, V_{i-1}^{q-1} \quad V_{i-1}^{q-1} \,] \odot (v_i z_i') \\
&= \frac{q}{n^{(q+1)/2}} \sum_{i=1}^{[ns]} [\, V_{i-1}^{q-1} \quad V_{i-1}^{q-1} \,] \odot e_i \\
&\quad + \frac{q}{n^{(q+1)/2}} \sum_{i=1}^{[ns]} [\, V_{i-1}^{q-1} \quad V_{i-1}^{q-1} \,] \odot \Lambda \\
&\Rightarrow q \int_0^s [\, B^{q-1} \quad B^{q-1} \,] dr \odot \Lambda
\end{aligned}$$

as $n \to \infty$. This follows since

$$\sup_{s \in [0,1]} \left| \frac{q}{n^{(q+1)/2}} \sum_{i=1}^{[ns]} [\, V_{i-1}^{q-1} \quad V_{i-1}^{q-1} \,] \odot e_i \right| \xrightarrow{p} 0$$

by Theorem 3.3 in Hansen (1992), and

$$\frac{q}{n^{(q+1)/2}} \sum_{i=1}^{[ns]} [\, V_{i-1}^{q-1} \quad V_{i-1}^{q-1} \,] \odot \Lambda \Rightarrow q \int_0^s [\, B^{q-1} \quad B^{q-1} \,] dr \odot \Lambda$$

as $n \to \infty$ by a CMT. Moreover, the term for $m = 2$ in (17) reads

$$\frac{c_2^q}{n^{(q+1)/2}} \sum_{i=1}^{[ns]} (V_{i-1}^{q-2} \odot v_i^2) z_i',$$

and by the triangle inequality and Hölder's inequality, we have that

$$\begin{aligned}
\mathrm{E} \sup_{s \in [0,1]} \left| \frac{c_2^q}{n^{(q+1)/2}} \sum_{i=1}^{[ns]} (V_{i-1}^{q-2} \odot v_i^2) z_i' \right| &\le \mathrm{E} \sup_{s \in [0,1]} \frac{c_2^q}{n^{(q+1)/2}} \left| \sum_{i=1}^{[ns]} (V_{i-1}^{q-2} \odot v_i^2) z_i' \right| \\
&\le \frac{c_2^q}{n^{(q+1)/2}} \sum_{i=1}^{n} \mathrm{E} \left| (V_{i-1}^{q-2} \odot v_i^2) z_i' \right| \\
&\le \frac{c_2^q}{n^{(q+1)/2}} \sum_{i=1}^{n} \| V_{i-1}^{q-2} \|_2 \| v_i^2 z_i \|_2 \\
&\le \frac{c_2^q}{n^{(q+1)/2}} \sum_{i=1}^{n} \| V_{i-1}^{q-2} \|_2 \| v_i \|_6^2 \| z_i \|_6 \to 0.
\end{aligned}$$

To see this, note that $\|v_i^2 z_i\|_2 \leq \|v_i\|_6^2 \|z_i\|_6 < \infty$ because $\|v_i\|_6 < \infty$ by assumption, $\|z_i\|_6 < 6\kappa \sum_{k=1}^{\infty} \alpha_k^{1/6-1/p} < \infty$ by Minkowski's inequality and McLeish's strong mixing inequality (McLeish 1975) and

$$
\begin{aligned}
\frac{1}{n^{(q+1)/2}} \sum_{i=1}^{n} \| V_{i-1}^{q-2} \|_2 &\leq \frac{1}{n^{(q+1)/2}} \sum_{i=1}^{n} \| V_n^{q-2} \|_2 \\
&= \frac{1}{n^{(q+1)/2}} \sum_{i=1}^{n} (\mathrm{E}U_n^{2q-4} + \mathrm{E}W_n^{2q-4})^{1/2} \\
&= n^{-1/2} \big(n^{-(q-2)} (\mathrm{E}U_n^{2q-4} + \mathrm{E}W_n^{2q-4})\big)^{1/2} \\
&= O\big(n^{-1/2}\big) \to 0
\end{aligned}
$$

as $n \to \infty$ since $\mathrm{E}V_n^{2q-4} = O(n^{q-2})$ by assumption.

Furthermore, it is straightforward to show that subsequent terms corresponding to $m = 3, 4, \ldots, q$ are also $o_p(1)$, but it is interesting to note that the term corresponding to $m = q - 1$ $(q > 1)$ comprises a sufficient moment condition for weak convergence in (14). More specifically, let $m = q - 1$, then

$$
\begin{aligned}
\mathrm{E} \sup_{s \in [0,1]} \left| \frac{c_{q-1}^q}{n^{(q+1)/2}} \sum_{i=1}^{[ns]} (V_{i-1} \odot v_i^{q-1}) z_i' \right| &\leq \mathrm{E} \sup_{s \in [0,1]} \frac{c_{q-1}^q}{n^{(q+1)/2}} \sum_{i=1}^{[ns]} \left| (V_{i-1} \odot v_i^{q-1}) z_i' \right| \\
&\leq \frac{c_{q-1}^q}{n^{(q+1)/2}} \sum_{i=1}^{n} \mathrm{E} \left| (V_{i-1} \odot v_i^{q-1}) z_i' \right| \\
&\leq \frac{c_{q-1}^q}{n^{(q+1)/2}} \sum_{i=1}^{n} \| V_{i-1} \|_2 \| v_i \|_{2q}^{q-1} \| z_i \|_{2q} \to 0
\end{aligned}
$$

as $n \to \infty$ if $\beta = 2q$ since then $\|v_i\|_{2q} < \infty$ and $\|z_i\|_{2q} < \infty$ (in particular, $\|z_i\|_{2q} \leq 6\kappa \sum_{k=1}^{\infty} \alpha_k^{1/2q-1/p} < \infty$ by once again applying the Minkowski's inequality and McLeish's strong mixing inequality).

The proof is complete. □

Proof of Theorem 1 The proof of (i). Scale (7) with n^{-1} to yield

$$
n^{-2} \sum_{t=1}^{n} \mathrm{sgn}(u_t) X_{t-1}^3 = 2 A_n n^2 \hat{\pi}_n,
$$

where

$$A_n = n^{-4} \sum_{t=1}^{n} \delta(u_t) X_{t-1}^6$$
$$+ \sum_{j=1}^{\infty} (-1)^{j+1} ((j+1)!)^{-1} \left(n^{-4} \sum_{t=1}^{n} \delta^{(j)}(u_t) X_{t-1}^6 (X_{t-1}^3 \hat{\pi}_n)^j \right).$$

Assuming that the second term of A_n is $o_p(1)$, it follows that

$$n^{-2} \sum_{t=1}^{n} \operatorname{sgn}(u_t) X_{t-1}^3 = 2 \left(n^{-4} \sum_{t=1}^{n} \delta(u_t) X_{t-1}^6 + o_p(1) \right) n^2 \hat{\pi}_n. \tag{18}$$

Furthermore,

$$n^{-4} \sum_{t=1}^{n} X_{t-1}^6 \Rightarrow \int B_1^6,$$

by a CMT, and noticing that $\mathrm{E}\delta(u_t) = f(0)$ we can apply Theorem 3.3 of Hansen (1992) to obtain

$$\sup_{s \in [0,1]} \left| n^{-4} \sum_{t=1}^{[ns]} \left(\delta(u_t) - f(0) \right) X_{t-1}^6 \right| \xrightarrow{p} 0,$$

and hence,

$$n^{-4} \sum_{t=1}^{n} \delta(u_t) X_{t-1}^6 \Rightarrow f(0) \int B_1^6. \tag{19}$$

By the $(1,2)$ element in (14) with $q = 3$, we also have that

$$n^{-2} \sum_{t=1}^{n} \operatorname{sgn}(u_t) X_{t-1}^3 \Rightarrow \int B_1^3 dB_2 + 3\Lambda_{12} \int B_1^2. \tag{20}$$

Combining the results in (18)–(20) gives the desirable result.

Finally, the claim that the second term of A_n is $o_p(1)$ as $n \to \infty$ follows by using arguments similar to those used in Phillips (1991, p. 455). More specifically, noticing $\mathrm{E}\delta^{(j)}(u_t) = (-1)^j f^{(j)}(0)$, we have, along the lines proving (16), that $n^{-4} \sum_{t=1}^{n} \delta^{(j)}(u_t) X_{t-1}^6 \Rightarrow (-1)^j f^{(j)}(0) \int B_1^6$. Finally, since $X_{t-1}^3 \hat{\pi}_n \xrightarrow{p} 0$ (uniformly in t), the claim follows.

The proof of (ii). Trivial. □

Proof of Corollary 2 Given consistent estimators of nuisance parameters and the density function, the proof of (i) and (ii) is trivial noticing that we can write $\int B_1^2 = \sigma_1^2 \int b_1^2$, $\int B_1^6 = \sigma_1^6 \int b_1^6$, and $\int B_1^3 dB_2 = \sigma_1^3 \sigma_2 \int b_1^3 db_2$. □

References

Andrews, D. W., & Ploberger, W. (1994). Optimal tests when a nuisance parameter is present only under the alternative. *Econometrica, 62*, 1383–1414.

Balke, N. S., & Fomby, T. B. (1997). Threshold cointegration. *International Economic Review, 38*, 627–645.

Bec, F., Ben Salem, M., & Carrasco, M. (2004). Tests for unit-root versus threshold specification with an application to the purchasing power parity relationsship. *Journal of Business & Economic Statistics, 22*, 382–395.

Billingsley, P. (1968). *Convergence of probability measures*. New York: Wiley.

Davidson, J. (1994). *Stochastic limit theory*. London: Oxford University Press.

Eitrheim, Ø., & Teräsvirta, T. (1996). Testing the adequacy of smooth transition autoregressive models. *Journal of Econometrics, 74*, 59–75.

Enders, W., & Granger, C. (1998). Unit root tests and asymmetric adjustment with an example using the term structure of interest rates. *Journal of Business & Economic Statistics, 16*, 304–311.

Franses, P. H., & Haldrup, N. (1994). The effects of additive outliers on tests for unit roots and cointegration. *Journal of Business & Economic Statistics, 12*, 471–478.

Gelfand, I. M., & Shilov, G. E. (1964). *Generalized functions* (Vol. 1). New York: Academic Press.

Granger, C. W., & Teräsvirta, T. (1993). *Modelling nonlinear economic relationships*. London: Oxford University Press.

Haldrup, N., Montañes, A., & Sansó, A. (2011). Detection of additive outliers in seasonal time series. *Journal of Time Series Econometrics, 3*, 1–18.

Hansen, B. E. (1992). Convergence to stochastic integrals for dependent heterogeneous processes. *Econometric Theory, 8*, 489–501.

Hansen, B. E. (1996). Inference when a nuisance parameter is not identified under the null hypothesis. *Econometrica, 64*, 413–430.

Herce, M. H. (1996). Asymptotic theory of LAD estimation in a unit root process with finite variance errors. *Econometric Theory, 12*, 129–153.

Huber, P. J. (1981). *Robust statistics*. New York: Wiley.

Kapetanios, G., & Shin, Y. (2006). Unit root tests in three-regime SETAR models. *Econometrics Journal, 9*, 252–278.

Kapetanios, G., Shin, Y., & Snell, A. (2003). Testing for a unit root in the nonlinear STAR framework. *Journal of Econometrics, 112*, 359–379.

Kiliç, R. (2011). Testing for a unit root in a stationary ESTAR process. *Econometric Reviews, 30*, 274–302.

Lucas, A. (1995a). Unit root tests based on M estimators. *Econometric Theory, 11*, 331–346.

Lucas, A. (1995b). An outlier robust unit root test with an application to the extended Nelson-Plosser data. *Journal of Econometrics, 66*, 153–173.

Lundbergh, S., Teräsvirta, T., & van Dijk, D. (2003). Time-varying smooth transition autoregressive models. *Journal of Business & Economic Statistics, 21*, 104–121.

Luukkonen, R., Saikkonen, P., & Teräsvirta, T. (1988). Testing linearity against smooth transition autoregressive models. *Biometrika, 75*, 491–499.

McLeish, D. L. (1975). A maximal inequality and dependent strong laws. *The Annals of Probability, 3*, 829–839.

Michael, P., Nobay, R. A., & Peel, D. A. (1997). Transaction costs and nonlinear adjustment in real exchange rates; an empirical investigation. *Journal of Political Economy, 105*, 862–879.

Newey, W. K., & West, K. D. (1987). A simple, positive definite, heteroskedasticity and autocorrelation consistent covariance matrix. *Econometrica, 55*, 277–301.

Park, J. Y., & Phillips, P. C. B. (1988). Statistical inference in regressions with integrated processes: part 1. *Econometric Theory, 4*, 468–497.

Phillips, P. C. B. (1987). Time series regression with a unit root. *Econometrica, 55*, 277–301.

Phillips, P. C. B. (1991). A shortcut to LAD estimator asymptotics. *Econometric Theory, 7*, 450–463.

Phillips, P. C. B., & Durlauf, S. N. (1986). Multiple time series regression with integrated processes. *Review of Economic Studies*, *53*, 473–496.

Phillips, P. C. B., & Perron, P. (1988). Testing for unit root in time series regression. *Biometrika*, *75*, 335–346.

Rothe, C., & Sibbertsen, P. (2006). Phillips-Perron-type unit root tests in the nonlinear ESTAR framework. *Allgemeines Statistisches Archiv*, *90*, 439–456.

Schwert, G. W. (1989). Tests for unit roots: a Monte Carlo investigation. *Journal of Business & Economic Statistics*, *7*, 147–159.

Shin, D. W., Sarkar, S., & Lee, J. H. (1996). Unit root tests for time series with outliers. *Statistics & Probability Letters*, *30*, 189–197.

Silverman, B. W. (1986). *Density estimation for statistics and data analysis*. London: Chapman & Hall.

Taylor, M. P., Peel, D. A., & Sarno, L. (2001). Nonlinear mean-reversion in real exchange rates: toward a solution to the purchasing power parity puzzles. *International Economic Review*, *42*, 1015–1042.

Teräsvirta, T., Tjøstheim, D., & Granger, C. (2010). *Modelling nonlinear economic time series*. New York: Oxford University Press.

Tong, H. (1990). *Non-linear time series*. Oxford: Oxford Sci.

Tsay, R. (1988). Outliers, level shifts, and variance changes in time series. *Journal of Forecasting*, *7*, 1–20.

van Dijk, D., Franses, P., & Lucas, A. (1999). Testing for smooth transition nonlinearity in the presence of outliers. *Journal of Business & Economic Statistics*, *17*, 217–235.

van Dijk, D., Teräsvirta, T., & Franses, P. (2002). Smooth transition autoregressive models-a survey of recent developments. *Journal of Business & Economic Statistics*, *21*, 1–47.

Vogelsang, T. J. (1999). Two simple procedures for testing for a unit root when there are additive outliers. *Journal of Time Series Analysis*, *20*, 237–252.

The Time-Varying Beveridge Curve

Luca Benati and Thomas A. Lubik

Abstract We use a Bayesian time-varying parameter structural VAR with stochastic volatility to investigate changes in both the reduced-form relationship between vacancies and the unemployment rate, and in their relationship conditional on permanent and transitory output shocks, in the post-WWII United States. Evidence points towards similarities and differences between the Great Recession and the Volcker disinflation, and widespread time variation along two key dimensions. First, the slope of the Beveridge curve exhibits a large extent of variation from the mid-1960s on. It is also notably pro-cyclical, whereby the gain is positively correlated with the transitory component of output. The evolution of the slope of the Beveridge curve during the Great Recession is very similar to its evolution during the Volcker recession in terms of both its magnitude and its time profile. Second, both the Great Inflation episode and the subsequent Volcker disinflation are characterized by a significantly larger negative correlation between the reduced-form innovations to vacancies and the unemployment rate than the rest of the sample period. Those years also exhibit a greater cross-spectral coherence between the two series at business-cycle frequencies. This suggests that they are driven by common shocks.

1 Introduction

The Beveridge curve describes the relationship between the unemployment rate and open positions, that is, vacancies, in the labor market. Plotting the former against the latter in a scatter diagram reveals a downward-sloping relationship that appears to be clustered around a concave curve (see Figure 1). The curve reflects the highly negative correlation between unemployment and vacancies that is a hallmark of labor markets in market economies.

L. Benati
Department of Economics, University of Bern, Schanzeneckstrasse 1, 3001 Bern, Switzerland
e-mail: luca.benati@vwi.unibe.ch

T.A. Lubik (✉)
Research Department, Federal Reserve Bank of Richmond, 701 E Byrd Street, Richmond, VA 23218, USA
e-mail: Thomas.Lubik@rich.frb.org

F. Schleer-van Gellecom (ed.), *Advances in Non-linear Economic Modeling*, Dynamic Modeling and Econometrics in Economics and Finance 17,
DOI 10.1007/978-3-642-42039-9_5,

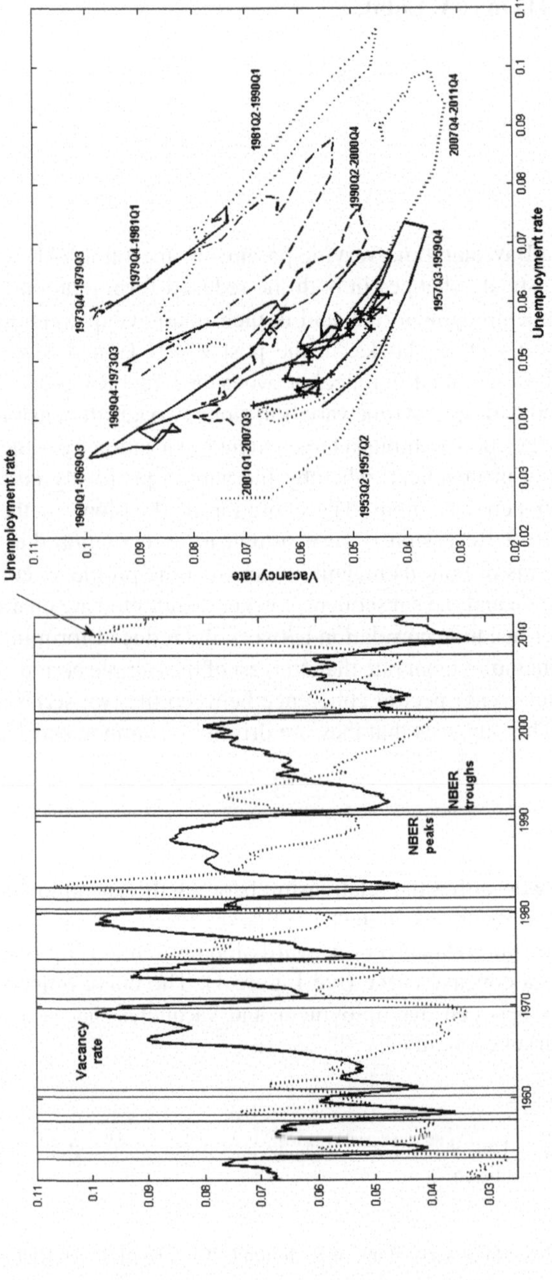

Fig. 1 The unemployment rate and the vacancy rate

Empirical work on the Beveridge curve has explored the relationship between vacancies and the unemployment rate under the maintained assumption that it can be regarded, as a first approximation, as time-invariant. The behavior of the two series during the Great Recession, with the unemployment rate seemingly stuck at high levels, even in the presence of a vacancy rate that has been progressively improving, has, however, raised doubts about the validity of the assumption of time-invariance. This suggests exploring the relationship between the two series allowing for the possibility that it may have evolved over time.

Our paper builds directly on the seminal contribution of Blanchard and Diamond (1989). These authors reintroduced the concept of the Beveridge curve as one of the key relationships in macroeconomic data. They conducted a vector autoregression (VAR) analysis of unemployment, vacancies, and the labor force in order to identify the driving forces behind movements in the Beveridge curve. We build upon their analysis by identifying both permanent and transitory structural shocks in a time-varying VAR context. By doing so, we are able to trace out the sources of movements, shifts, and tilts in the Beveridge curve over time.

The theoretical background for our study, and one that we use for identifying the structural shocks, is the simple search and matching approach to modeling labor markets (see Shimer 2005). The Beveridge curve encapsulates the logic of this model. In times of economic expansions, unemployment is low and vacancies, that is, open positions offered by firms, are plentiful. Firms want to expand their workforce, but they are unable to do so since the pool of potential employees (that is, the unemployed) is small. As economic conditions slow down and demand slackens, firms post fewer vacancies and unemployment rises, consistent with a downward move along the Beveridge curve. At the trough of the business cycle, firms may have expectations of a future uptick in demand and start posting open positions. This decision is amplified by the large pool of unemployed, which guarantees firms high chances of finding suitable candidates and thus outweighs the incurred search costs. As the economy improves, unemployment falls and vacancy postings rise in an upward move along the Beveridge curve.

We approach the Beveridge curve empirically by specifying a time-varying parameter VAR with stochastic volatility. Our choice is informed by the observation that there are patterns in the Beveridge-curve relationship that are ill described by a linear framework. Specifically, the data suggest that the slope of the Beveridge curve is different for each business cycle episode, that the curve shifts over time, and that the pattern of driving forces change in a nonlinear fashion as well. Naturally, nonlinearity can take many forms, as do the framework to capture this. We utilize a time-varying parameter framework since it is a reasonably straightforward extension of linear VARs. Moreover, and perhaps more importantly, it can capture and approximate a wide range of underlying nonlinear behavior. To this point, we will introduce time variation in a nonlinear theoretical model of the labor market in order to relate it to the results from the VAR.

Our empirical analysis starts by documenting the presence of time variation in the relationship between vacancies and the unemployment rate by means of Stock and

Watson's (1996, 1998) time-varying parameter median-unbiased estimation, which allows us to test for the presence of random-walk time variation in the data. Having detected evidence of time variation in the bivariate relationship between vacancies and the unemployment rate, we then use a Bayesian time-varying parameter structural VAR with stochastic volatility to characterize changes over time in the relationship. Evidence points towards both similarities and differences between the Great Recession and the Volcker disinflation, and widespread time variation along two key dimensions.

First, the slope of the Beveridge curve, which we capture by the average gain of the unemployment rate onto vacancies at business cycle frequencies is strongly negatively correlated with the Congressional Budget Office's (CBO) estimate of the output gap. The evolution of the slope of the Beveridge curve during the Great Recession is very similar to its evolution during the Volcker recession in terms of both its magnitude and its time profile. This suggests that the seemingly anomalous behavior of the Beveridge curve during the Great Recession, which has attracted much attention in the literature, may not have been that unusual. Second, both the Great Inflation episode and the subsequent Volcker disinflation, are characterized by a significantly larger (in absolute value) negative correlation between the reduced-form innovations to vacancies and the unemployment rate than the rest of the sample period. These years also show a greater cross-spectral coherence between the two series at business-cycle frequencies. This suggests that they are driven, to a larger extent than the rest of the sample, by common shocks.

Having characterized changes over time in the relationship between vacancies and the unemployment rate, we then proceed to interpret these stylized facts based on an estimated search and matching model. Specifically, we explore within a simple theoretical model how changes in individual parameter values affect the relationship between vacancies and the unemployment rate in order to gauge the origin of the variation in the Beveridge relationship.

The paper is organized as follows. The next section presents preliminary evidence on the presence of (random-walk) time variation in the bivariate relationship between the vacancy rate and the unemployment rate. Section 3 describes the Bayesian methodology we use to estimate the time-varying parameter VAR with stochastic volatility, whereas Section 4 discusses the evidence of changes over time in the Beveridge relationship. Section 5 details our structural identification procedure based on insights from a simple search and matching model, where we discuss implementation of both long-run and sign restrictions. We present the results of the structural identification procedure in Section 6. We also explore how changes in individual structural parameters of the search and matching model map into corresponding changes in the relationship between vacancies and the unemployment rate. Section 7 concludes. Appendices A–D contain a detailed description of the econometric methods and the theoretical model we use for identification purposes.

2 Searching for Time Variation in the Beveridge Relationship

Figure 1 presents a time series plot of the unemployment rate and vacancies from 1949 to 2011. The negative relationship between the two series is readily apparent. At the peak of the business cycle unemployment is low and vacancies are high. Over the course of a downturn the former rises and the latter declines as fewer and fewer workers are employed and firms have fewer and fewer open positions. Volatility and serial correlation of both series appear of similar magnitude. The second panel in Figure 1 depicts the same series in a scatter plot of vacancies against unemployment, resulting in the well-known downward-sloping relationship that has come to be known as the Beveridge curve. In the graph, we plot individual Beveridge curves for each NBER business cycle. Each episode starts at the business cycle peak and ends with the period before the next peak. Visual inspection reveals two observations. First, all curves are downward-sloping, but with different slopes. Second, there is substantial lateral movement in the individual Beveridge curves, ranging from the innermost cycle, the 1953–1957 episode, to the outermost, 1982–1990. We take these observations as motivating evidence that the relationship between unemployment and vacancies exhibits substantial variation over time, which a focus on a single aggregate Beveridge curve obscures.

Time variation in data and in theoretical models can take many forms, from continuous variations in unit-root time-varying parameter models to discrete parameter shifts such as regime-switching. We regard both discrete and continuous changes as a priori plausible. In this paper, we focus on the latter. We thus provide evidence of time variation in the bivariate relationship between vacancies and unemployment. We apply the methodology developed by Stock and Watson (1996, 1998) to test for the presence of random-walk time-variation in the two-equation VAR representation for the two variables.[1]

The regression model we consider is:

$$x_t = \mu + \alpha(L)V_{t-1} + \beta(L)U_{t-1} + \epsilon_t \equiv \theta' Z_t + \epsilon_t, \tag{1}$$

where $x_t = V_t, U_t$, with V_t and U_t being the vacancy rate and the unemployment rate, respectively. $\alpha(L)$ and $\beta(L)$ are lag polynomials; $\theta = [\mu, \alpha(L), \beta(L)]'$ and $Z_t = [1, V_{t-1}, \dots, U_{t-p}]'$. We select the lag order as the maximum of the lag orders individually chosen by the Akaike, Schwartz, and Hannan-Quinn criteria. Letting $\theta_t = [\mu_t, \alpha_t(L), \beta_t(L)]'$, the time-varying parameter version of (1) is given by:

$$x_t = \theta_t' Z_t + \epsilon_t, \tag{2}$$

[1]From an empirical perspective, we prefer their methodology over, for instance, structural break tests for reasons of robustness to uncertainty regarding the specific form of time-variation present in the data. While time-varying parameter models can successfully track processes subject to structural breaks, Cogley and Sargent (2005) and Benati (2007) show that break tests possess low power when the true data-generating process (DGP) is characterized by random walk time variation. Generally speaking, break tests perform well if the DGP is subject to discrete structural breaks, while time-varying parameter models perform well under both scenarios.

Table 1 Results based on the Stock-Watson TVP-MUB methodology: *exp*- and *sup*-Wald test statistics, simulated *p*-values, and median-unbiased estimates of λ

Equation for:	*exp*-Wald (*p*-value)	$\hat{\lambda}$	*sup*-Wald (*p*-value)	$\hat{\lambda}$
Newey and West (1987) correction				
Vacancy rate	9.40 (0.0053)	0.0286	28.91 (0.0028)	0.0327
Unemployment rate	4.97 (0.1661)	0.0153	16.17 (0.1770)	0.0153
Andrews (1991) correction				
Vacancy rate	7.65 (0.0195)	0.0235	25.40 (0.0086)	0.0286
Unemployment rate	4.68 (0.1987)	0.0133	14.61 (0.2594)	0.0122

$$\theta_t = \theta_{t-1} + \eta_t, \tag{3}$$

with $\eta_t \sim$ iid $\mathcal{N}(0_{4p+1}, \lambda^2\sigma^2 Q)$, where 0_{4p+1} is a $(4p+1)$-dimensional vector of zeros. σ^2 is the variance of ϵ_t, Q a covariance matrix, and $E[\eta_t\epsilon_t] = 0$. Following Stock and Watson (1996, 1998), we set $Q = [E(Z_t Z_t')]^{-1}$. Under this normalization, the coefficients on the transformed regressors, $[E(Z_t Z_t')]^{-1/2} Z_t$, evolve according to a $(4p+1)$-dimensional standard random walk, where λ^2 is the ratio between the variance of each transformed innovation and the variance of ϵ_t. We estimate the matrix Q as $\hat{Q} = [T^{-1}\sum_{t=1}^{T} z_t z_t']^{-1}$.

We estimate the specification (1) by OLS, from which we obtain an estimate of the innovation variance, $\hat{\sigma}^2$. We then perform an exp- and a sup-Wald joint test for a single unknown break in μ and in the sums of the α's and β's, using either Newey and West (1987) or Andrews (1991) HAC covariance matrix estimator to control for possible autocorrelation and/or heteroskedasticity in the residuals. Following Stock and Watson (1996), we compute the empirical distribution of the test statistic by considering a 100-point grid of values for λ over the interval [0, 0.1]. For each element of the grid we compute the corresponding estimate of the covariance matrix of η_t as $\hat{Q}_j = \lambda_j^2\hat{\sigma}^2\hat{Q}$; conditional on $\hat{Q}_j$ we simulate the model (2)–(3) 10,000 times, drawing the pseudo innovations from pseudo-random iid $N(0, \hat{\sigma}^2)$. We compute the median-unbiased estimate of λ as that particular value for which the median of the simulated empirical distribution of the test is closest to the test statistic previously computed based on the actual data. Finally, we compute the p-value based on the empirical distribution of the test conditional on $\lambda_j = 0$, which we compute based on Benati's (2007) extension of the Stock and Watson (1996, 1998) methodology.

We report the estimation results in Table 1. We find strong evidence of random-walk time variation in the equation for the vacancy rate. The p-values for the null of no time variation range from 0.0028 to 0.0195, depending on the specific test statistic. The median-unbiased estimates of λ are comparatively large, between 0.0235 and 0.0327. On the other hand, the corresponding p-values for the unemployment rate are much larger, ranging from 0.1661 to 0.2594, which suggests time-invariance. However, the densities of the median-unbiased estimates of λ in Figure 2 paint a more complex picture. A substantial fraction of the probability mass are clearly above zero, whereas median-unbiased estimates of λ range between 0.0122 and 0.0153. Although, strictly speaking, the null hypothesis of no

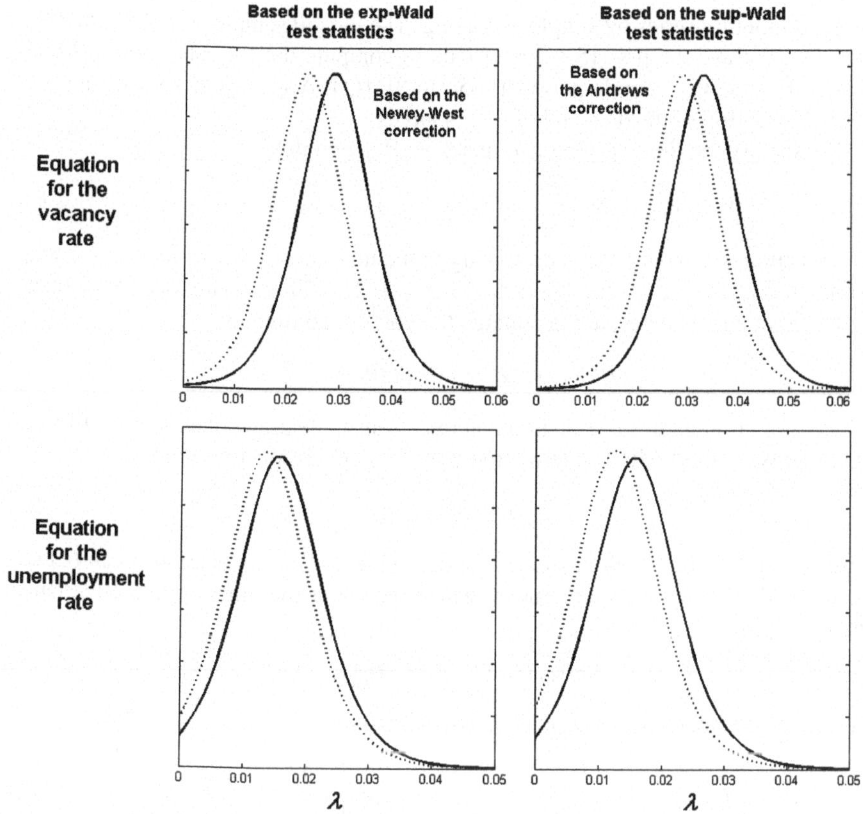

Fig. 2 Deconvoluted PDFs of γ

time variation cannot be rejected at conventional significance levels in a frequentist sense, the evidence reported in Figure 2 suggests more caution. In what follows, we will proceed under the assumption that both equations feature random-walk time variation. We now investigate the changing relationship between the vacancy rate and the unemployment rate based on a Bayesian time-varying parameter VAR.

3 A Bayesian Time-Varying Parameter VAR with Stochastic Volatility

We define the data vector $Y_t \equiv [\Delta y_t, V_t, U_t]'$, where Δy_t is real GDP growth, computed as the log-difference of real GDP; V_t is the vacancy rate based the Conference Board's Help-Wanted Index and Barnichon's (2010) extension; and U_t is the unemployment rate. The data are all quarterly. The vacancies and unemployment series are both normalized by the labor force, seasonally adjusted, and converted from the original monthly series by simple averaging. The overall sample period is 1951Q1–2011Q4. We use the first 15 years of data to compute the Bayesian priors, which makes the effective sample period 1965Q1–2011Q4. Appendix A contains a complete description of the data and of their sources.

We specify the time-varying parameter VAR(p) model:

$$Y_t = B_{0,t} + B_{1,t}Y_{t-1} + \cdots + B_{p,t}Y_{t-p} + \epsilon_t \equiv X_t'\theta_t + \epsilon_t. \tag{4}$$

The notation is standard. As is customary in the literature on Bayesian time-varying parameter VARs, we set the lag order to $p = 2$. The time-varying lag coefficients, collected in the vector θ_t, are postulated to evolve according to:

$$p(\theta_t|\theta_{t-1}, Q) = I(\theta_t) f(\theta_t|\theta_{t-1}, Q), \tag{5}$$

where $I(\theta_t)$ is an indicator function that rejects unstable draws and thereby enforces stationarity on the VAR. The transition $f(\theta_t|\theta_{t-1}, Q)$ is given by:

$$\theta_t = \theta_{t-1} + \eta_t, \tag{6}$$

with $\eta_t \sim$ iid $\mathcal{N}(0, Q)$. We assume that the reduced-form innovations ϵ_t in (4) are normally distributed with zero mean, where we factor the time-varying covariance matrix Ω_t as:

$$\mathrm{Var}(\epsilon_t) \equiv \Omega_t = A_t^{-1} H_t \left(A_t^{-1}\right)'. \tag{7}$$

The time-varying matrices H_t and A_t are defined as:

$$H_t \equiv \begin{bmatrix} h_{1,t} & 0 & 0 \\ 0 & h_{2,t} & 0 \\ 0 & 0 & h_{3,t} \end{bmatrix}, \qquad A_t \equiv \begin{bmatrix} 1 & 0 & 0 \\ \alpha_{21,t} & 1 & 0 \\ \alpha_{31,t} & \alpha_{32,t} & 1 \end{bmatrix}. \tag{8}$$

We assume that the $h_{i,t}$ evolve as geometric random walks:

$$\ln h_{i,t} = \ln h_{i,t-1} + \nu_{i,t}, \qquad i = 1, 2, 3. \tag{9}$$

For future reference, we define $h_t \equiv [h_{1,t}, h_{2,t}, h_{3,t}]'$ and $\nu_t = [\nu_{1,t}, \nu_{2,t}, \nu_{3,t}]'$, with $\nu_t \sim$ iid $\mathcal{N}(0, Z)$ and Z diagonal. We assume, as in Primiceri (2005), that the non-zero and non-unity elements of the matrix A_t, which we collect in the vector $\alpha_t \equiv [\alpha_{21,t}, \alpha_{31,t}, \alpha_{32,t}]'$, evolve as random walks:

$$\alpha_t = \alpha_{t-1} + \tau_t, \tag{10}$$

where $\tau_t \sim$ iid $\mathcal{N}(0, S)$. Finally, we assume that the innovations vector $[u_t', \eta_t', \tau_t', \nu_t']'$ is distributed as:

$$\begin{bmatrix} u_t \\ \eta_t \\ \tau_t \\ \nu_t \end{bmatrix} \sim N(0, V), \quad \text{with } V = \begin{bmatrix} I_3 & 0 & 0 & 0 \\ 0 & Q & 0 & 0 \\ 0 & 0 & S & 0 \\ 0 & 0 & 0 & Z \end{bmatrix}, \tag{11}$$

where u_t is such that $\epsilon_t \equiv A_t^{-1} H_t^{\frac{1}{2}} u_t$.

We follow the literature in imposing a block-diagonal structure for V, mainly for parsimony, since the model is already quite heavily parameterized. Allowing for a completely generic correlation structure among different sources of uncertainty would also preclude any structural interpretation of the innovations. Finally, following Primiceri (2005) we adopt the additional simplifying assumption of a block-diagonal structure for S:

$$S \equiv \mathrm{Var}(\tau_t) = \mathrm{Var}(\tau_t) = \begin{bmatrix} S_1 & 0_{1\times 2} \\ 0_{2\times 1} & S_2 \end{bmatrix}, \tag{12}$$

with $S_1 \equiv \mathrm{Var}(\tau_{21,t})$, and $S_2 \equiv \mathrm{Var}([\tau_{31,t}, \tau_{32,t}]')$. This implies that the non-zero and non-one elements of A_t, which belong to different rows, evolve independently. This assumption simplifies inference substantially, since it allows Gibbs sampling on the non-zero and non-one elements of A_t equation by equation. We estimate (4)–(12) using standard Bayesian methods. Appendix B discusses our choices for the priors, and the Markov-Chain Monte Carlo algorithm we use to simulate the posterior distribution of the hyperparameters and the states conditional on the data.

4 Reduced-Form Evidence

Figure 3 presents the first set of reduced-form results. It shows statistics of the estimated time-varying innovations in the VAR (4). The first panel depicts the median posterior estimate of the correlation coefficient of the innovations to vacancies and the unemployment rate and associated 68 % and 90 % coverage regions. The plot shows substantial time variation in this statistic. From the late 1960s to the early 1980s the correlation strengthens from −0.4 to −0.85 before reaching a low of −0.25. Over the course of the last decade, the correlation has strengthened again, settling close to the average median value of −0.55. This suggests that the unemployment-vacancy correlation strengthens during periods of broad downturns and high volatility, whereas it weakens in general upswings with low economic turbulence. The evidence over the last decade also supports the impression that the U.S. economy is in a period of a prolonged downswing.[2]

The impression of substantial time variation is strongly supported by the second panel, which shows the fraction of draws from the posterior distribution for which the correlation coefficient is greater than the average median value over the sample. The fraction of draws sinks toward zero at the end of the Volcker disinflation, while it oscillates for much of the Great Moderation between 0.6 and 0.9. Similarly to the results in the first panel, the period since the beginning of the financial crisis in August 2007 is characterized by a substantial decrease in the fraction of draws.

[2]We also note that at the same time the coverage regions are tightly clustered around the median estimate during the period of highest instability, namely the last 1970s and the Volcker disinflation, whereas they are more spread out in the beginning and towards the end of the sample.

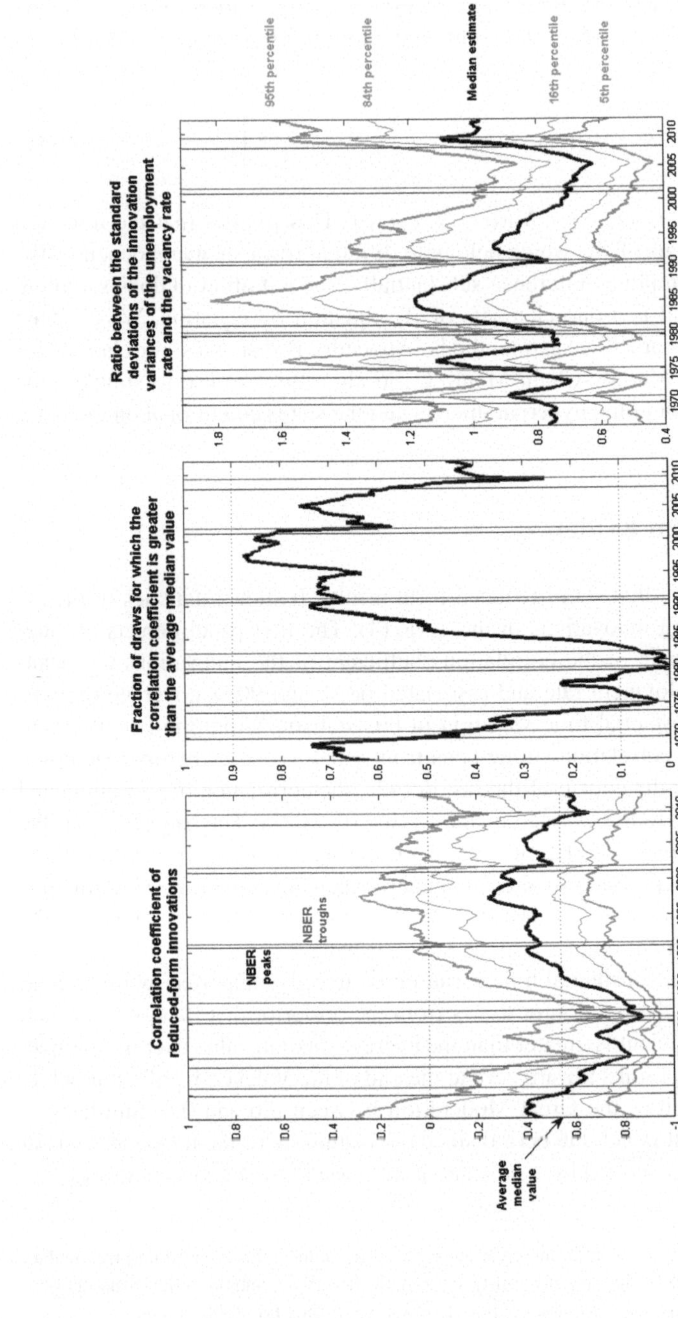

Fig. 3 Correlation coefficient of reduced-form innovations to vacancies and the unemployment rate, and ratio between the standard deviations of reduced-form innovations to the two variables (Color figure online)

In the third panel of Figure 3, we highlight the ratio of the estimated standard deviations of the unemployment and vacancy innovations. The graph shows substantial time variation in this ratio, although overall both innovation variances are of roughly equal size. While the innovation variance of the vacancy rate appears overall dominant, unemployment innovations play a relatively larger role at the end of the Great Inflation, the Volcker disinflation, and the Great Recession. All of these are periods during which the unemployment rate shot up sharply. This suggests a dominant role of specific shocks, namely those tied closely to reduced-form innovations to the unemployment rate, at the onset of an economic downturn. We attempt to identify the sources for this behavior in the following section.

We now narrow our focus to the behavior of unemployment and vacancies at the business-cycle frequencies between six quarters and eight years. We report these results using statistics from the frequency domain. Figure 4 shows median posterior estimates (and associated coverage regions) of the average cross-spectral gain and coherence between the two variables. The gain of a variable x_t onto another variable y_t at the frequency ω is defined as the absolute value of the OLS-coefficient in the regression of y_t on x_t at that frequency, whereas the coherence is the R^2 in that regression. Consequently, the gain has a natural interpretation in terms of the slope of the Beveridge curve, while the coherence measures the fraction of the vacancy-rate's variance at given frequencies that is accounted for by the variation in the unemployment rate. We find it convenient to express time variation in the Beveridge curve in terms of the frequency domain since it allows us to isolate the fluctuations of interest, namely policy-relevant business cycles, and therefore abstract from secular movements.

Overall, evidence of time variation is significantly stronger for the gain than for the coherence. The coherence between the two series appears to have remained broadly unchanged since the second half of the 1960s, except for a brief run-up during the Great Inflation of the 1970s, culminating in the tight posterior distribution during the Volcker disinflation of the early 1980s. Moreover, average coherence is always above 0.8, with 0.9 contained in the 68 % coverage region. The high explanatory power of one variable for the other at the business cycle frequencies thus suggests that unemployment and vacancies are driven by a set of common shocks over the sample period.

The gain is large during the same periods in which the relative innovation variance of reduced-form shocks to the unemployment rate is large, namely during the first oil shock, the Volcker recession, and the Great Recession; that is, during these recessionary episodes movements in the unemployment rate are relatively larger than those in the vacancy rate. This points towards a flattening of the Beveridge curve in downturns, when small movements in vacancies are accompanied by large movements in unemployment. Time variation in the gain thus captures the shifts and tilts in the individual Beveridge curves highlighted in Figure 1 in one simple statistic.

As a side note, our evidence does not indicate fundamental differences between the Volcker disinflation and the Great Recession, that is, between the two deepest recessions in the post-war era. This is especially apparent from the estimated

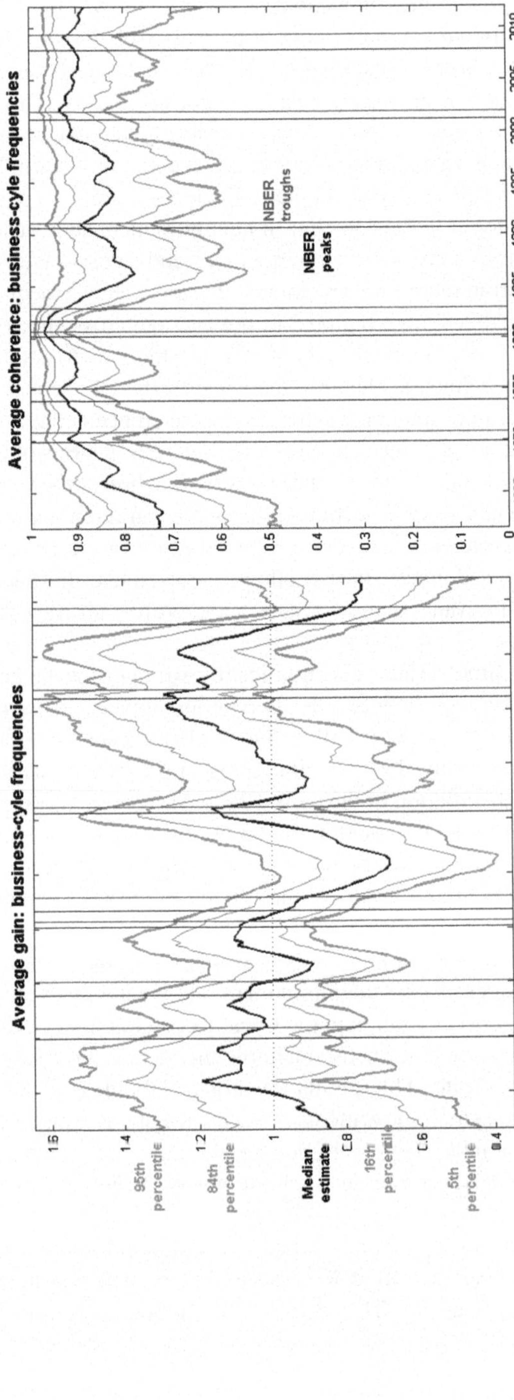

Fig. 4 Average gain of the unemployment rate onto vacancies, and average coherence between vacancies and the unemployment rate, at the business-cycle frequencies

gain in Figure 4, which shows a similar time profile during both episodes. The relationship between vacancies and unemployment, although clearly different from the years leading up to the financial crisis, is broadly in line with that of the early 1980s.

We can now summarize our findings from the reduced-form evidence as follows. The correlation pattern between unemployment and vacancies shows a significant degree of time variation. It strengthens during downturns and weakens in upswings. This is consistent with the idea that over the course of a business cycle, as the economy shifts from a peak to a trough, labor market equilibrium moves downward along the Beveridge curve. This movement creates a tight negative relationship between unemployment and vacancies. As the economy recovers, however, vacancies start rising without much movement in unemployment. Hence, the correlation weakens. The economy thus goes off the existing Beveridge curve, in the manner of a counter-clockwise loop, as identified by Blanchard and Diamond (1989), or it moves to a new Beveridge curve, as suggested by the recent literature on mismatch, e.g. Furlanetto and Groshenny (2012), Lubik (2013), and Sahin et al. (2012). Evidence from the frequency domain suggests that the same shocks underlie movements in the labor market, but that over the course of the business cycle shocks change in their importance. During recessions movements in unemployment dominate, while in upswings vacancies play a more important role. We now try to identify the structural factors determining this reduced-form behavior.

5 Identification

A key focus of our analysis is to identify the underlying sources of the movements in the Beveridge curve. In order to do so we need to identify the structural shocks behind the behavior of unemployment and vacancies. Our data set contains a nonstationary variable, GDP, and two stationary variables, namely the unemployment and vacancy rates. This allows us to identify one permanent and two transitory shocks from the reduced-form innovation covariance matrix. While the permanent shock has no effect on the two labor market variables in the long run, it can still lead to persistent movements in these variables, and thus in the Beveridge curve, in the short to medium run.[3] More specifically, we are interested in which shocks can be tied to the changing slope and the shifts in the Beveridge curve. We let our identification strategy be guided by the implications of the simple search and matching model, which offers predictions for the effects of permanent and transitory productivity shocks as well as for other transitory labor market disturbances.

[3]While this rules out strict hysteresis effects, in the sense that temporary shocks can have permanent effects, it can still lead to behavior that looks over typical sample periods as hysteresis-induced. Moreover, the empirical evidence concerning hysteresis is decidedly mixed.

5.1 A Simple Theoretical Framework

We organize the interpretation of our empirical findings around the predictions of the standard search and matching model of the labor market model as described in Shimer (2005). The model is a data-generating process for unemployment and vacancies that is driven by a variety of fundamental shocks. The specific model is taken from Lubik (2013). The specification and derivation is described in more detail in Appendix A–D.

The model can be reduced to three key equations that will guide our thinking about the empirics. The first equation describes the law of motion for employment:

$$N_{t+1} = (1 - \rho_{t+1})\left[N_t + m_t U_t^{\xi} V_t^{1-\xi}\right]. \tag{13}$$

The stock of existing workers N_t is augmented by new hires $m_t U_t^{\xi} V_t^{1-\xi}$, which are the result of a matching process between open positions V_t and job seekers U_t via the matching function. The matching process is subject to exogenous variation in the match efficiency m_t, which affects the size of the workforce with a one-period lag. Similarly, employment is subject to exogenous variations in the separation rate ρ_t, which affects employment contemporaneously. It is this timing convention that gives rise to an identifying restriction.

The second equation is the job-creation condition, which describes the optimal vacancy posting decision by a firm:

$$\frac{\kappa_t}{m_t}\theta_t^{\xi} = \beta E_t(1 - \rho_{t+1})\left[(1 - \eta)(A_{t+1} - b_{t+1}) - \eta\kappa_{t+1}\theta_{t+1} + \frac{\kappa_{t+1}}{m_{t+1}}\theta_{t+1}^{\xi}\right]. \tag{14}$$

$\theta_t = V_t/U_t$ is labor market tightness and a crucial statistic in the search and matching model. Effective vacancy creation costs $\frac{\kappa_t}{m_t}\theta_t^{\xi}$ are increasing in tightness since firms have to compete with other firm's hiring efforts given the size of the applicant pool. Hiring costs are subject to exogenous variations in the component κ_t and have to be balanced against the expected benefits, namely the right-hand side of the above equation. This consists of surplus of a worker's marginal product A_t over his outside options (unemployment benefits b_t), net of a hold-up term, $\eta\kappa_t\theta_t$, accruing to workers and extracted from the firm on account of the latter's costly participation in the labor market, and the firm's implicit future cost savings $\frac{\kappa_t}{m_t}\theta_t^{\xi}$ when having already hired someone. η is a parameter indicating the strength of worker's bargaining power. Finally, production is assumed to be linear in employment, but subject to both permanent and temporary productivity shocks, A_t^P and A_t^T, respectively:

$$Y_t = A_t N_t = \left(A_t^P A_t^T\right)N_t. \tag{15}$$

The permanent shock is a pure random walk, while the temporary shock is serially correlated but stationary.

Figure 5 depicts the theoretical impulse response functions of the unemployment and vacancy rate to each of the shocks. We can categorize the shocks in two groups,

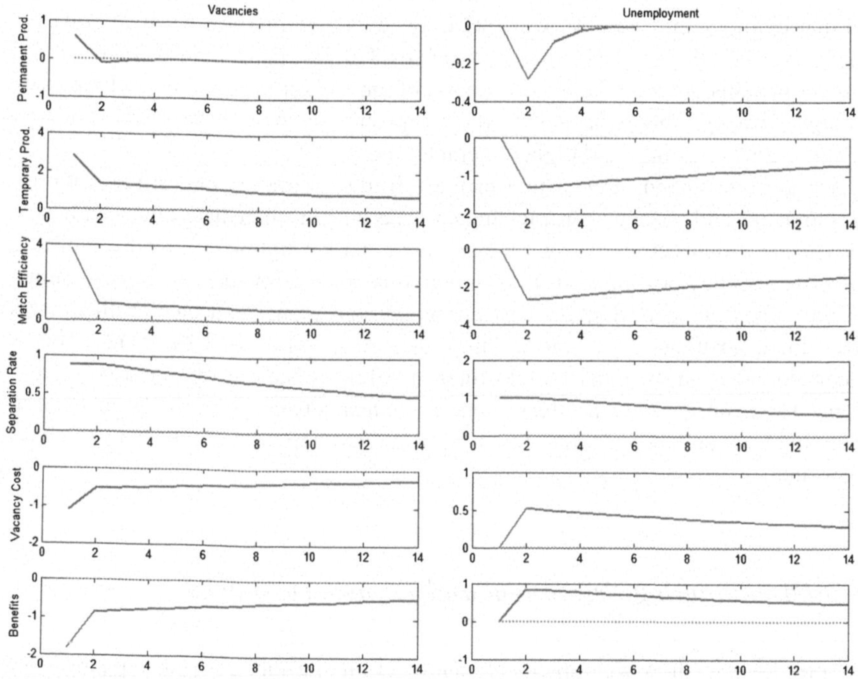

Fig. 5 Impulse response functions of the calibrated search and matching model

namely into shocks that move unemployment and vacancies in the same direction, and those that imply opposite movements of these variables. This classification underlies the identification by sign restrictions that we use later on. Both productivity shocks increase vacancies on impact and lower unemployment over the course of the adjustment period. The effect of the temporary shock is much more pronounced since it is calibrated at a much higher level of persistence than the productivity growth rate shock. Persistent productivity shocks increase vacancy posting because they raise the expected value of a filled position. As more vacancies get posted, new employment relationships are established and the unemployment rate falls. We note that permanent shocks have a temporary effect on the labor market because they tilt the expected profit profile in a manner similar to temporary shocks. However, they are identified by their long-run effect on output, which by definition no other shock can muster.

Shocks to match efficiency, vacancy posting costs and unemployment benefits lead to negative comovement between unemployment and benefits. Increases in match efficiency and decreases in the vacancy costs both lower effective vacancy creation cost $\frac{\kappa_t}{m_t}\theta_t^{\xi}$ ceteris paribus and thereby stimulate initial vacancy creation. These vacancies then lead to lower unemployment over time. In the case of κ_t there is additional feedback from wage setting since the hold-up term $\kappa_t\theta_t$ can rise or fall. Similarly, increases in match efficiency have an additional effect via the matching function as the higher level of vacancies is now turned into even more new hires, so that employment rises. Movements in benefits also produce negative comovements between the key labor market variables, but the channel is via wage setting. Higher benefits increase the outside option of the worker in bargaining which leads to higher wages. This reduces the expected profit stream to the firm and results in fewer vacancy postings and higher unemployment.

On the other hand, a persistent increase in the separation rate drives both unemployment and vacancy postings higher. There is an immediate effect on unemployment, which ceteris paribus lowers labor market tightness, thereby reducing effective vacancy posting cost. In isolation, this effect stimulates vacancy creation. At the same time, persistent increases in separations reduce expected profit streams from filled positions which has a dampening effect on desired vacancies. This is balanced, however, by persistent declines in tightness because of increased separations. The resulting overall effect is that firms take advantage of the larger pool of potential hires and increase vacancy postings to return to the previous long-run level over time.

5.2 Disentangling Permanent and Transitory Shocks

We now describe how we implement identification of a single permanent shock and two transitory shocks in our time-varying parameter VAR model, based on the theoretical insights derived in the previous section. The permanent shock is identified from a long-run restriction as originally proposed by Blanchard and Quah (1989). We label a shock as permanent if it affects only GDP in the long run, but not the labor market variables. The short- and medium-run effects on all variables is left unrestricted. In terms of the simple model, the identified permanent shock is consistent with the permanent productivity shock A_t^P which underlies the stochastic trend in output. We follow the procedure proposed by Galí and Gambetti (2009) for imposing long-run restrictions within a time-varying parameter VAR model.

Let $\Omega_t = P_t D_t P_t'$ be the eigenvalue-eigenvector decomposition of the VAR's time-varying covariance matrix Ω_t in each time period and for each draw from the ergodic distribution. We compute a local approximation to the matrix of the cumulative impulse-response functions (IRFs) to the VAR's structural shocks as:

$$\bar{C}_{t,\infty} = \underbrace{[I_N - B_{1,t} - \cdots - B_{p,t}]^{-1}}_{C_0} \bar{A}_{0,t}, \tag{16}$$

where I_N is the $N \times N$ identity matrix. The matrix of the cumulative impulse-response functions is then rotated via an appropriate Householder matrix H in order to introduce zeros in the first row of $\bar{C}_{t,\infty}$, which corresponds to GDP, except for the $(1, 1)$ entry. Consequently, the first row of the cumulative impulse-response functions,

$$C_{t,\infty}^P = \bar{C}_{t,\infty} H = C_0 \bar{A}_{0,t} H = C_0 A_{0,t}^P, \tag{17}$$

is given by [x 0 0], with x being a non-zero entry. By definition, the first shock identified by $A^P_{0,t}$ is the only one exerting a long-run impact on the level of GDP. We therefore label it the permanent output shock.

5.3 Identifying the Transitory Shocks Based on Sign Restrictions

We identify the two transitory shocks by assuming that they induce a different impact pattern on vacancies and the unemployment rate. Our theoretical discussion of the search and matching model has shown that a host of shocks, e.g., temporary productivity, vacancy cost, or match efficiency shocks, imply negative comovement for the two variables, while separation rate shocks increase vacancies and unemployment on impact. We transfer these insights to the structural VAR identification scheme.

Let $u_t \equiv [u^P_t, u^{T_1}_t, u^{T_2}_t]'$ be the vector of the structural shocks in the VAR: u^P_t is the permanent output shock, $u^{T_1}_t$ and $u^{T_2}_t$ are the two transitory shocks; let $u_t = A^{-1}_{0,t}\epsilon_t$, with $A_{0,t}$ being the VAR's structural impact matrix. Our sign restriction approach postulates that $u^{T_1}_t$ induces the opposite sign on vacancies and the unemployment rate contemporaneously, while $u^{T_2}_t$ induces an impact response of the same sign. We compute the time-varying structural impact matrix $A_{0,t}$ by combining the methodology proposed by Rubio-Ramirez et al. (2005) for imposing sign restrictions, and the procedure proposed by Galí and Gambetti (2009) for imposing long-run restrictions in time-varying parameter VARs.

Let $\Omega_t = P_t D_t P_t'$ be the eigenvalue-eigenvector decomposition of the VAR's time-varying covariance matrix Ω_t, and let $\tilde{A}_{0,t} \equiv P_t D_t^{\frac{1}{2}}$. We draw an $N \times N$ matrix K from a standard-normal distribution and compute the QR decomposition of K; that is, we find matrices Q and R such that $K = Q \cdot R$. A proposal estimate of the time-varying structural impact matrix can then be computed as $\bar{A}_{0,t} = \tilde{A}_{0,t} \cdot Q'$. We then compute the local approximation to the matrix of the cumulative IRFs to the VAR's structural shocks, $\bar{C}_{t,\infty}$, from (16). In order to introduce zeros in the first row of $\bar{C}_{t,\infty}$, we rotate the matrix of the cumulative IRFs via an appropriate Householder matrix H. The first row of the matrix of the cumulative IRFs, $C^P_{t,\infty}$ as in (17), is given by [x 0 0]. If the resulting structural impact matrix $A_{0,t} = \bar{A}_{0,t} H$ satisfies the sign restrictions, we store it; otherwise it is discarded. We then repeat the procedure until we obtain an impact matrix that satisfies both the sign restrictions and the long-run restriction at the same time.

6 Structural Evidence

Our identification strategy discussed in Section 5 allows us to distinguish between one permanent and two transitory shocks. The permanent shock is identified as having a long-run effect on GDP, while the transitory shocks are identified from sign

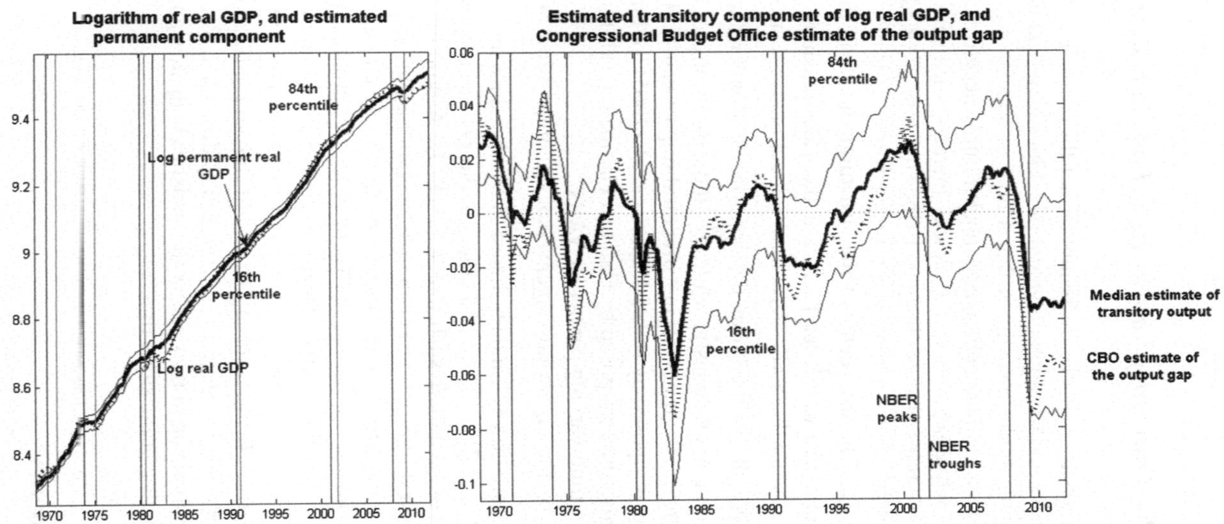

Fig. 6 Logarithm of real GDP and estimated permanent component, estimated transitory component of log real GDP, and Congressional Budget Office estimate of the output gap

restrictions derived from a simple search and matching model. A side product of our strategy is that we can identify the natural rate of output as its permanent component. Figure 6 shows real GDP in logs together with the median of the posterior distribution of the estimated permanent component and the 68 % coverage region. We also report the corresponding transitory component together with the output gap estimate from the CBO.

Our estimate of the transitory component is most of the time quite close to the CBO output gap, which is produced from a production function approach to potential output, whereas our estimate is largely atheoretical. The main discrepancy between the two estimates is in the wake of the Great Recession, particularly the quarters following the collapse of Lehman Brothers. Whereas the CBO estimate implies a dramatic output shortfall of around 7.5 % of potential output in the first half of 2009, our estimated gap is much less at between 3–4 % with little change since then. The reason behind our smaller estimate of the current gap is a comparatively large role played by permanent output shocks in the Great Recession. As the first panel shows, the time profile of the permanent component of log real GDP is estimated to have been negatively affected in a significant way by the Great Recession, with a downward shift in the trend path; that is, natural output is now permanently lower. The question we now investigate is whether and to what extent these trend shifts due to permanent output shocks seep into the Beveridge curve.

6.1 *Impulse Response Functions*

As a first pass, we report IRFs to unemployment and vacancies for each of the three shocks in Figures 7, 8 and 9. Because of the nature of the time-varying parameter VAR, there is not a single IRF for each shock-variable combination. We therefore represent the IRFs by collecting the time-varying coefficients on impact, two quarters ahead, one year ahead, and five years ahead in individual graphs in order to track how the dynamic behavior of the labor market variables changes over time. An IRF for a specific period can then be extracted by following the impulse response coefficient over the four panels. The IRFs are normalized such that the long-run effect is attained at a value of one, while transitory shocks eventually return the responses to zero.

In Figure 7, an innovation to the permanent component of output raises GDP on impact by one-half of the long-run effect, which is obtained fairly quickly after around one year in most periods. A permanent shock tends to raise the vacancy rate on impact, after which it rises for a few quarters before falling to its long-run level. The unemployment rate rises on impact, but then quickly settles around zero. The initial, seemingly counterfactual response is reminiscent of the finding by Galí (1999) that positive productivity shocks have negative employment consequences, which in our model translates into an initial rise in the unemployment rate. Furthermore, the behavior of the estimated impulse responses is broadly consistent with the results from the calibrated theoretical model, both in terms of direction and size of

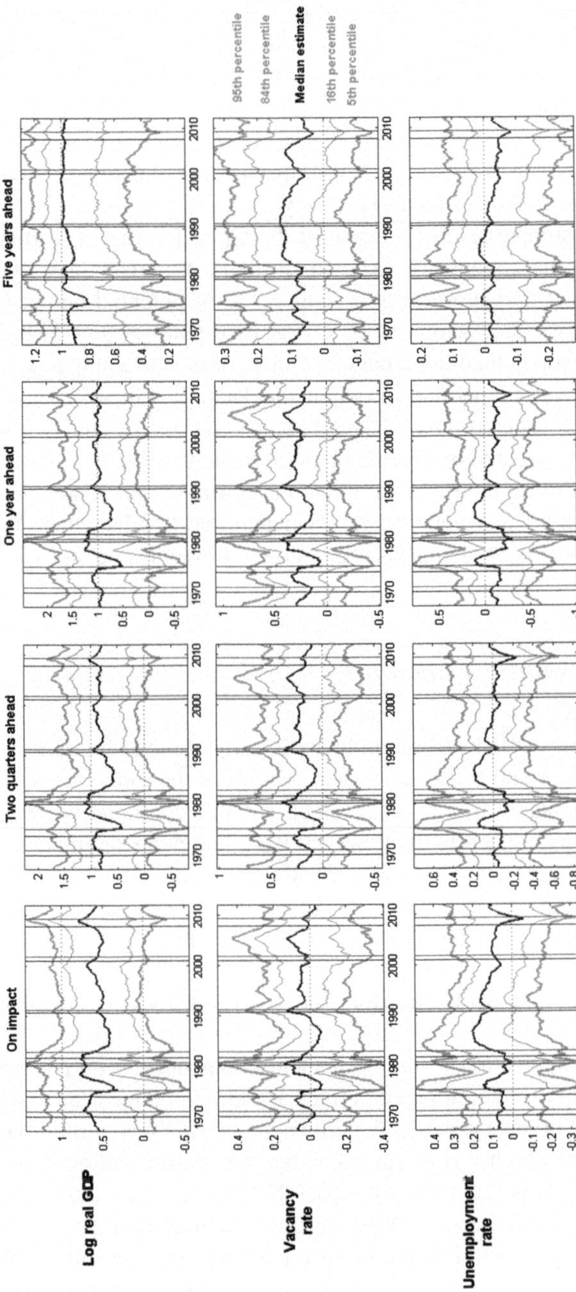

Fig. 7 Impulse response functions to a permanent output shock (Color figure online)

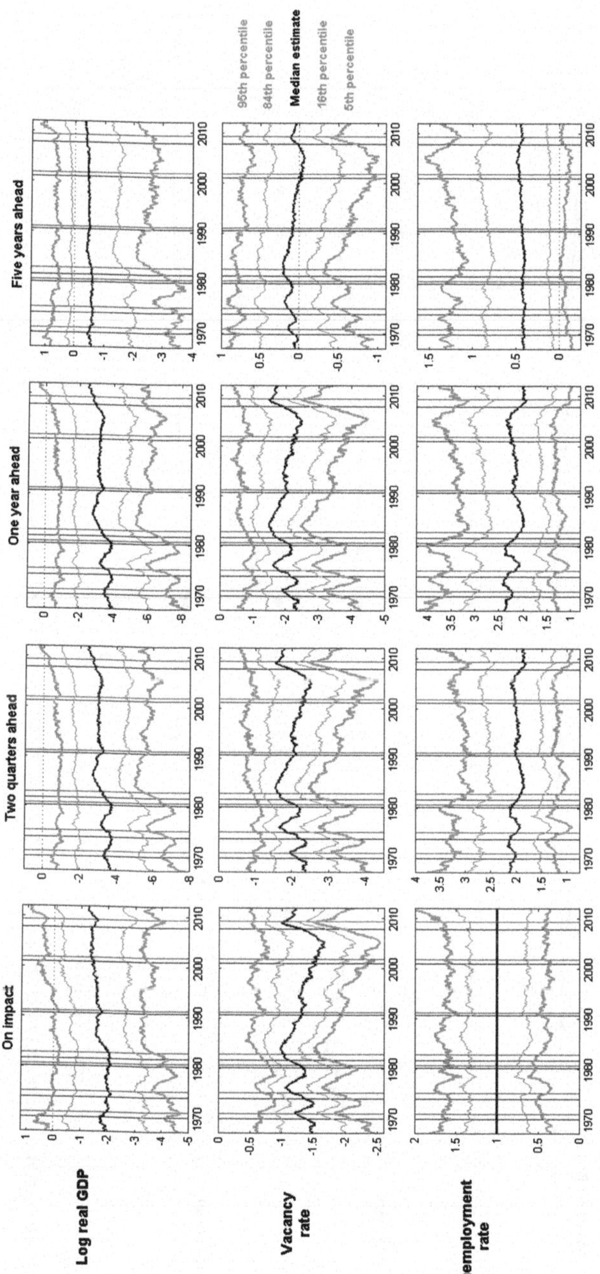

Fig. 8 Impulse response functions to the first transitory shock (Color figure online)

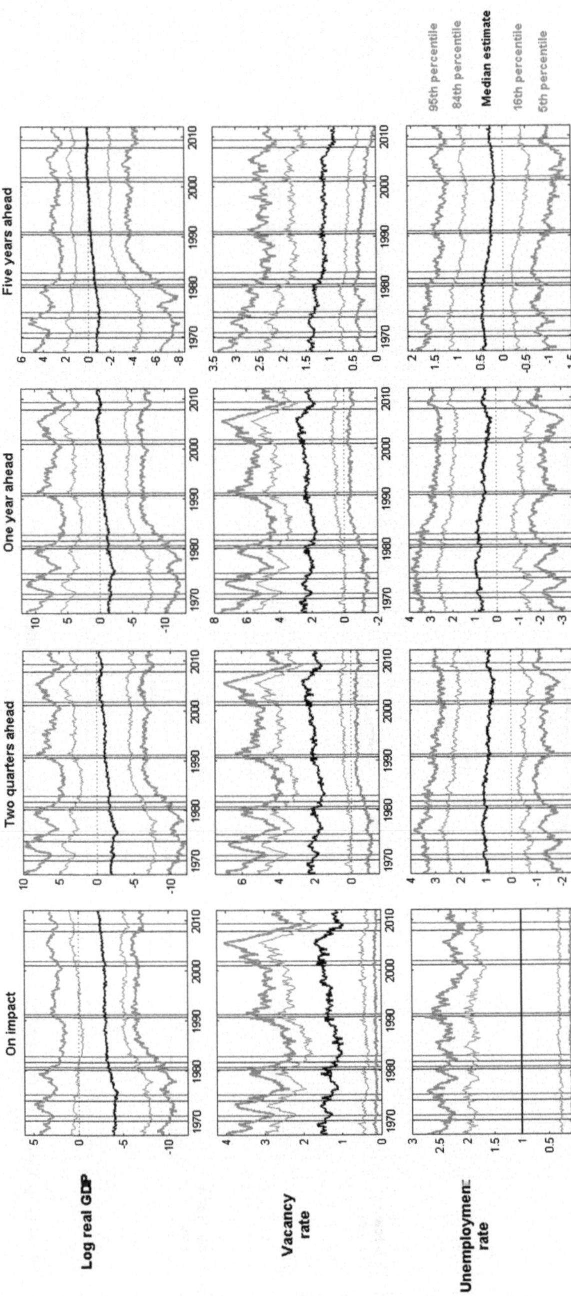

Fig. 9 Impulse response functions to the second transitory shock (Color figure online)

the responses. As we will see below, compared to the transitory shocks the permanent productivity shock, which in the theoretical model takes the form of a growth rate variation, exerts only a small effect on unemployment and vacancy rates. Notably, the coverage regions for both variables include zero at all horizons. Overall, the extent of time variation in the IRFs appears small. It is more pronounced at shorter horizons than in the long run.

We report the IRFs to the first transitory shock in Figure 8. This shock is identified as inducing an opposite response of the vacancy and the unemployment rate on impact. In the theoretical model, this identified empirical shock is associated with a transitory productivity shock, variations in match efficiency, hiring costs, or benefit movements. The IRFs of all three variables in the VAR are hump-shaped, with a peak response after one year. Moreover, the amplitudes of the responses are much more pronounced than in the previous case. The vacancy rate is back at its long-run level after 5 years, while there is much more persistence in the unemployment rate and GDP. We also note that our simple theoretical framework cannot replicate this degree of persistence.

The vacancy rate exhibits the highest degree of time variation. What stands out is that its response is asymmetric over the business cycle, but only in the pre-1984 period. During the recessions of the early and mid-1970s, and the deep recession of the early 1980s culminating in the Volcker disinflation, the initial vacancy response declines (in absolute value) over the course of the downturn before increasing in the recovery phase. That is, the vacancy rate responds less elastically to the first transitory shock during downturns than in expansions—which is not the case for the unemployment rate. This pattern is visible at all horizons. Between the Volcker disinflation and shortly before the onset of the Great Recession the impact response of the vacancy rate declines gradually from -1 % to almost -2 % before rising again sharply during a recession.

The second transitory shock is identified by imposing the same sign response on unemployment and vacancies. In the context of the theoretical model, such a pattern is due to movements in the separation rate. The IRFs in Figure 9 show that the vacancy rate rises on impact, then reaches a peak four quarters out before returning gradually over the long run. The unemployment rate follows the same pattern, while the shock induces a large negative response of GDP. None of the responses exhibits much time variation, at best there are slow-moving changes in the IRF-coefficients towards less elastic responses. Interestingly, the impact behavior of the vacancy rate declines over the course of the Great Recession. We note, however, that the coverage regions are very wide and include zero for the unemployment rate and GDP at all horizons.

6.2 *Variance Decompositions*

Figure 10 provides evidence on the relative importance of permanent and transitory shocks for fluctuations in vacancies and the unemployment rate. We report the median of the posterior distributions of the respective fractions of innovation variance

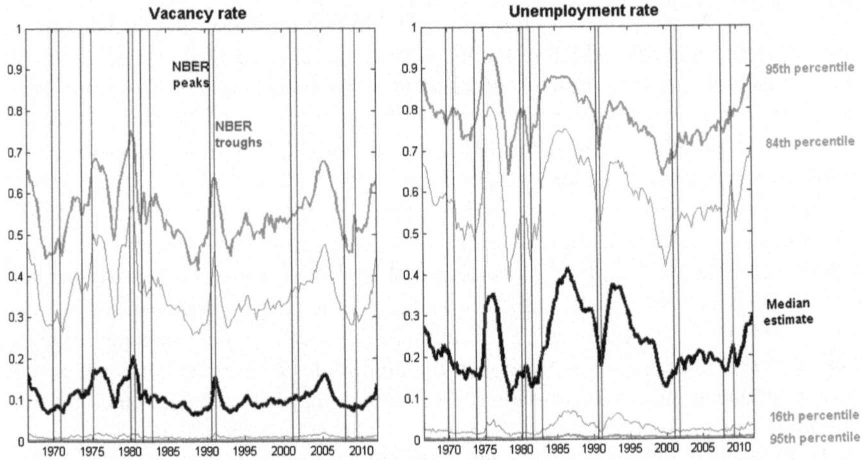

Fig. 10 Fractions of innovation variance due to the permanent output shock (Color figure online)

due to the permanent shock and the associated coverage regions. For the vacancy rate, permanent shocks appear to play a minor role, with a median estimate of between 10 % and 20 %. The median estimate for the unemployment rate exhibits a greater degree of variation, oscillating between 10 % and 40 %. Despite this large extent of time variation, it is difficult to relate fluctuations in the relative importance of permanent shocks to key macroeconomic events. Possible candidates are the period after the first oil shock, when the contribution of permanent shocks increased temporarily, and the long expansion of the 1980s until the late 1990s, which was temporarily punctured by the recession in 1991. Moreover, there is no consistent behavior of the permanent shock contribution over the business cycle. Their importance rises both in downturns and in upswings. On the other hand, this observation gives rise to the idea that all business cycles, at least in the labor market, are different along this dimension.

We now turn to the relative contribution of the two transitory shocks identified by sign restrictions. The evidence is fairly clear-cut. Given the strongly negative unconditional relationship between vacancies and the unemployment rate, we would expect the contribution of $u_t^{T_2}$, that is, the shock that induces positive contemporaneous comovement between the two variables, to be small. This is, in fact, borne out by the second column of the graph in Figure 11. The median estimate of the fraction of innovation variance of the two series due to $u_t^{T_2}$ is well below 20 %. Correspondingly, the first transitory shock appears clearly to be dominant for both variables. Based on the theoretical model, we can associate this shock with either temporary productivity disturbances or with stochastic movements in hiring costs, match efficiency, or unemployment benefits. Given the parsimonious nature of both the theoretical and empirical model, however, we cannot further disentangle this.

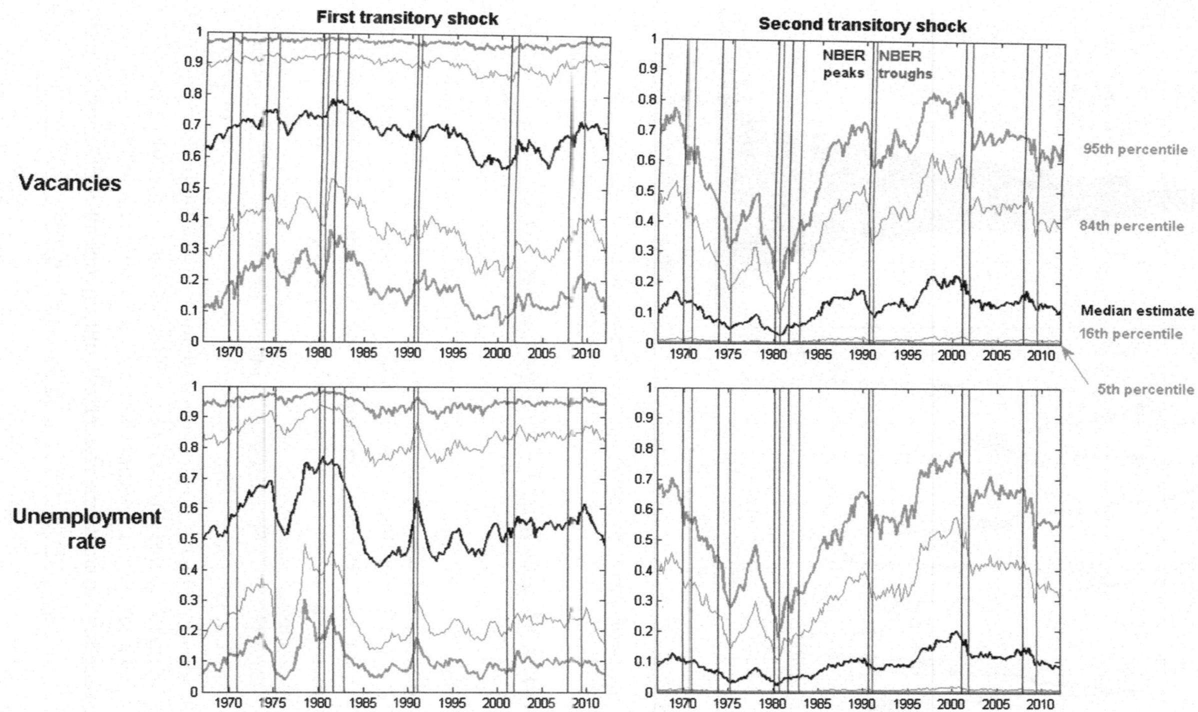

Fig. 11 Fractions of innovation variance due to the two transitory shocks identified from sign restrictions (Color figure online)

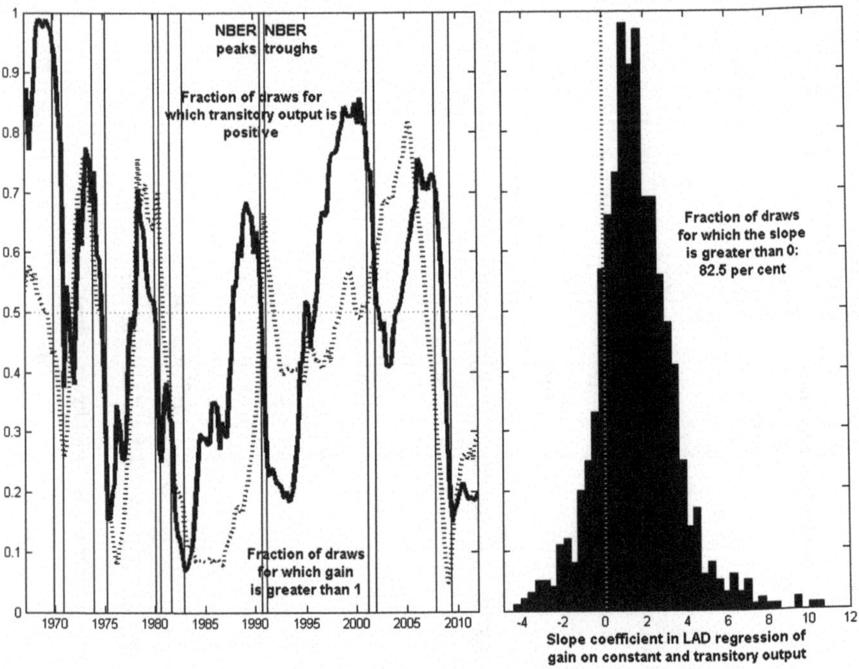

Fig. 12 Evidence on the pro-cyclicality of the Beveridge curve

6.3 Structural Shocks and Beveridge Curve Shifts

We now turn to one of the main results of the paper, namely the structural sources of time variation in the Beveridge curve. We first discuss the relationship between the business cycle, as identified by the transitory component in GDP and measures of the Beveridge curve. We then decompose the estimated gain and coherence of unemployment and vacancies into their structural components based on the identification scheme discussed above.

Figure 12 reports two key pieces of evidence on the cyclical behavior of the slope of the Beveridge curve. The left panel shows the fraction of draws from the posterior distribution for which the transitory component of output is positive. This is plotted against the fraction of draws for which the average cross-spectral gain between vacancies and the unemployment rate at the business-cycle frequencies is greater than one. The graph thus gives an indication of how the slope of the Beveridge curve moves with aggregate activity over the business cycle.

We can differentiate two separate time periods. During the 1970s and the early 1980s, that is, during the Great Inflation, the slope of the Beveridge curve systematically comoves contemporaneously with the state of the business cycle. It is comparatively larger (in absolute value) during business-cycle upswings, and comparatively smaller during periods of weak economic activity. Similarly, the Great Recession is characterized by very strong comovement between the slope of the Beveridge curve and the transitory component of output, but this time the slope slightly leads the business cycle.

On the other hand, in the long expansion period from 1982 to 2008, labelled the Great Moderation that was only marred by two minor recessions, the slope of the Beveridge curve comoves less clearly with the business cycle, a pattern that is especially apparent during the 1990s. In the early and late part of this period the Beveridge curve appears to lag the cycle. This is consistent with the notion of jobless recoveries after the two mild recessions. Despite upticks in economic activity, the labor market did not recover quickly after 1992 and, especially, after 2001. In the data, this manifests itself in a large gain between unemployment and vacancies (see Figure 13). Moreover, this is also consistent with the changing impulse response patterns to structural shocks discussed above. In a sense, the outlier is the Great Recession, which resembles more the recessions of the Great Inflation rather than those of the Great Moderation.

The second panel reports additional evidence on the extent of cyclicality of the slope of the Beveridge curve. It shows the distribution of the slope coefficient in the LAD (Least Absolute Deviations) regression of the cross-spectral gain on a constant and the transitory component of output. Overall, the LAD coefficient is greater than zero for 82.5 % of the draws from the posterior distribution, which points towards the pro-cyclicality of the slope of the Beveridge curve.

Figure 13 shows how the two types of shocks shape the evolution of the Beveridge curve. We plot the average gain and coherence between vacancies and the unemployment rate at business-cycle frequencies over time together with the fraction of draws for which the average gain is greater than one. The upper row of the panel reports the statistics conditional on the permanent shock, the lower panel contains those conditional on the two transitory shocks. Whereas the coherence conditional on the permanent shock does not show much time variation, conditioning on transitory shocks reveals a pattern that is broadly similar to the reduced-form representation. This suggests that the comparatively greater coherence between the two series around the time of the Great Inflation and of the Volcker disinflation is mostly due to transitory shocks.

Time variation in the gain, on the other hand, appears to be due to both types of shocks. Although the middle column suggests that the extent of statistical significance of the fluctuations in the gain is similar, the first column shows a different magnitude. In particular, fluctuations in the gain conditional on permanent output shocks, which accounted for a comparatively minor fraction of the innovation variance of the two series, is significantly wider than the corresponding fluctuations conditional on transitory shocks. Moreover, and unsurprisingly in the light of the previously discussed evidence on the relative importance of the two types of shocks, both the magnitude and the time-profile of the fluctuations of the gain conditional on transitory shocks are very close to the reduced-form evidence.

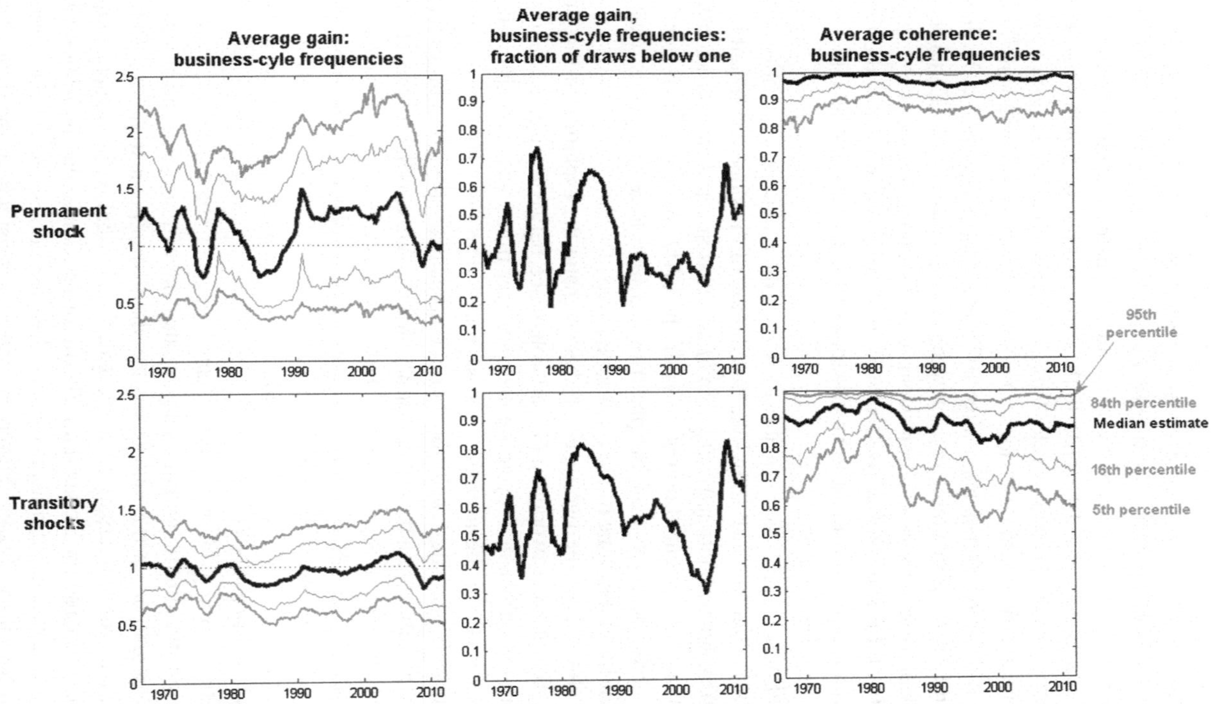

Fig. 13 Business-cycle frequencies: average gain and coherence between vacancies and the unemployment rate conditional on the permanent and the transitory output shocks (Color figure online)

6.4 *Interpreting Changes in the Beveridge Curve Based on an Estimated DSGE Model*

One of our key contributions is to have demonstrated the presence and the extent of non-linearity in the Beveridge-curve relationship over time in U.S. data. We did so in a structural VAR, where the identification restrictions were derived from the behavior of a simple search and matching model for the labor market. Nevertheless, VARs are by their very nature largely atheoretical, in the sense that they represent the reduced form of a potentially much richer underlying dynamic stochastic general equilibrium (DSGE) model. The best that we can do is to identify structural innovations, but they do not necessarily reveal much about the structure of the theoretical model. However, researchers may want to go further than this. One particular point of interest is the source of the non-linearity in the data. We now make some forays in this direction.

Following Fernández-Villaverde and Rubio-Ramirez (2007) we assume that all parameters in the DSGE model in Section 4 are first-order autoregressive processes. The set of parameters thus includes the original primitive parameters, their first-order autocorrelation coefficients, and their innovation variances. We then estimate the model using Bayesian methods.[4] We draw from the posterior distribution for each individual parameter and compute the associated gain of the unemployment rate onto vacancies at business cycle frequencies. Figure 14 contains a graph that shows, for parameter intervals around the modal estimates generated by the Random Walk Metropolis algorithm, the average gain of the unemployment rate onto vacancies at the business-cycle frequencies, as a function of each individual parameter.

The parameters with the largest impact on the gain, as a measure of the slope of the Beveridge curve, are the separation rate ρ, the match efficiency m, and the match elasticity ξ. The estimated gain is independent of the vacancy creation cost κ and the bargaining share η. It is also largely inelastic to variations in the parameters' autocorrelation coefficients, with the exception of the match efficiency and separation rate parameters and the serial correlation of the permanent productivity shock. The innovation variances do not affect the gain, the exception being the separation rate. These results are in line with the observation of Lubik (2013) that the key driver of shifts in the Beveridge curve are variations in the matching function parameters. While productivity shocks can generate movements along the Beveridge curve, movements of the Beveridge curve have to come through changes in the law of motion for employment. In contrast to his findings, our exercise puts additional weight on variations in the separation rate. As Figure 14 suggests, variations in the level, the persistence and the volatility of the separation rate are the main factor underlying the non-linearity and the time variation in the Beveridge curve relationship. We regard this as a crucial starting point for future research.

---

[4]The specification of the prior follows Lubik (2013). Posterior estimates and additional results are available from the authors upon request.

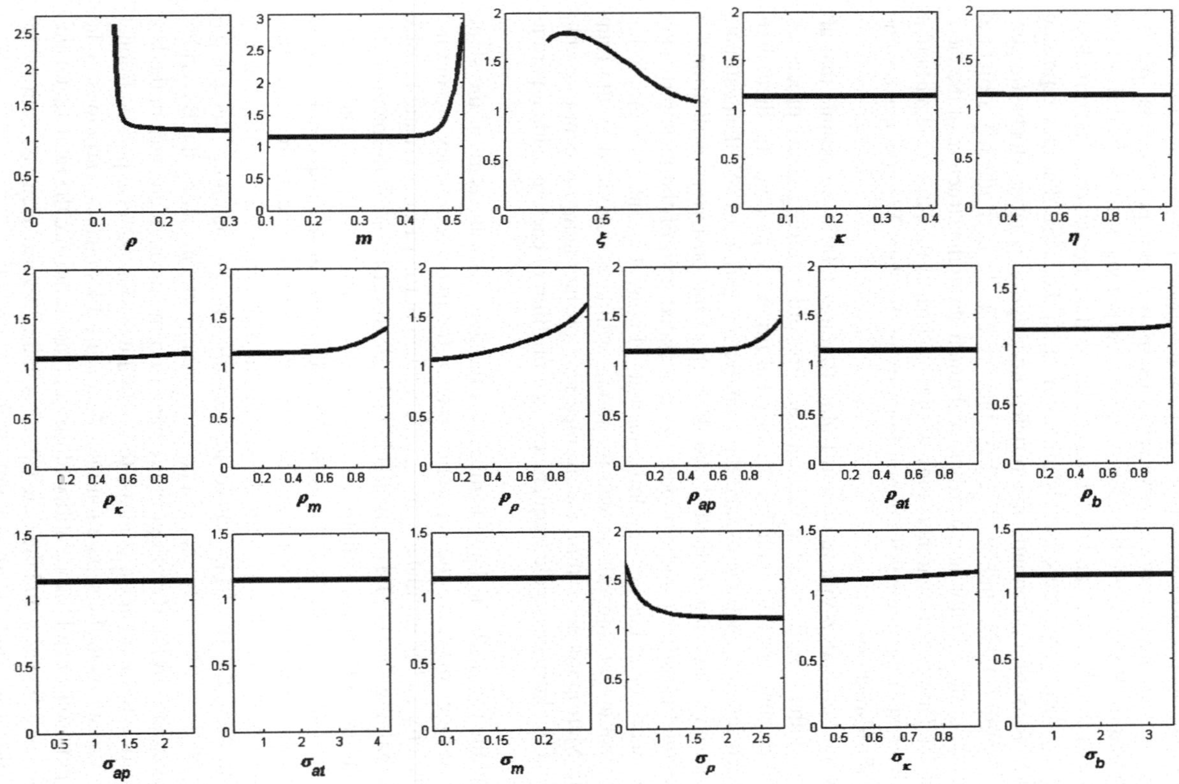

Fig. 14 Average gain of the unemployment rate onto vacancies at the business-cycle frequencies, as a function of individual parameters of the DSGE model (for parameter intervals around the modal estimates generated by the Random Walk Metropolis)

7 Conclusion

We have used a Bayesian time-varying parameter structural VAR with stochastic volatility to investigate changes in both the reduced-form relationship between vacancies and the unemployment rate, and in their relationship conditional on permanent and transitory output shocks, for the post-WWII United States. Evidence points towards both similarities and differences between the Great Recession and the Volcker disinflation, and a widespread time-variation along two key dimensions. First, the slope of the Beveridge curve, as captured by the average cross-spectral gain between vacancies and the unemployment rate at business-cycle frequencies, exhibits a large extent of variation since the second half of the 1960s. Moreover, it is broadly pro-cyclical, with the gain being positively correlated with the transitory component of output. The evolution of the slope of the Beveridge curve during the Great Recession appears to be very similar to its evolution during the Volcker recession in terms of its magnitude and its time profile. Second, both the Great Inflation episode, and the subsequent Volcker disinflation, are characterized by a significantly larger (in absolute value) negative correlation between the reduced-form innovations to vacancies and the unemployment rate than the rest of the sample period. Those years also exhibit a greater cross-spectral coherence between the two series at the business-cycle frequencies, thus pointing towards them being driven, to a larger extent than the rest of the sample, by common shocks.

Acknowledgements The views in this paper are those of the authors and should not be interpreted as those of the Federal Reserve Bank of Richmond, the Board of Governors, or the Federal Reserve System. We are grateful to participants at the Applied Time Series Econometrics Workshop at the Federal Reserve Bank of St. Louis and the Midwest Macroeconomics Meetings at the University of Colorado Boulder for useful comments and suggestions.

Appendix A: The Data

The series for real GDP ('GDPC96, Real Gross Domestic Product, 3 Decimal, Seasonally Adjusted Annual Rate, Quarterly, Billions of Chained 2005 Dollars') is from the U.S. Department of Commerce: Bureau of Economic Analysis. It is collected at quarterly frequency and seasonally adjusted. A quarterly seasonally adjusted series for the unemployment rate has been computed by converting the series UNRATE ('Civilian Unemployment Rate, Seasonally Adjusted, Monthly, Percent, Persons 16 years of age and older') from the U.S. Department of Labor: Bureau of Labor Statistics to quarterly frequency (by taking averages within the quarter). A monthly seasonally adjusted series for the vacancy rate has been computed as the ratio between the 'Help Wanted Index' (HWI) and the civilian labor force. The HWI index is from the Conference Board up until 1994Q4, and from Barnichon (2010) after that. The labor force series is from the U.S. Department of Labor: Bureau of Labor Statistics ('CLF16OV, Civilian Labor Force, Persons 16 years of age and older, Seasonally Adjusted, Monthly, Thousands of Persons'). The monthly seasonally adjusted series for the vacancy rate has been converted to the quarterly frequency by taking averages within the quarter.

Appendix B: Deconvoluting the Probability Density Function of $\hat{\lambda}$

This appendix describes the procedure we use in Section 2 to deconvolute the probability density function of $\hat{\lambda}$. We consider the construction of a $(1-\alpha)$ % confidence interval for $\hat{\lambda}$, $[\hat{\lambda}^L_{(1-\alpha)}, \hat{\lambda}^U_{(1-\alpha)}]$. We assume for simplicity that λ_j and $\hat{\lambda}$ can take any value over $[0; \infty)$. Given the duality between hypothesis testing and the construction of confidence intervals, the $(1-\alpha)$ % confidence set for $\hat{\lambda}$ comprises all the values of λ_j that cannot be rejected based on a two-sided test at the α % level. Given that an increase in λ_j automatically shifts the probability density function (pdf) of $\hat{L}_j$ conditional on λ_j upwards, $\hat{\lambda}^L_{(1-\alpha)}$ and $\hat{\lambda}^U_{(1-\alpha)}$ are therefore such that:

$$P\left(\hat{L}_j > \hat{L}|\lambda_j = \hat{\lambda}^L_{(1-\alpha)}\right) = \alpha/2, \tag{18}$$

and

$$P\left(\hat{L}_j < \hat{L}|\lambda_j = \hat{\lambda}^U_{(1-\alpha)}\right) = \alpha/2. \tag{19}$$

Let $\phi_{\hat{\lambda}}(\lambda_j)$ and $\Phi_{\hat{\lambda}}(\lambda_j)$ be the pdf and, respectively, the cumulative pdf of $\hat{\lambda}$, defined over the domain of λ_j. The fact that $[\hat{\lambda}^L_{(1-\alpha)}, \hat{\lambda}^U_{(1-\alpha)}]$ is a $(1-\alpha)$ % confidence interval automatically implies that $(1-\alpha)$ % of the probability mass of $\phi_{\hat{\lambda}}(\lambda_j)$ lies between $\hat{\lambda}^L_{(1-\alpha)}$ and $\hat{\lambda}^U_{(1-\alpha)}$. This, in turn, implies that $\Phi_{\hat{\lambda}}(\hat{\lambda}^L_{(1-\alpha)}) = \alpha/2$ and $\Phi_{\hat{\lambda}}(\hat{\lambda}^U_{(1-\alpha)}) = 1-\alpha/2$. Given that this holds for any $0 < \alpha < 1$, we therefore have that:

$$\Phi_{\hat{\lambda}}(\lambda_j) = P(\hat{L}_j > \hat{L}|\lambda_j). \tag{20}$$

Based on the *exp*-Wald test statistic, $\hat{L}$, and on the simulated distributions of the $\hat{L}_j$'s conditional on the λ_j's in Λ, we thus obtain an estimate of the cumulative pdf of $\hat{\lambda}$ over the grid Λ, $\hat{\Phi}_{\hat{\lambda}}(\lambda_j)$. Finally, we fit a logistic function to $\hat{\Phi}_{\hat{\lambda}}(\lambda_j)$ via nonlinear least squares and we compute the implied estimate of $\phi_{\hat{\lambda}}(\lambda_j)$, $\hat{\phi}_{\hat{\lambda}}(\lambda_j)$, whereby we scale its elements so that they sum to one.

Appendix C: Details of the Markov-Chain Monte Carlo Procedure

We estimate (4)–(12) using Bayesian methods. The next two subsections describe our choices for the priors, and the Markov-Chain Monte Carlo algorithm we use to simulate the posterior distribution of the hyperparameters and the states conditional on the data. The third section discusses how we check for convergence of the Markov chain to the ergodic distribution.

C.1 Priors

The prior distributions for the initial values of the states, θ_0 and h_0, which we postulate to be normally distributed, are assumed to be independent both from each other and from the distribution of the hyperparameters. In order to calibrate the prior distributions for θ_0 and h_0 we estimate a time-invariant version of (4) based on the first 15 years of data. We set:

$$\theta_0 \sim \mathcal{N}\big[\hat{\theta}_{OLS}, 4 \cdot \hat{V}(\hat{\theta}_{OLS})\big], \tag{21}$$

where $\hat{V}(\hat{\theta}_{OLS})$ is the estimated asymptotic variance of $\hat{\theta}_{OLS}$. As for h_0, we proceed as follows. Let $\hat{\Sigma}_{OLS}$ be the estimated covariance matrix of ϵ_t from the time-invariant VAR, and let C be its lower-triangular Cholesky factor, $CC' = \hat{\Sigma}_{OLS}$. We set:

$$\ln h_0 \sim \mathcal{N}(\ln \mu_0, 10 \times I_N), \tag{22}$$

where μ_0 is a vector collecting the logarithms of the squared elements on the diagonal of C. As stressed by Cogley and Sargent (2002), this prior is weakly informative for h_0.

Turning to the hyperparameters, we postulate independence between the parameters corresponding to the two matrices Q and A for convenience. Further, we make the following standard assumptions. The matrix Q is postulated to follow an inverted Wishart distribution:

$$Q \sim \mathcal{IW}(\bar{Q}^{-1}, T_0), \tag{23}$$

with prior degrees of freedom T_0 and scale matrix $T_0\bar{Q}$. In order to minimize the impact of the prior, we set T_0 equal to the minimum value allowed, the length of θ_t plus one. As for $\bar{Q}$, we calibrate it as $\bar{Q} = \gamma \times \hat{\Sigma}_{OLS}$, setting $\gamma = 1.0 \times 10^{-4}$, as in Cogley and Sargent (2002). This is a comparatively conservative prior in the sense of allowing little random-walk drift. We note, however, that it is smaller than the median-unbiased estimates of the extent of random-walk drift discussed in Section 2, ranging between 0.0235 and 0.0327 for the equation for the vacancy rate, and between 0.0122 and 0.0153 for the equation for the unemployment rate. As for α, we postulate it to be normally distributed with a large variance:

$$f(\alpha) = \mathcal{N}(0, 10000 \cdot I_{N(N-1)/2}). \tag{24}$$

Finally, we follow Cogley and Sargent (2002, 2005) and postulate an inverse-Gamma distribution for $\sigma_i^2 \equiv \text{Var}(\nu_{i,t})$ for the variances of the stochastic volatility innovations:

$$\sigma_i^2 \sim \mathcal{IG}\left(\frac{10^{-4}}{2}, \frac{1}{2}\right). \tag{25}$$

C.2 Simulating the Posterior Distribution

We simulate the posterior distribution of the hyperparameters and the states conditional on the data using the following MCMC algorithm (see Cogley and Sargent 2002). x^t denotes the entire history of the vector x up to time t, that is, $x^t \equiv [x_1', x_2', x_t']'$, while T is the sample length.

1. *Drawing the elements of* θ_t: Conditional on Y^T, α, and H^T, the observation equation (4) is linear with Gaussian innovations and a known covariance matrix. Following Carter and Kohn (1994), the density $p(\theta^T|Y^T, \alpha, H^T)$ can be factored as:

$$p(\theta^T|Y^T, \alpha, H^T) = p(\theta_T|Y^T, \alpha, H^T) \prod_{t=1}^{T-1} p(\theta_t|\theta_{t+1}, Y^T, \alpha, H^T). \tag{26}$$

   Conditional on α and H^T, the standard Kalman filter recursions determine the first element on the right hand side of (26), $p(\theta_T|Y^T, \alpha, H^T) = N(\theta_T, P_T)$, with P_T being the precision matrix of θ_T produced by the Kalman filter. The remaining elements in the factorization can then be computed via the backward recursion algorithm found in Cogley and Sargent (2005). Given the conditional normality of θ_t, we have:

$$\theta_{t|t+1} = \theta_{t|t} + P_{t|t} P_{t+1|t}^{-1} (\theta_{t+1} - \theta_t), \tag{27}$$

   and

$$P_{t|t+1} = P_{t|t} - P_{t|t} P_{t+1|t}^{-1} P_{t|t}, \tag{28}$$

   which provides, for each t from $T-1$ to 1, the remaining elements in (4), $p(\theta_t|\theta_{t+1}, Y^T, \alpha, H^T) = N(\theta_{t|t+1}, P_{t|t+1})$. Specifically, the backward recursion starts with a draw from $\mathcal{N}(\theta_T, P_T)$, call it $\tilde{\theta}_T$. Conditional on $\tilde{\theta}_T$, (27)–(28) give us $\theta_{T-1|T}$ and $P_{T-1|T}$, thus allowing us to draw $\tilde{\theta}_{T-1}$ from $N(\theta_{T-1|T}, P_{T-1|T})$, and so on until $t = 1$.
2. *Drawing the elements of* H_t: Conditional on Y^T, θ^T, and α, the orthogonalized innovations $u_t \equiv A(Y_t - X_t'\theta_t)$, with $\text{Var}(u_t) = H_t$, are observable. Following Cogley and Sargent (2002), we then sample the $h_{i,t}$'s by applying the univariate algorithm of Jacquier et al. (1994) element by element.
3. *Drawing the hyperparameters*: Conditional on Y^T, θ^T, H^T, and α, the innovations to θ_t and to the $h_{i,t}$'s are observable, which allows us to draw the hyperparameters, namely the elements of Q and the σ_i^2, from their respective distributions.
4. *Drawing the elements of* α: Finally, conditional on Y^T and θ^T, the ϵ_t's are observable. They satisfy:

$$A\epsilon_t = u_t, \tag{29}$$

with the u_t being a vector of orthogonalized residuals with known time-varying variance H_t. Following Primiceri (2005), we interpret (29) as a system of unrelated regressions. The first equation in the system is given by $\epsilon_{1,t} \equiv u_{1,t}$, while the following equations can be expressed as transformed regressions:

$$\begin{aligned}\left(h_{2,t}^{-\frac{1}{2}}\epsilon_{2,t}\right) &= -\alpha_{2,1}\left(h_{2,t}^{-\frac{1}{2}}\epsilon_{1,t}\right) + \left(h_{2,t}^{-\frac{1}{2}}u_{2,t}\right),\\ \left(h_{3,t}^{-\frac{1}{2}}\epsilon_{3,t}\right) &= -\alpha_{3,1}\left(h_{3,t}^{-\frac{1}{2}}\epsilon_{1,t}\right) - \alpha_{3,2}\left(h_{3,t}^{-\frac{1}{2}}\epsilon_{2,t}\right) + \left(h_{3,t}^{-\frac{1}{2}}u_{3,t}\right),\end{aligned} \tag{30}$$

where the residuals are independently standard normally distributed. Assuming normal priors for each equation's regression coefficients the posterior is also normal and can be computed as in Cogley and Sargent (2005).

Summing up, the MCMC algorithm simulates the posterior distribution of the states and the hyperparameters, conditional on the data, by iterating on (1)–(4). In what follows, we use a burn-in period of 50,000 iterations to converge to the ergodic distribution. After that, we run 10,000 more iterations sampling every 10th draw in order to reduce the autocorrelation across draws.

Appendix D: A Simple Search and Matching Model of the Labor Market

The model specification follows Lubik (2013). Time is discrete and the time period is a quarter. The model economy is populated by a continuum of identical firms that employ workers, each of whom inelastically supplies one unit of labor. Output Y_t of a typical firm is linear in employment N_t:

$$Y_t = A_t N_t. \tag{31}$$

A_t is a stochastic aggregate productivity process. It is composed of a permanent productivity shock, A_t^P, which follows a random walk, and a transitory productivity shock, A_t^T, which is an $AR(1)$-process. Specifically, we assume that $A_t = A_t^P A_t^T$.

The labor market matching process combines unemployed job seekers U_t with job openings (vacancies) V_t. This can be represented by a constant returns matching function, $M_t = m_t U_t^{\xi} V_t^{1-\xi}$, where m_t is stochastic match efficiency, and $0 < \xi < 1$ is the match elasticity. Unemployment is defined as those workers who are not currently employed:

$$U_t = 1 - N_t, \tag{32}$$

where the labor force is normalized to one. Inflows to unemployment arise from job destruction at rate $0 < \rho_t < 1$, which can vary over time. The dynamics of employment are thus governed by the following relationship:

$$N_t = (1 - \rho_t)\left[N_{t-1} + m_{t-1} U_{t-1}^{\xi} V_{t-1}^{1-\xi}\right]. \tag{33}$$

This is a stock-flow identity that relates the stock of employed workers N_t to the flow of new hires $M_t = m_t U_t^{\xi} V_t^{1-\xi}$ into employment. The timing assumption is such that once a worker is matched with a firm, the labor market closes. This implies that if a newly hired worker and a firm separate, the worker cannot re-enter the pool of searchers immediately and has to wait one period before searching again.

The matching function can be used to define the job finding rate, i.e., the probability that a worker will be matched with a firm:

$$p(\theta_t) = \frac{M_t}{U_t} = m_t \theta_t^{1-\xi}, \tag{34}$$

and the job matching rate, i.e., the probability that a firm is matched with a worker:

$$q(\theta_t) = \frac{M_t}{V_t} = m_t \theta_t^{-\xi}, \tag{35}$$

where $\theta_t = V_t/U_t$ is labor market tightness. From the perspective of an individual firm, the aggregate match probability $q(\theta_t)$ is exogenous and unaffected by individual decisions. Hence, for individual firms new hires are linear in the number of vacancies posted: $M_t = q(\theta_t)V_t$.

A firm chooses the optimal number of vacancies V_t to be posted and its employment level N_t by maximizing the intertemporal profit function:

$$E_0 \sum_{t=0}^{\infty} \beta^t [A_t N_t - W_t N_t - \kappa_t V_t], \tag{36}$$

subject to the employment accumulation equation (33). Profits are discounted at rate $0 < \beta < 1$. Wages paid to the workers are W_t, while $\kappa_t > 0$ is a firm's time-varying cost of opening a vacancy. The first-order conditions are:

$$N_t: \quad \mu_t = A_t - W_t + \beta E_t\big[(1-\rho_{t+1})\mu_{t+1}\big], \tag{37}$$

$$V_t: \quad \kappa_t = \beta q(\theta_t) E_t\big[(1-\rho_{t+1})\mu_{t+1}\big], \tag{38}$$

where μ_t is the multiplier on the employment equation.

Combining these two first-order conditions results in the *job creation condition* (JCC):

$$\frac{\kappa_t}{q(\theta_t)} = \beta E_t\left[(1-\rho_{t+1})\left(A_{t+1} - W_{t+1} + \frac{\kappa_{t+1}}{q(\theta_{t+1})}\right)\right]. \tag{39}$$

This captures the trade-off faced by the firm: the marginal effective cost of posting a vacancy, $\frac{\kappa_t}{q(\theta_t)}$, that is, the per-vacancy cost κ adjusted for the probability that the position is filled, is weighed against the discounted benefit from the match. The latter consists of the surplus generated by the production process net of wage payments to the workers, plus the benefit of not having to post a vacancy again in the next period.

In order to close the model, we assume in line with the existing literature that wages are determined based on the Nash bargaining solution: surpluses accruing to the matched parties are split according to a rule that maximizes their weighted average. Denoting the workers' weight in the bargaining process as $\eta \in [0, 1]$, this implies the sharing rule:

$$\mathcal{W}_t - \mathcal{U}_t = \frac{\eta}{1-\eta}(\mathcal{J}_t - \mathcal{V}_t), \tag{40}$$

where $\mathcal{W}_t$ is the asset value of employment, $\mathcal{U}_t$ is the value of being unemployed, $\mathcal{J}_t$ is the value of the marginal worker to the firm, and $\mathcal{V}_t$ is the value of a vacant job. By free entry, $\mathcal{V}_t$ is assumed to be driven to zero.

The value of employment to a worker is described by the following Bellman equation:

$$\mathcal{W}_t = W_t + E_t\beta\big[(1-\rho_{t+1})\mathcal{W}_{t+1} + \rho_{t+1}\mathcal{U}_{t+1}\big]. \tag{41}$$

Workers receive the wage W_t, and transition into unemployment next period with probability ρ_{t+1}. The value of searching for a job, when the worker is currently unemployed, is:

$$\mathcal{U}_t = b_t + E_t\beta\big[p_t(1-\rho_{t+1})\mathcal{W}_{t+1} + \big(1 - p_t(1-\rho_{t+1})\big)\mathcal{U}_{t+1}\big]. \tag{42}$$

An unemployed searcher receives stochastic benefits b_t and transitions into employment with probability $p_t(1-\rho_{t+1})$. Recall that the job finding rate p_t is defined as $p(\theta_t) = M(V_t, U_t)/U_t$ which is decreasing in tightness θ_t. It is adjusted for the probability that a completed match gets dissolved before production begins next period. The marginal value of a worker $\mathcal{J}_t$ is equivalent to the multiplier on the employment equation, $\mathcal{J}_t = \mu_t$, so that the respective first-order condition defines the Bellman-equation for the value of a job. Substituting the asset equations into the sharing rule (40) results in the wage equation:

$$W_t = \eta(A_t + \kappa_t\theta_t) + (1-\eta)b_t. \tag{43}$$

Wage payments are a weighted average of the worker's marginal product A_t, which the worker can appropriate at a fraction η, and the outside option b_t, of which the firm obtains the portion $(1-\eta)$. Moreover, the presence of fixed vacancy posting costs leads to a hold-up problem where the worker extracts an additional $\eta\kappa_t\theta_t$ from the firm.

Finally, we can substitute the wage equation (43) into (39) to derive an alternative representation of the job creation condition:

$$\frac{\kappa_t}{m_t}\theta_t^{\xi} = \beta E_t(1-\rho_{t+1})\left[(1-\eta)(A_{t+1} - b_t) - \eta\kappa_t\theta_{t+1} + \frac{\kappa_t}{m_{t+1}}\theta_{t+1}^{\xi}\right]. \tag{44}$$

References

Andrews, D. K. (1991). Heteroskedasticity and autocorrelation-consistent covariance matrix estimation. *Econometrica*, *59*, 817–858.

Barnichon, R. (2010). Building a composite help-wanted index. *Economics Letters*, *109*, 175–178.

Benati, L. (2007). Drifts and breaks in labor productivity. *Journal of Economic Dynamics & Control*, *31*, 2847–2877.

Blanchard, O. J., & Diamond, P. (1989). The Beveridge curve. *Brookings Papers on Economic Activity*, *1*, 1–60.

Blanchard, O. J., & Quah, D. (1989). The dynamic effects of aggregate demand and supply disturbances. *The American Economic Review*, *79*, 655–673.

Carter, C. K., & Kohn, R. (1994). On Gibbs sampling for state space models. *Biometrika*, *81*, 541–553.

Cogley, T., & Sargent, T. J. (2002). Evolving post-World War II U.S. inflation dynamics. *NBER Macroeconomics Annual*, *16*, 331–388.

Cogley, T., & Sargent, T. J. (2005). Drifts and volatilities: monetary policies and outcomes in the post-WWII U.S. *Review of Economic Dynamics*, *8*, 262–302.

Fernández-Villaverde, J., & Rubio-Ramirez, J. (2007). How structural are structural parameter values? *NBER Macroeconomics Annual*, *22*, 83–132.

Furlanetto, F., & Groshenny, N. (2012). *Mismatch shocks and unemployment during the Great Recession*. Manuscript, Norges Bank.

Galí, J. (1999). Technology, employment, and the business cycle: do technology shocks explain aggregate fluctuations? *The American Economic Review*, *89*, 249–271.

Galí, J., & Gambetti, L. (2009). On the sources of the great moderation. *American Economic Journal: Macroeconomics*, *1*, 26–57.

Jacquier, E., Polson, N. G., & Rossi, P. E. (1994). Bayesian analysis of stochastic volatility models. *Journal of Business & Economic Statistics*, *12*, 371–389.

Lubik, T. A. (2013). *The shifting and twisting Beveridge curve: an aggregate perspective*. Manuscript, Fęderal Reserve Bank of Richmond.

Newey, W., & West, K. (1987). A simple positive semi-definite heteroskedasticity and autocorrelation-consistent covariance matrix. *Econometrica*, *55*, 703–708.

Primiceri, G. (2005). Time varying structural vector autoregressions and monetary policy. *Review of Economic Studies*, *72*, 821–852.

Rubio-Ramirez, J., Waggoner, D., & Zha, T. (2005). Structural vector autoregressions: theory of identification and algorithms for inference. *Review of Economic Studies*, *77*, 665–696.

Sahin, A., Song, J., Topa, G., & Violante, G. L. (2012). *Mismatch unemployment* (Staff Reports No. 566). Federal Reserve Bank of New York.

Shimer, R. (2005). The cyclical behavior of equilibrium unemployment and vacancies. *The American Economic Review*, *95*, 25–49.

Stock, J., & Watson, M. M. (1996). Evidence of structural instability in macroeconomic time series relations. *Journal of Business & Economic Statistics*, *14*, 11–30.

Stock, J., & Watson, M. M. (1998). Median-unbiased estimation of coefficient variance in a time-varying parameter model. *Journal of the American Statistical Association*, *93*, 349–358.

Bilinear Forecast Risk Assessment for Non-systematic Inflation: Theory and Evidence

Wojciech W. Charemza, Yuriy Kharin, and Vladislav Maevskiy

Abstract The paper aims at assessing the forecast risk and the maximum admissible forecast horizon for the non-systematic component of inflation modeled autoregressively, where a distortion is caused by a simple nonlinear (bilinear) process. The concept of the guaranteed upper risk of forecasting and the δ-admissible distortion level is defined. In order to make this concept operational we propose a method of evaluation of the p-maximum admissible forecast risk, on the basis of the maximum likelihood estimates of the bilinear coefficient. It has been found that for the majority of developed countries (in terms of average GDP per capita) the maximum admissible forecast horizon is between 5 and 12 months, while for the poorer countries it is either shorter than 5 or longer than 12. There is also a negative correlation of the maximum admissible forecast horizon with the average GDP growth.

1 Introduction

The literature on inflation forecasting has, so far, focused on identification and further analysis of its systematic part, often described as the core or underlying inflation. This component of inflation is loosely defined as the dynamics of prices being neutral regarding to output in the medium and long-run. The literature on this subject is huge (see e.g. the seminal works by Eckstein 1981; Cecchetti 1996; Quah and Vahey 1995; Cristadoro et al. 2005, current critical reviews and advances by Silver

W.W. Charemza (✉)
University of Leicester, Leicester, UK
e-mail: wch@le.ac.uk

W.W. Charemza
AFiBV, Warsaw, Poland

Y. Kharin
Belarusian State University, Minsk, Belarus
e-mail: kharin@bsu.by

V. Maevskiy
EPAM-Systems, Minsk, Belarus

F. Schleer-van Gellecom (ed.), *Advances in Non-linear Economic Modeling*,
Dynamic Modeling and Econometrics in Economics and Finance 17,
DOI 10.1007/978-3-642-42039-9_6,

2007; Rich and Steindel 2007, Bodenstein et al. 2008; Siviero and Veronese 2011; Wynne 2008 and Bermingham 2010 and more recent developments by Blankmeyer 2012 and Charemza and Shah 2013). In fact forecasting core inflation has become a common practice at many central banks and other financial institutions. There is, however, a limited interest in investigation of the non-systematic part of inflation, described as the difference between the headline (or observed) inflation and its core component. It is usually acknowledged that the non-systematic inflation is stationary and short-term forecastable. Nevertheless, the specific forecasting techniques have not been researched so far and, in particular, the length of the admissible forecasting horizon has usually been defined here rather vaguely.

In this paper we aim to assess the forecast risk and the maximum admissible forecast horizon for the non-systematic component of inflation, where there is a certain type of non-linearity in the process, defined (or approximated) by a simple first-order bilinear process. The presence in such component creates misspecification in forecasting for periods longer than one. This prompts the question about the maximum admissible forecast horizon, for which distortions caused to the forecast due to such misspecification are not substantial. Obviously, what is 'substantial' here is arbitrary and has to be defined prior to any investigation. Here we start with the concept of the guaranteed upper risk of forecasting. The δ-admissible distortion level defined as the maximal value of the bilinear coefficient for which the forecast instability does not exceed a priori given admissible risk level δ (Section 2). In order to make the concept of the admissible risk level operational, in Section 3 we propose a method for evaluation of the p-maximum admissible forecast risk, which corresponds to the pth fractile of the distribution of a statistic used for evaluation of the null hypothesis of no bilinearity. After a series of Monte Carlo experiments, we suggest to use, as such statistic, a Student-t ratio for the maximum likelihood estimates of the bilinear coefficient. After computing the p-maximum admissible forecast risk, it is possible to evaluate the maximum forecast horizon for which, under such level of risk, the estimated bilinear coefficient is equal to its maximal admissible value.

Section 4 contains a description of empirical results for time series of monthly data on inflation for 122 countries. The maximum span of the series is from 1957 to April 2011 (some series are shorter). This section also discusses the relationship between GDP (in terms of levels and growths) and the maximum admissible forecast horizon.

2 Risk Assessment Problem

Suppose that the non-systematic inflation, π_t, that is a difference between the headline and core inflations at time t, $t = 0, 1, \ldots, T$, is described by a simple stationary bilinear autoregressive BL$(1, 0, 1, 1)$ process:

$$\pi_t = \alpha\pi_{t-1} + \beta\pi_{t-1}u_{t-1} + u_t, \tag{1}$$

where α and β are the parameters and $\{u_t\}$ is a sequence of i.i.d. random variables with zero expected value (both unconditional and conditional on past information) and finite higher moments. The rationale for the existence of the bilinear term in (1) can be grounded, for instance, within the theory of speculative inflation (see e.g. Schmitt-Grohé 2005; Sims 2005 for economies with inflationary targeting) and within the modern hyperinflation theories (see Vázquez 1998; Jha et al. 2002; Adam et al. 2006 and Arce 2009).

Let us consider forecasting from (1) outside T, initially assuming the knowledge of α but not β. In this case the forecasting scheme is analogous to that from a linear $AR(1)$ model, that is:

$$\pi^f_{T+\tau} = \alpha\pi^f_{T+\tau-1} = \alpha^\tau \pi_T, \quad \text{for } \tau = 1, 2, 3, \ldots. \tag{2}$$

The absence of information regarding β leads in the above forecasting scheme causes a distortion and creates a forecasting risk. Let us define such risk as the mean-square error (MSE) of the forecast, that is:

$$MSE(\tau) = E\big(\pi_{T+\tau} - \pi^f_{T+\tau}\big)^2.$$

Theorem 1 (see Appendix A) gives the asymptotic expansion of $MSE(\tau)$ in terms of model parameters. In order to evaluate a possible impact of the bilinear distortion on $MSE(\tau)$, let us define the guaranteed upper risk of forecast as the maximum admissible mean square forecast error for a given set of the bilinear parameters, that is (see Kharin 1996):

$$MSE_+(\tau) = \sup_{\beta\in[-\beta_+,\beta_+]} MSE(\tau).$$

Let us also define the forecast instability coefficient $\kappa(\tau)$ as:

$$\kappa(\tau) = \frac{MSE_+(\tau) - MSE_0(\tau)}{MSE_0(\tau)}, \quad \text{where } MSE_0(\tau) = \frac{\sigma^2(1-\alpha^{2\tau})}{1-\alpha^2}$$

is the minimally admissible risk value for the situation without bilinear distortions, and σ^2 is variance of u_t. Following Kharin (1996), we can define the δ-admissible distortion level $\beta^+(\delta, \tau)$ as the maximal distortion level β_+ for which the instability coefficient $\kappa(\tau)$ does not exceed a priori given admissible risk level δ. It can be shown (see Theorem 2 in Appendix A) that, under additional assumption of normality for u_t, the following asymptotic expansions are true:

$$MSE_+(\tau) = \frac{\sigma^2(1-\alpha^{2\tau})}{1-\alpha^2} + \beta_+^2\sigma^4\Upsilon + o\big(\beta_+^2\big), \tag{3}$$

$$\kappa(\tau) = \beta_+^2\sigma^2\Upsilon + o\big(\beta_+^2\big), \tag{4}$$

$$\beta_+(\delta,\tau) = \frac{\sqrt{\delta}}{\sigma\sqrt{\Upsilon}} + o\big(\beta_+^2\big), \quad \text{where: } \Upsilon = 2\frac{1-\alpha^\tau}{(1-\alpha)^2} + \frac{1-\alpha^{2\tau}}{(1-\alpha^2)^2} - 2\frac{\alpha^{2\tau-1}}{(1-\alpha)}. \tag{5}$$

With the use of the formula above one might evaluate the potential distortion to the means square error of forecast due to omitted nonlinearity. Figures 1a, 1b and 1c show the results of a numerical evaluation of $MSE_{+}(\tau)$, $\kappa(\tau)$ and $\beta^{+}(\delta,\tau)$ for values of α varying from -0.99 to 0.99, $\tau = 1, 2, 3$ and 5, $\sigma^2 = 1$ and $\delta = 1$.

Figures 1a–1c suggest that nonlinear and asymmetric responses of the guaranteed forecast risk and instability coefficients might cause practical problems in establishing the admissible risk level δ. The fact that for large α's (typical for inflationary processes), the $MSE_{+}(\tau)$ rapidly approaching infinity makes it particularly cumbersome.

This is illustrated by relating the admissible risk level δ to a range of $AR(1)$ α coefficients corresponding to a certain level of the nonstationarity which is defined as:

$$\Phi = \alpha^2 + \sigma^2\beta_{+}^2(\tau). \tag{6}$$

If $\Phi = 1$, (6) constitutes the stationarity limit for (1) (see e.g. Granger and Anderson 1978). For $0 \leq \Phi < 1$ it is a general measure of time-dependence (predictability) of a stationary process (1) so that $\Phi = 0$ refers to a purely random unpredictable process (white noise). For a given Φ, σ^2, τ and α, values of δ can be solved out from (5) and plotted against α. Figure 2 shows δ as a function of $\alpha \in [-\Phi, \Phi]$ for Φ equal respectively to $0.25, 0.75$ and 0.95, $\sigma^2 = 1$ and $\tau = 2$, where $\beta_{+}^2(\tau) = (\Phi - \alpha^2)/\sigma^2$.

Figure 2 shows that the increase in the admissible risk for a given predictability is not linear and not even monotonous if regarded as function of the degree of predictability. For large predictability and large α (in excess of 0.8) the level of admissible risk falls. So that, establishing the appropriate admissible risk level in inflation forecasting might be difficult.

3 Econometric Problem

The problem with establishing the admissible risk level, outlined in Section 2, might be to some extent relaxed if it is possible to estimate the parameters of (1) econometrically. Let us assume that there exists statistical data on inflation for the period $t = 0, 1, \ldots, T$, and, prior to forecasting for the periods $T + \tau$, it is possible to estimate the parameters α and β. Denoting these estimates respectively by $\hat{\alpha}$ and $\hat{\beta}$ and using some initial values π_0 and u_0, it is possible to obtain the estimates of u_t recursively as:

$$\hat{u}_t = \pi_t - \hat{\alpha}\pi_{t-1} - \hat{\beta}\pi_{t-1}\hat{u}_{t-1}.$$

This might help in constructing a one-step ahead forecast as:

$$\pi_{T+1}^{f} = \alpha\pi_T + \beta\pi_t u_T.$$

However, for forecast horizons longer than one, there is no possibility of recovering $u_{T+\tau}, \tau = 2, 3 \ldots$. In this case forecast from the estimated Eq. (1) coincides with

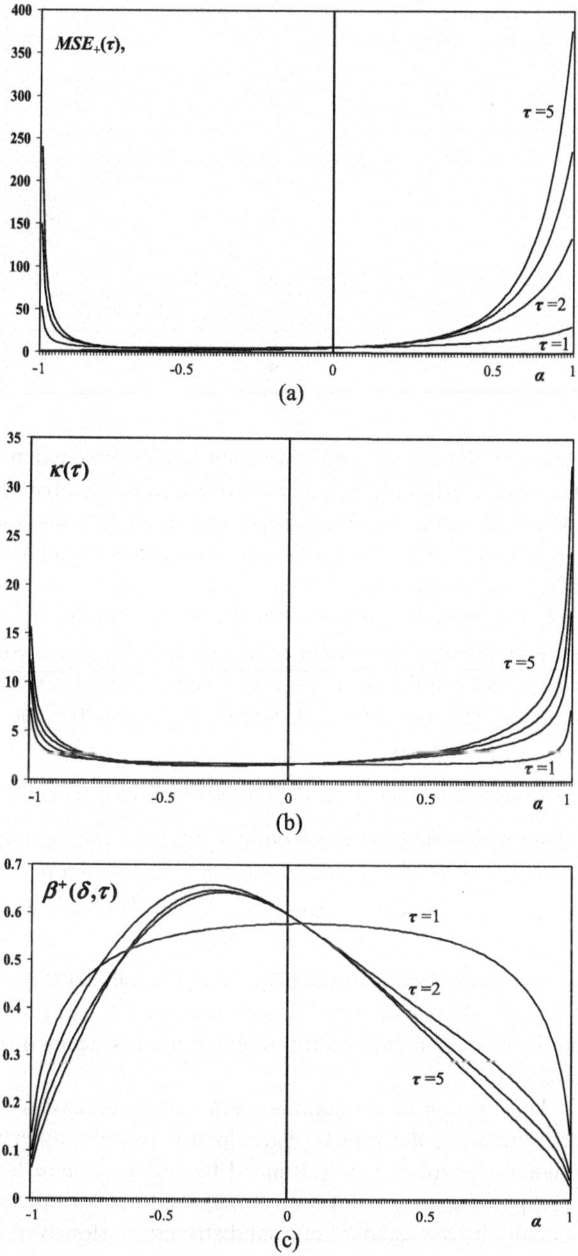

Fig. 1 (**a**) Dependence of $MSE_{+}(\tau)$ on autoregressive parameter and forecast horizon. (**b**) Dependence of $\kappa(\tau)$ on autoregressive parameter and forecast horizon. (**c**) Dependence of $\beta^{+}(\delta,\tau)$ on autoregressive parameter and forecast horizon

the forecast from a simple $AR(1)$ model and is based upon information on a single parameter α, that is on:

$$\pi^{f}_{T+\tau} = \hat{\alpha}\pi^{f}_{T+\tau-1} = \alpha^{\tau}\pi_{T} + \alpha^{\tau-1}\hat{\beta}\pi_{T}\hat{u}_{T}.$$

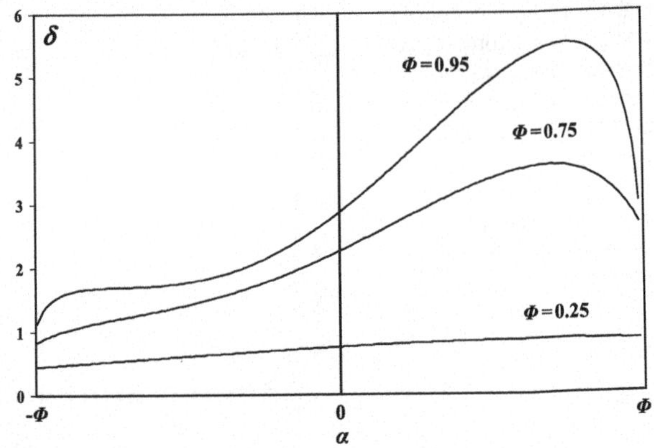

Fig. 2 Dependence of δ on the degree of predictability

However, the econometric estimates can, to some extent, help with establishing the admissible risk level. In turn, this can lead to establishing the maximum admissible forecast horizon (MAF), that is the maximum value of τ for which, given δ, the absence of $\hat{\beta}$ in the forecasting process does not lead to the increase of the expected $MSE(\tau)$ over the $MSE_+(\tau)$.

Let ξ_β be a well-defined statistic for β with the argument $\hat{\beta}$. In particular it can be the Student-t statistic for β, that is $(\hat{\beta}-\beta)/S(\hat{\beta})$, where $S(\hat{\beta})$ is the standard deviation of $\hat{\beta}$, or the normalised estimate of β, that is $\hat{\beta}\cdot S(u_t)$. $S(u_t)$ is the estimated standard deviation of u_t. Denote by $\hat{\beta}^p_{\xi|\beta=0}$ such value of $\hat{\beta}$ which corresponds to the pth fractile of the distribution of ξ_β for $\beta=0$. Knowing $\hat{\beta}^p_{\xi|\beta=0}$ and $\hat{\alpha}$, it is possible to find the p-maximum admissible forecast risk $\delta^p_{\hat{\beta}}(\tau)$ which can be obtained by solving (5) for δ with $\beta_+(\delta,\tau)=\hat{\beta}^p_{\xi|\beta=0}$ and $S(u_t)$. Since, in practice, a normalisation for a unitary variance of u_t is required, it can be achieved by using $\tilde{\beta}_+(\delta,\tau)=\hat{\beta}^p_{\xi|\beta=0}\cdot S(u_t)$ rather than $\beta_+(\delta,\tau)$. It is convenient to interpret the p-maximum admissible forecast risk $\delta^p_{\hat{\beta}}(\tau)$ as the risk which is associated with ignoring, in the forecasting scheme, the β parameter if it is equal to the unusually high (or low) estimate of β, in the case where the hypothesis that $\beta=0$ is true. Whether the value of $\hat{\beta}$ is 'unusually' high (or low) is decided by using tail percentiles like 0.05 or 0.95.

The concept of p-maximum admissible forecast risk requires knowledge of the distribution of the statistic ξ_β, which is usually either the distribution of $\hat{\beta}$, or its Student-t ratio. If β is estimated by the maximum likelihood (ML) method, the asymptotic normality of the estimates allows for approximation of the normalised statistics by the standard normal distribution. However, the behavior of the statistics in finite samples depends on the speed of convergence.

In order to investigate the finite sample properties of the statistics, the following Monte Carlo experiments have been performed. The data generating process is (1) with $\beta=0$ and $u_t\sim$ i.i.d. $N(0,1)$, $t=0,1,\ldots,T$, which reduces it to a simple $AR(1)$ process with a random initial value. The parameter α varies as 0.25, 0.5 and 0.75, T varies as 75, 100 and 250 and, for each sets of parameters and

Table 1 Bera-Jarque statistics for the ML estimates of β and their t ratios

T	$\alpha = 0.25$	$\alpha = 0.50$	$\alpha = 0.75$
For $\hat{\beta}$			
75	326.6 (0.00)	777.6 (0.00)	341.8 (0.00)
100	71.12 (0.00)	86.42 (0.00)	127.3 (0.00)
250	1.64 (0.44)	1.76 (0.41)	3.00 (0.22)
500	2.20 (0.33)	1.03 (0.60)	0.25 (0.88)
For $t(\hat{\beta})$			
75	61.32 (0.00)	7843 (0.00)	839900 (0.00)
100	8.25 (0.02)	40.26 (0.00)	1317 (0.00)
250	0.57 (0.75)	0.86 (0.65)	0.90 (0.64)
500	3.36 (0.18)	1.91 (0.36)	0.56 (0.75)

each T, 10,000 replications are generated. In each replication the parameters α and β are estimated by the constrained maximum likelihood method used for the Kalman Filter representation of (1), where the constraint is the stationarity condition.[1]

Table 1 shows the Bera-Jarque measures of normality for the empirical distributions of the estimates of β and their Student-t statistics, $t(\hat{\beta})$ with p-values in the parentheses. It indicates that the convergence to normality is relatively slow here. This prompts the question whether the percentiles of the standard normal distribution can be used as critical values for the t-ratios of the estimated β parameters. Table 2 shows the empirical percentiles of the simulated distributions of the t-ratios for the ML β estimates in comparison with the percentiles of the standard normal distribution, which is the asymptotic distribution for the ML estimates.

Results in Table 2 suggest that, although the finite sample distributions of the Student-t ratios are not normal and the tails of the distributions are heavy, especially for the large values of α and small samples, the differences are not very substantial. With some caution, percentiles of normal distribution can be used here for testing the significance of the estimates of β.

[1] Computations were performed in Aptech GAUSS using the constrained maximum likelihood package (CML) and Roncalli (1995) Kalman Filter routines.

Table 2 Simulated percentiles of $t(\hat{\beta})$

		Percentiles				
		99 %	97.5 %	95 %	90 %	50 %
$T = 75$	$\alpha = 0.25$	2.39	1.97	1.67	1.28	0.00
	$\alpha = 0.50$	2.46	2.01	1.68	1.29	0.00
	$\alpha = 0.75$	2.54	2.11	1.75	1.33	0.00
$T = 100$	$\alpha = 0.25$	2.35	1.95	1.61	1.27	0.00
	$\alpha = 0.50$	2.35	1.98	1.65	1.27	0.00
	$\alpha = 0.75$	2.44	2.05	1.68	1.30	0.00
$T = 250$	$\alpha = 0.25$	2.29	1.95	1.65	1.26	−0.02
	$\alpha = 0.50$	2.33	1.85	1.60	1.25	0.01
	$\alpha = 0.75$	2.39	1.89	1.61	1.25	−0.01
∞		*2.33*	*1.96*	*1.64*	*1.28*	*0.00*

Figures 3a–3c show the computed values of $\delta_{\hat{\beta}}^{0.95}(\tau)$ obtained by solving (5) for δ, that is:

$$\delta_{\hat{\beta}}^{p}(\tau) = \left(\hat{\beta}_{\xi|\beta=0}^{p}\right)^2 \sigma^2 \Upsilon,$$

where $\hat{\beta}_{\xi|\beta=0}^{p}$ has been selected alternatively by three criteria: percentiles of $\hat{\beta}$ (Figure 3a), percentiles of $t(\hat{\beta})$ (Figure 3b) and percentiles of normalised $\hat{\beta}$, that is $\tilde{\beta} = \hat{\beta} \cdot S(u_t)$, where $S(u_t)$ is the estimated standard deviation of u_t (Figure 3c). These are compared with their sample estimates $\hat{\delta}_{\hat{\beta}}^{p}(\tau)$, that is:

$$\hat{\delta}_{\hat{\beta}}^{p}(\tau) = \left(\hat{\beta}_{\xi|\beta=0}^{p}\right)^2 S^2(u) \hat{\Upsilon}, \tag{7}$$

where $\hat{\Upsilon}$ is computed as Υ in (3)–(5), except that the estimates $\hat{\alpha}$ are used here rather than α.

Figures 3a–3c indicate that, although the estimates of the 5 % admissible forecast risk are biased (either negatively, as in Figure 3a, or positively, as in Figures 3b and 3c), its values stabilizes quickly with the increase of forecast horizon and, for the horizons greater than 9, they are virtually constant. Similar results are observed for different sample sizes and different values of α. Generally, it appears that the criterion of selecting $\hat{\beta}_{\xi|\beta=0}^{p}$ according to the percentiles of $t(\hat{\beta})$ is most advisable, since the bias of the estimates is usually the smallest.

4 Risk Assessment and Forecast Horizon for Worldwide Inflation

The concept of p-maximum admissible forecast risk can be applied in practice for assessing the rationale of forecasting of the non-systematic part of inflation and, in

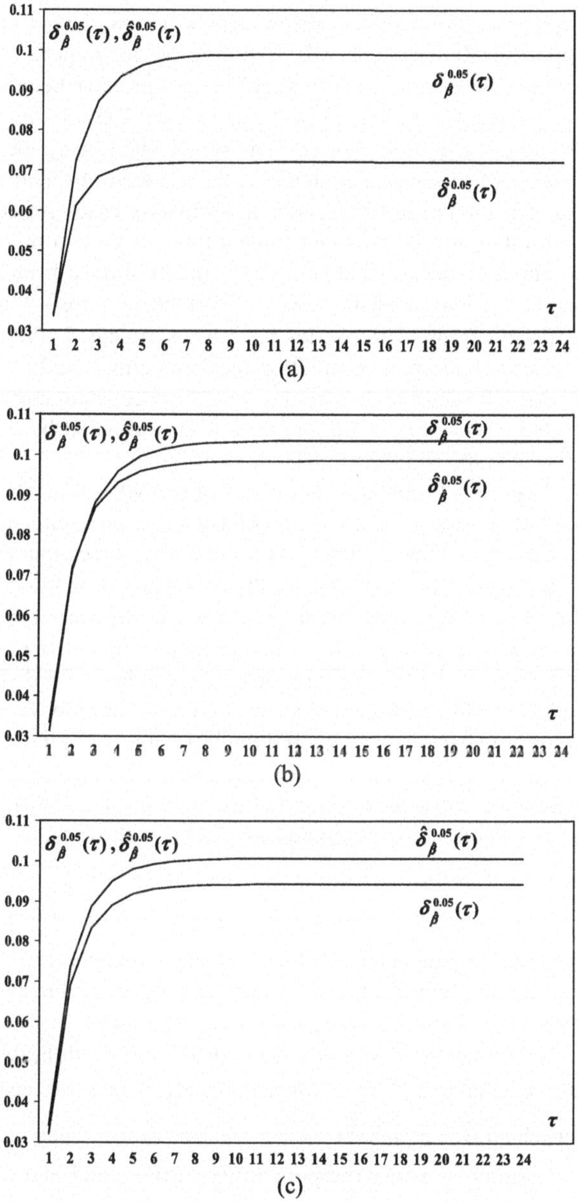

Fig. 3 True and estimated 5 % maximum admissible forecast risk $T = 100$, $\alpha = 0.5$, 10,000 replications. (**a**) Criterion: percentiles of $\hat{\beta}$; (**b**) criterion: percentiles of $t(\hat{\beta})$; (**c**) criterion: percentiles of $\tilde{\beta}$

particular, evaluating the maximum forecast horizon for which the bilinear distortions do not cause the risk in excess of the admissible value. For the empirical analysis a panel of monthly time series of annual inflation rates (that is, on the basis of the corresponding month of the previous year) for a wide number of countries have been used. The data are taken from the International Monetary Fund database (see http://www.imfstatistics.org/imf, can be accessed e.g. through the ESDC database at http://esds80.mcc.ac.uk/wds_ifs/ReportFolders/reportFolders.aspx). Out of the data set for 170 countries, series for 122 countries have been selected with the maximum time coverage of the data set from January 1957 to April 2011 (for most countries the series have been shorter). The series which were incomplete, with a substantial number of missing or systematically repeated observations, have been eliminated. For the remaining series, in a few obvious cases infrequent missing values have been interpolated and some evident typos in data corrected. From the original data the monthly series of annual (y/y) inflation have been computed which gives the maximum length of the series of 591 observations. Outliers greater than 5 standard deviations of the series have been truncated (there were very few of them). The systematic part of inflation has been eliminated by smoothing the data by the Hodrick-Prescott filter with the smoothing constant equal to 16,000. For each country the parameters of Eq. (1) have been estimated by the constrained ML Kalman Filter method (see Section 3).

Appendix B contains the results of the ML estimates of coefficients α and β for individual series. Tables 4 shows the estimation results. In columns (1)–(4), after the country codes and number of observations, the estimates of the $AR(1)$ coefficients, $\hat{\alpha}$, are given and followed by their t ratios. In column (5) the significance of the $AR(1)$ the $AR(1)$ coefficients which are significant at the 0.01 level are marked by (3) and those with p-values smaller than 0.01 by (0). Columns (6)–(9) describe the estimates of the bilinear coefficient; columns (6) and (7) give the non-normalised and normalised estimates correspondingly, column (8) shows the t-ratios for the non-normalised estimators and the last column (9) indicates the significance.

Table 5 present the forecast risk assessment characteristics. Column (3) gives the stationarity measures computed as: $\hat{\Phi} = \hat{\alpha}^2 + \hat{\beta}^2 S^2(u_t)$. Column (4) presents the $\hat{\beta}^{0.90}_{\xi|\beta=0}$ coefficients computed as in (7), with the selection criteria being the 90th percentile of the $t(\hat{\beta})$ statistic. The corresponding $\hat{\delta}^{0.90}_{\hat{\beta}}(\tau^*)$ values, where $\tau^* = 24$ and represents the most remote forecast horizon, for which the values of $\delta^{0.90}_{\hat{\beta}}(\tau)$ are virtually independent from τ, are shown in column (5). These values are halved, in order to allow for the symmetry of positive and negative bilinearity. Column (6) shows the estimates of the maximum admissible forecast horizon for which the effect of bilinearity does not exceed the maximal admissible distortion level computed at risk equal to $\hat{\delta}^{0.90}_{\hat{\beta}}(\tau^*)$. More precisely, $\tau_{\max}$ is defined as such forecast horizon τ for which $\tilde{\beta} \approx \beta_{+}[\hat{\delta}^{0.90}_{\hat{\beta}}(\tau^*), \tau]$.

In order to assess the poolability of the panel and to decide whether particular series in the panel can be analysed separately, a simple correlation analysis between the pairs of ML residuals of the estimated Eq. (1) has been performed. For 7,381 correlations the percentage of significant correlations at 5 % equal to 8.89 %. Although this is more than the expected 5 %, nevertheless this percentage rate is not very high, so that the possible distortions to the estimates for the individual countries due to interdependence within the panel are likely not substantial. The estimated bilinear coefficients are, in most cases, insignificant; there are only 24 significant (at the 5 % level of significance) bilinear coefficients.

Table 3 Distribution of τ_{max} for non-systematic inflation

τ_{max}	No. of countries
Smaller than 6	29
Between 6 and 9	45
Between 10 and 14	36
Greater than 14	12

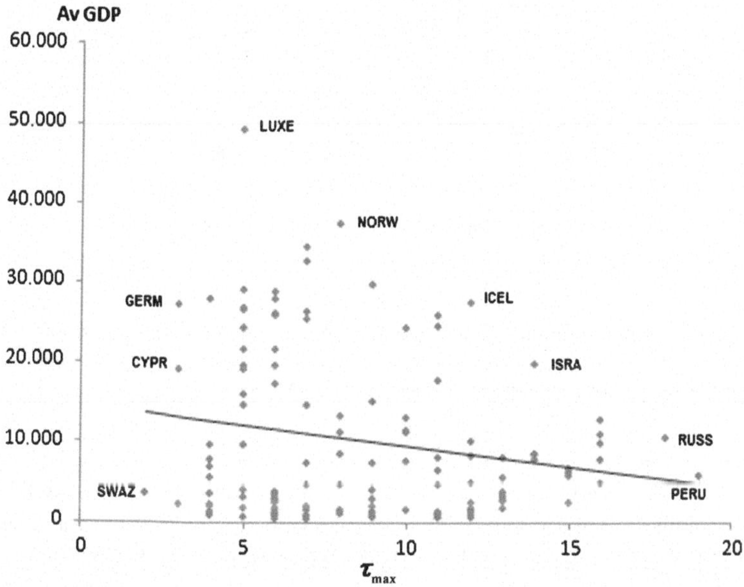

Fig. 4 Average levels of GDP and the maximum admissible inflation forecast horizons. Legend: GDP level per capita adjusted for purchasing power disparities measured at constant Geary-Khamis 2005 international dollars (see http://esds80.mcc.ac.uk/WDS_WB/)

The distribution of countries according to the maximum admissible forecast horizon is given in Table 3.

There is an interesting regularity between the World Bank estimates of the annual GDP level per capita adjusted for purchasing power disparities measured at constant 2005 international dollars (see http://esds80.mcc.ac.uk/WDS_WB/) and τ_{max}. Figure 4 shows a scatter diagram of the average GDP per capita and τ_{max}. The periods for which means of the GDP have been computed correspond to the periods used for computing τ_{max}. Some visible outliers on the diagrams have been marked by country symbols. There is also a linear regression line presented at this figure.

There is a visible, albeit not very strong, negative relationship between the maximum admissible forecast horizon and the average GDP level. The correlation coefficient is equal to -0.194, with Student-t ratio equal to 2.143 and a one-sided

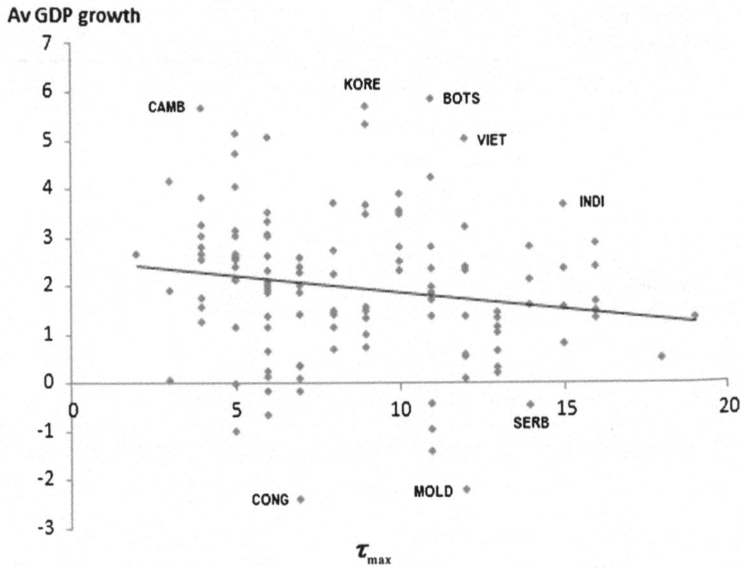

Fig. 5 Average GDP growth and the maximum admissible inflation forecast horizons. Legend: GDP level per capita adjusted for purchasing power disparities measured at constant Geary-Khamis 2005 international dollars (see http://esds80.mcc.ac.uk/WDS_WB/)

p-value 0.01606. The triangular shape of the scatter points suggests a nonlinearity of the dependence pattern. Out of 12 countries with $\tau_{max} < 5$, 9 have average per capita GDP level below the level of 10,000 International Geary-Khamis \$. Similarly, out of 19 countries with $\tau_{max} > 12$, 8 have the average per capita GDP below 10,000 international Geary-Khamis \$. For countries with τ_{max} between 5 and 12, the proportion of richer countries is greater. If there is a relation between the level of development of a country measured by its GDP per capita and the maximum admissible forecast horizon it can be stated that the developed countries have usually the linearly forecastable inflation with the moderate forecast horizons, while the poorer countries usually have inflation linearly forecastable for either very short, or very long periods.

The idea of the maximum admissible forecast horizon might also add to the empirical evidence of GDP convergence. Figure 5 depicts the relationship between τ_{max} and the average rate of growth of the 122 countries analysed here. The data for growth have been obtained from the World Bank sources at http://esds80.mcc.ac.uk/WDS_WB/.

There is a significant negative correlation between τ_{max} and the average GDP growth (the correlation coefficient is equal to -0.1648, with Student t-ratio equal to 1.820 and one-sided p-value 0.03438). Detailed interpretation is beyond the scope of this paper, but it seems possible that it may contribute to a further discussion on the empirical evidence for convergence in growth.

5 Concluding Remarks

The paper presents a relatively simple method of assessing the maximal admissible forecast horizon for non-systematic inflation when an autoregressive forecasting model is used. The empirical results indicate the plausibility of the method which might be implemented in practice by monetary policy authorities and forecasting institutions. It can also be used as an auxiliary tool for evaluation the rationale of inflation smoothing and for assessing the quality of linear autoregressive forecasting models. However, the bilinear model used here is relatively simple and its extension (for instance, by allowing for more complicated lags structure) is likely to increase the practical relevance of the method proposed.

Acknowledgements Financial support of the ESRC/ORA project RES-360-25-0003 *Probabilistic Approach to Assessing Macroeconomic Uncertainties* is gratefully acknowledged. We are grateful to Svetlana Makarova for her helpful comments on an earlier draft of the paper and to Sun Qi for help with data preparation. We are solely responsible for all remaining deficiencies.

Appendix A: Proofs of Theorems 1 and 2 and Corollaries

Lemma *If the time series π_t satisfies the bilinear model* (1), $\alpha^2+\beta^2\sigma^2<1$, $\tau\in N$, $\beta\to 0$, *then the following asymptotic expansions for the second order moments hold*:

$$\begin{aligned}E\{\pi_t^2\} &= \sigma^2\frac{1}{1-\alpha^2}+2\beta\mu_3\frac{\alpha}{1-\alpha^2}\\ &\quad+\beta^2\sigma^4\left(\frac{\alpha^2}{(1-\alpha^2)^2}+\frac{4\alpha}{(1-\alpha)(1-\alpha^2)}\right)\\ &\quad+\beta^2\mu_4\frac{1}{1-\alpha^2}+o(\beta^2),\end{aligned}$$

$$\begin{aligned}E\{\pi_T\pi_{T+\tau}\} &= \sigma^2\frac{\alpha^\tau}{1-\alpha^2}+\beta\mu_3\alpha^{\tau-1}\frac{1+\alpha^2}{1-\alpha^2}\\ &\quad+\beta^2\sigma^4\left(\frac{3\alpha^{\tau+2}+\alpha^{\tau+1}+\alpha^{\tau-1}}{(1-\alpha^2)^2}+\frac{1}{(1-\alpha)^2}\right)\\ &\quad+\beta^2\mu_4\frac{\alpha^\tau}{1-\alpha^2}+o(\beta^2).\end{aligned}$$

Proof (1) Using the decomposition of (1), analogous to the moving average decomposition of the $AR(1)$ process, that is:

$$\pi_t = u_t+\sum_{i=1}^{\infty}u_{t-i}\prod_{k=1}^{i}(\alpha+\beta u_{t-k})$$

and applying the assumption of independence of u_t and u_{t-i} at $i \geq 1$, we have:

$$E\{\pi_t^2\} = \sigma^2 + E\left\{\sum_{i=1}^{\infty}\sum_{j=1}^{\infty} u_{t-i}u_{t-j} \prod_{k=1}^{i}(\alpha + \beta u_{t-k}) \prod_{m=1}^{j}(\alpha + \beta u_{t-m})\right\}$$

$$= \sigma^2 + E\left\{\sum_{i=1}^{\infty}\sum_{j=1}^{\infty} u_{t-i}u_{t-j}\left(\alpha^{i+j} + \beta\alpha^{i+j-1}\sum_{m=1}^{j} u_{t-m}\right.\right.$$

$$+ \beta\alpha^{j+i-1}\sum_{k=1}^{i} u_{t-k} + \beta^2\alpha^{i+j-2}\sum_{m=1}^{j-1}\sum_{p=m+1}^{j} u_{t-m}u_{t-p}$$

$$+ \beta^2\alpha^{j+i-2}\sum_{k=1}^{i-1}\sum_{l=k+1}^{i} u_{t-k}u_{t-l} + \beta^2\alpha^{i-1+j-1}\sum_{k=1}^{i}\sum_{m=1}^{j} u_{t-k}u_{t-m}$$

$$\left.\left. + \alpha^{i+j-3}o(\beta^2)\right)\right\}.$$

Considering that $E\{u_t\} = 0$, $\text{Var}\{u_t\} = \sigma^2 < +\infty$, and u_{t-i} are independent at $i \neq j$, and using the fact that $E\{u_{t_1}u_{t_2}u_{t_3}u_{t_4}\} \neq 0$ only for the situations where either $t_1 = t_2 = t_3 = t_4$ or where these four indices are pairwise equal), we get:

$$E\{\pi_t^2\} = \sigma^2 + E\left\{\sum_{i=1}^{\infty} u_{t-i}^2\alpha^{2i}\right\} + 2\beta E\left\{\sum_{i=1}^{\infty} u_{t-i}^3\alpha^{2i-1}\right\}$$

$$+ \beta^2 E\left\{\sum_{j=2}^{\infty}\sum_{i=1}^{j-1} \alpha^{i+j-2}u_{t-i}^2u_{t-j}^2\right\} + \beta^2 E\left\{\sum_{i=2}^{\infty}\sum_{j=1}^{i-1} \alpha^{i+j-2}u_{t-i}^2u_{t-j}^2\right\}$$

$$+ \beta^2 E\left\{\sum_{i=2}^{\infty}\sum_{k=1}^{i-1} \alpha^{2i-2}u_{t-i}^2u_{t-k}^2 + \sum_{i=1}^{\infty}\sum_{j=1, j\neq i}^{\infty} \alpha^{i+j-2}u_{t-i}^2u_{t-j}^2\right.$$

$$\left. + \sum_{i=1}^{\infty} \alpha^{2i-2}u_{t-i}^4\right\} + \sum_{i=1}^{\infty}\sum_{j=1}^{\infty} \alpha^{i+j-3}o(\beta^2)$$

$$= \sigma^2 + \sigma^2\sum_{i=1}^{\infty}\alpha^{2i} + 2\beta\mu_3\sum_{i=1}^{\infty}\alpha^{2i-1} + 2\beta^2\sigma^4\sum_{j=2}^{\infty}\sum_{i=1}^{j-1}\alpha^{i+j-2}$$

$$+ \beta^2\sigma^4\sum_{i=2}^{\infty}(i-1)\alpha^{2i-2} + \beta^2\sigma^4\sum_{i=1}^{\infty}\sum_{j=1, j\neq i}^{\infty}\alpha^{i+j-2}$$

$$+ \beta^2\mu_4\sum_{i=1}^{\infty}\alpha^{2i-2} + o(\beta^2)$$

$$=\sigma^2+\sigma^2\frac{\alpha^2}{1-\alpha^2}+2\beta\mu_3\frac{\alpha}{1-\alpha^2}+2\beta^2\sigma^4\frac{\alpha}{(1-\alpha)(1-\alpha^2)}$$

$$+\beta^2\sigma^4+\beta^2\sigma^4\left(-\frac{1}{1-\alpha^2}\right)+\beta^2\mu_4\frac{1}{1-\alpha^2}+o(\beta^2)$$

$$=\sigma^2\frac{1}{}1-\alpha^2+2\beta\mu_3\frac{\alpha}{1-\alpha^2}$$

$$+\beta^2\sigma^4\left(\frac{\alpha^2}{(1-\alpha^2)^2}+\frac{4\alpha}{(1-\alpha)(1-\alpha^2)}\right)+\beta^2\mu_4\frac{1}{1-\alpha^2}+o(\beta^2).$$

(2) From (5), as$\tau \geq 1$, we have:

$$E\{\pi_T\pi_{T+\tau}\}=E\left\{u_T\sum_{j=1}^{\infty}u_{T+\tau-j}\prod_{m=1}^{j}(\alpha+\beta u_{T+\tau-m})\right\}$$

$$+E\left\{\sum_{i=1}^{\infty}\sum_{j=1}^{\infty}u_{T-i}u_{T+\tau-j}\prod_{k=1}^{i}(\alpha+\beta u_{T-k})\prod_{m=1}^{j}(\alpha+\beta u_{T+\tau-m})\right\}.$$

Using independence of $\{u_t\}$ and selecting nonlinear elements in the first summand, we find:

$$E\left\{u_T\sum_{j=1}^{\infty}u_{T+\tau-j}\prod_{m=1}^{j}(\alpha+\beta u_{T+\tau-m})\right\}$$

$$=E\left\{u_T\sum_{j=1}^{\infty}u_{T+\tau-j}\left(\alpha^j+\alpha^{j-1}\beta\sum_{m=1}^{j}u_{T+\tau-m}\right)+\alpha^{j-3}o(\beta^2)\right\}$$

$$+u_T\sum_{j=2}^{\infty}u_{T+\tau-j}\alpha^{j-2}\beta^2\sum_{m=1}^{j-1}\sum_{p=m+1}^{j}u_{T+\tau-m}u_{T+\tau-p}$$

$$=\sigma^2\alpha^\tau+\alpha^{\tau-1}\beta\mu_3+\beta^2E\left\{u_T^2\sum_{j=\tau+1}^{\infty}u_{T+\tau-j}^2\alpha^{j-2}\right\}+o(\beta^2)$$

$$=\sigma^2\alpha^\tau+\alpha^{\tau-1}\beta\mu_3+\beta^2\sigma^4\alpha^{\tau-1}(1-\alpha)^{-1}+o(\beta^2).$$

Selecting nonlinear elements in the second summand, we get:

$$E\left\{\sum_{i=1}^{\infty}\sum_{j=1}^{\infty}u_{T-i}u_{T+\tau-j}\prod_{k=1}^{i}(\alpha+\beta u_{T-k})\prod_{m=1}^{j}(\alpha+\beta u_{T+\tau-m})\right\}$$

$$=E\left\{\sum_{i=1}^{\infty}\sum_{j=1}^{\infty}u_{T-i}u_{T+\tau-j}\left(\alpha^{i+j}+\beta\alpha^{i+j-1}\sum_{m=1}^{j}u_{T+\tau-m}\right.\right.$$

$$+\beta\alpha^{j+i-1}\sum_{k=1}^{i}u_{T-k}+\beta^2\alpha^{i+j-2}\sum_{m=1}^{j-1}\sum_{p=m+1}^{j}u_{T+\tau-m}u_{T+\tau-p}$$

$$+\beta^2\alpha^{j+i-2}\sum_{k=1}^{i-1}\sum_{l=k+1}^{i}u_{T-k}u_{T-l}+\alpha^{i+j-3}o(\beta^2)$$

$$+\beta^2\alpha^{i+j-2}\sum_{k=1}^{i}\sum_{m=1}^{j}u_{T-k}u_{T+\tau-m}\Bigg)\Bigg\}$$

$$=\sigma^2\sum_{i=1}^{\infty}\alpha^{2i+\tau}+2\beta\mu_3\sum_{i=1}^{\infty}\alpha^{2i+\tau-1}+o(\beta^2)+\beta^2\sigma^4\sum_{i=1}^{\infty}\sum_{j=i+\tau+1}^{\infty}\alpha^{i+j-2}$$

$$+\beta^2\sigma^4\sum_{i=2}^{\infty}\sum_{j=\tau+1}^{i+\tau-1}\alpha^{i+j-2}+\beta^2E\left\{\sum_{i=2}^{\infty}\sum_{k=1}^{i-1}u_{T-i}^2u_{T-k}^2\alpha^{2i+\tau-2}\right.$$

$$\left.+\sum_{i=1}^{\infty}\sum_{j=1,j\neq i+\tau}^{\infty}u_{T-i}^2u_{T+\tau-j}^2\alpha^{i+j-2}+\sum_{i=1}^{\infty}u_{T-i}^4\alpha^{2i+\tau-2}\right\}$$

$$=\sigma^2\frac{\alpha^{2+\tau}}{1-\alpha^2}+\beta\mu_3+\beta^2\sigma^4\frac{\alpha^{2+\tau}}{(1-\alpha^2)^2}+\beta^2\sigma^4\frac{2\alpha^{1+\tau}}{(1-\alpha)(1-\alpha^2)}$$

$$+\beta^2\sigma^4\left(\frac{1}{(1-\alpha)^2}-\frac{\alpha^\tau}{1-\alpha^2}\right)+\beta^2\mu_4\frac{\alpha^\tau}{1-\alpha^2}+o(\beta^2).$$

Then:

$$E\{\pi_T\pi_{T+\tau}\}=\sigma^2\alpha^\tau\left(1+\frac{\alpha^2}{1-\alpha^2}\right)+\alpha^{\tau-1}\beta\mu_3\frac{1+\alpha^2}{1-\alpha^2}$$

$$+\beta^2\sigma^4\alpha^{\tau-1}\left(\frac{1}{1-\alpha}+\frac{2\alpha^2}{(1-\alpha)(1-\alpha^2)}+\frac{\alpha^3}{(1-\alpha^2)^2}-\frac{\alpha}{1-\alpha^2}\right)$$

$$+\beta^2\sigma^4\frac{1}{(1-\alpha)^2}+\beta^2\mu_4\frac{\alpha^\tau}{1-\alpha^2}+o(\beta^2)$$

$$=\sigma^2\frac{\alpha^\tau}{1-\alpha^2}+\beta\mu_3\alpha^{\tau-1}\frac{1+\alpha^2}{1-\alpha^2}+\beta^2\sigma^4\left(\frac{3\alpha^{\tau+2}+\alpha^{\tau+1}+\alpha^{\tau-1}}{(1-\alpha^2)^2}\right.$$

$$\left.+\frac{1}{(1-\alpha)^2}\right)+\beta^2\mu_4\frac{\alpha^\tau}{1-\alpha^2}+o(\beta^2).$$

□

Theorem 1 *If the time series π_t satisfies the bilinear model* (1) *with*$\beta\to 0$, $\alpha^2+\beta^2\sigma^2<1$, $\tau\in N$, *and the forecasting procedure* (2) *is used, then the mean square* *risk satisfies the asymptotic expansion*:

$$MSE(\tau) = \sigma^2 \frac{1-\alpha^{2\tau}}{1-\alpha^2} + 2\beta\mu_3 \frac{\alpha(1-\alpha^{2(\tau-1)})}{1-\alpha^2}$$
$$+ \beta^2\sigma^4 \left(\frac{4\alpha + 5\alpha^2 - \alpha^{2\tau+2}}{(1-\alpha^2)^2} - 2\frac{\alpha^{2\tau-1}}{1-\alpha^2} - 2\frac{\alpha^\tau}{(1-\alpha)^2} \right)$$
$$+ \beta^2\mu_4 \frac{1-\alpha^{2\tau}}{1-\alpha^2} + o(\beta^2). \tag{8}$$

Proof Using (2), we have $MSE(\tau) = \alpha^{2\tau} E\{x_T^2\} - 2\alpha^\tau E\{x_T x_{T+\tau}\} + E\{x_{T+\tau}^2\}$.
By Lemma we get:

$$MSE(\tau) = \sigma^2 \frac{\alpha^{2\tau}}{1-\alpha^2} + \beta\mu_3 \frac{2\alpha^{2\tau+1}}{1-\alpha^2} + \beta^2\sigma^4 \left(\frac{\alpha^{2\tau+2}}{(1-\alpha^2)^2} + \frac{4\alpha^{2\tau+1}}{(1-\alpha)(1-\alpha^2)} \right)$$
$$+ \beta^2\mu_4 \frac{\alpha^{2\tau}}{1-\alpha^2} - 2\left(\sigma^2 \frac{\alpha^{2\tau}}{1-\alpha^2} + \beta\mu_3 \alpha^{2\tau-1} \frac{1+\alpha^2}{1-\alpha^2} \right.$$
$$\left. + \beta^2\sigma^4 \left(\frac{3\alpha^{2\tau+2} + \alpha^{2\tau+1} + \alpha^{2\tau-1}}{(1-\alpha^2)^2} + \frac{\alpha^\tau}{(1-\alpha)^2} \right) + \beta^2\mu_4 \right)$$
$$+ \sigma^2 \frac{1}{1-\alpha^2} + 2\beta\mu_3 \frac{\alpha}{1-\alpha^2} + \beta^2\sigma^4 \left(\frac{\alpha^2}{(1-\alpha^2)^2} \right.$$
$$\left. + \frac{4\alpha}{(1-\alpha)(1-\alpha^2)} \right) + \beta^2\mu_4 (1-\alpha^2)^{-1} + o(\beta^2)$$
$$= \sigma^2 \frac{1-\alpha^{2\tau}}{1-\alpha^2} + \beta\mu_3 \frac{2\alpha(1-\alpha^{2(\tau-1)})}{1-\alpha^2}$$
$$+ \beta^2\sigma^4 \left(\frac{4\alpha + 5\alpha^2 - 2\alpha^{2\tau-1} + 2\alpha^{2\tau+1} - \alpha^{2\tau+2}}{(1-\alpha^2)^2} - \frac{2\alpha^\tau}{(1-\alpha)^2} \right)$$
$$+ \beta^2\mu_4 \frac{1-\alpha^{2\tau}}{1-\alpha^2} + o(\beta^2).$$ □

Corollary 1 *If the random errors* $\{u_t\}$ *in* (1) *have the Gaussian probability distribution* $N_1(0, \sigma^2)$, *then*:

$$MSE(\tau) = \sigma^2 \frac{1-\alpha^{2\tau}}{1-\alpha^2} + \beta^2\sigma^4 \left(2\frac{1-\alpha^\tau}{(1-\alpha)^2} + \frac{1-\alpha^{2\tau}}{(1-\alpha^2)^2} - \frac{2\alpha^{2\tau-1}}{1-\alpha} \right) + o(\beta^2). \tag{9}$$

Proof For the Gaussian probability distribution $N_1(0, \sigma^2)$ we have$\mu_3 = 0$, $\mu_4 = 3\sigma^4$. Then (9) follows from (8). □

Note, that the risk functional in (9), (8) has an additive form: the first summand is the risk value for the non-distorted model ($\beta = 0$), i.e. for the autoregression model; the second term proportional to β^2 is generated by the bilinear distortion.

Corollary 2 *Under Theorem* 1 *the condition at* $\tau = 1$ *is*:

$$MSE(\tau) = \sigma^2 + \mu_3 \frac{2\alpha}{1-\alpha^2}\beta + \left(\sigma^4 \frac{\alpha^2}{1-\alpha^2} + \mu_4\right)\beta^2 + o(\beta^2).$$

Theorem 2 *If the time series* π_t *satisfies the bilinear model* (1), $\beta \in [-\beta_+, \beta_+]$, $\beta_+ \to 0$, $\alpha^2 + \beta_+^2\sigma^2 < 1$, $\tau \in N$, *random errors* $\{u_t\}$ *have the Gaussian probability distribution* $N_1(0, \sigma^2)$, *and the forecasting procedure* (2) *is used, then the guaranteed upper risk, the instability coefficient and the* δ*-admissible distortion level satisfy the asymptotic expansions*:

$$MSE_+(\tau) = \sigma^2 \frac{1-\alpha^{2\tau}}{1-\alpha^2} + \beta_+^2\sigma^4\left(2\frac{1-\alpha^\tau}{(1-\alpha)^2} + \frac{1-\alpha^{2\tau}}{(1-\alpha^2)^2} - \frac{2\alpha^{2\tau-1}}{1-\alpha}\right) + o(\beta_+^2),$$

$$\kappa(\tau) = \beta_+^2\sigma^2\left(2\frac{1-\alpha^\tau}{(1-\alpha)^2} + \frac{1-\alpha^{2\tau}}{(1-\alpha^2)^2} - \frac{2\alpha^{2\tau-1}}{1-\alpha}\right) + o(\beta_+^2), \qquad (10)$$

$$\beta^+(\delta, \tau) = \delta^{\frac{1}{2}}\sigma^{-1}\left(2\frac{1-\alpha^\tau}{(1-\alpha)^2} + \frac{1-\alpha^{2\tau}}{(1-\alpha^2)^2} - \frac{2\alpha^{2\tau-1}}{1-\alpha}\right)^{-\frac{1}{2}} + o(\beta_+^2).$$

Proof (1) The coefficient at $\beta^2\sigma^4$ in (9) equals to:

$$K_{\beta^2} = 2\frac{1-\alpha^\tau}{(1-\alpha)^2} + \frac{1-\alpha^{2\tau}}{(1-\alpha^2)^2} - 2\frac{\alpha^{2\tau-1}}{1-\alpha}.$$

It can be shown that this coefficient is positive: if $\alpha = 0$, then $K_{\beta^2} = 3$; if $-1 < \alpha < 0$, then for $\tau \in N$ we have $\frac{1-\alpha^\tau}{(1-\alpha)^2} > 0$, $\frac{1-\alpha^{2\tau}}{(1-\alpha^2)^2} > 0$, $\frac{\alpha^{2\tau-1}}{1-\alpha} < 0$. This is why $K_{\beta^2} > 0$; if $0 < \alpha < 1$, then for $\tau \in N$ we have: $\frac{1-\alpha^\tau}{(1-\alpha)^2} - \frac{\alpha^{2\tau-1}}{1-\alpha} = \frac{(1-\alpha^{\tau-1})+\alpha^{\tau-1}(1-\alpha^\tau)(1-\alpha)}{(1-\alpha)^2} > 0$, therefore $K_{\beta^2} > 0$.

(2) From (9) and the definition of $MSE_+(\tau)$ it can be shown that:

$$MSE_+(\tau) = \sigma^2 \frac{1-\alpha^{2\tau}}{1-\alpha^2} + \max_{\beta \in [-\beta_+, \beta_+]} \left(K_{\beta^2}\beta^2\sigma^4 + o(\beta^2)\right)$$
$$= \sigma^2 \frac{1-\alpha^{2\tau}}{1-\alpha^2} + K_{\beta^2}\beta_+^2\sigma^4 + o(\beta_+^2).$$

(3) The second and the third expansions in (10) follow from the expansion of the guaranteed risk. □

Corollary 3 *Under Theorem 2 conditions at* $\tau = 1$:

$$MSE_{+}(\tau) = \sigma^2 + \beta_{+}^2\sigma^4\frac{3-2\alpha^2}{1-\alpha^2} + o\left(\beta_{+}^2\right), \qquad \kappa(\tau) = \beta_{+}^2\sigma^2\frac{3-2\alpha^2}{1-\alpha^2} + o\left(\beta_{+}^2\right),$$

$$\beta^{+}(\delta,\tau) = \sigma^{-1}\sqrt{\delta\frac{1-\alpha^2}{3-2\alpha^2}} + o\left(\beta_{+}^2\right).$$

Appendix B

Table 4 ML Kalman Filter estimates

Country	No. obs.	$AR(1)$ coefficient			Bilinear coefficient			
		$\hat{\alpha}$	$t(\alpha)$	signif	$\hat{\beta}$	$\tilde{\beta}$	$t(\beta)$	signif
(1)	(2)	(3)	(4)	(5)	(6)	(7)	(8)	(9)
ALBA	183	0.793	16.97	(3)	0.720	0.007	0.213	(0)
ARGE	591	0.951	72.07	(3)	−3.067	−0.012	−2.870	(3)
ARME	156	0.855	21.49	(3)	0.387	0.006	0.459	(0)
AUST	591	0.701	23.29	(3)	−33.362	−0.044	−3.131	(3)
BARB	483	0.753	24.89	(3)	−0.534	−0.002	−0.222	(0)
BELG	591	0.755	27.95	(3)	7.781	0.008	0.747	(0)
BENI	171	0.771	15.68	(3)	−0.279	−0.003	−0.095	(0)
BOLI	590	0.911	53.25	(3)	0.766	0.003	0.385	(0)
BOTS	370	0.884	29.22	(3)	−21.670	−0.039	−2.373	(3)
BRAZ	315	0.943	49.89	(3)	1.604	0.012	2.256	(3)
BULG	183	0.860	22.43	(3)	0.328	0.009	0.630	(0)
BURK	566	0.693	22.72	(3)	2.355	0.022	1.441	(0)
BURU	383	0.789	24.57	(3)	−1.512	−0.012	−0.836	(0)
CAMB	137	0.703	11.44	(3)	−1.283	−0.007	−0.199	(0)
CAME	455	0.853	33.05	(3)	5.980	0.035	2.267	(3)
CANA	590	0.789	31.02	(3)	−4.014	−0.005	−0.444	(0)
CAPE	171	0.706	12.76	(3)	5.553	0.039	1.017	(0)
CENT	299	0.781	21.65	(3)	0.000	0.000	0.000	(0)
CHAD	274	0.812	23.01	(3)	1.954	0.029	1.034	(0)
CHHK	304	0.731	18.52	(3)	−5.509	−0.013	−0.479	(0)
CHMC	218	0.780	18.38	(3)	0.000	0.000	0.000	(0)
COLO	591	0.884	43.85	(3)	0.336	0.000	0.369	(0)
CONG	496	0.802	29.30	(3)	0.947	0.011	1.796	(3)
COTE	550	0.796	30.53	(3)	0.888	0.004	0.156	(0)
CROA	243	0.948	45.53	(3)	0.934	0.011	1.875	(3)
CYPR	591	0.653	20.98	(3)	−2.530	−0.007	−0.431	(0)
CZEC	159	0.884	23.75	(3)	4.309	0.010	0.459	(0)

Table 4 (Continued)

Country	No. obs.	AR(1) coefficient			Bilinear coefficient			
		$\hat{\alpha}$	$t(\alpha)$	signif	$\hat{\beta}$	$\tilde{\beta}$	$t(\beta)$	signif
(1)	(2)	(3)	(4)	(5)	(6)	(7)	(8)	(9)
CZEC	159	0.884	23.75	(3)	4.309	0.010	0.459	(0)
DENM	471	0.774	26.51	(3)	−10.670	−0.015	−0.924	(0)
DOMR	591	0.940	65.23	(3)	−0.483	−0.002	−0.109	(0)
EQUA	591	0.932	60.62	(3)	−2.568	−0.006	−1.405	(0)
EGYP	591	0.790	31.04	(3)	5.898	0.019	1.795	(3)
ELSA	591	0.835	36.50	(3)	4.853	0.011	0.903	(0)
ESTO	171	0.859	23.59	(3)	−13.709	−0.055	−2.032	(3)
ETHI	482	0.846	34.26	(3)	−1.126	−0.009	−0.531	(0)
FIJI	446	0.788	26.85	(3)	2.363	0.007	0.408	(0)
FINL	591	0.742	26.07	(3)	−4.947	−0.006	−0.108	(0)
FRAN	591	0.803	32.78	(3)	−12.762	−0.011	−0.565	(0)
GAMB	542	0.826	33.61	(3)	3.806	0.015	1.108	(0)
GEOR	147	0.723	9.06	(3)	1.949	0.034	2.577	(3)
GERM	183	0.639	11.17	(3)	−24.200	−0.030	−1.014	(0)
CHAN	517	0.895	47.29	(3)	0.563	0.002	0.115	(0)
GREE	591	0.741	23.77	(3)	−7.579	−0.009	−0.262	(0)
GREN	359	0.716	19.13	(3)	7.837	0.025	1.048	(0)
GUAT	591	0.908	50.64	(3)	−7.763	−0.015	−2.050	(3)
GUIN	241	0.766	18.47	(3)	−0.277	−0.004	−0.215	(0)
GUYA	143	0.660	10.34	(3)	1.447	0.006	0.147	(0)
HAIT	586	0.879	44.34	(3)	7.445	0.022	1.865	(3)
HOND	591	0.902	49.75	(3)	0.000	0.000	0.001	(0)
HUNG	363	0.871	33.27	(3)	1.369	0.002	0.192	(0)
ICEL	279	0.895	34.67	(3)	−1.265	−0.003	−0.223	(0)
INDI	589	0.860	40.12	(3)	−3.707	−0.008	−0.499	(0)
INDI	459	0.938	57.46	(3)	0.504	0.002	0.376	(0)
IREL	111	0.865	16.92	(3)	−20.044	−0.029	−0.322	(0)
ISRA	591	0.914	54.48	(3)	0.423	0.001	0.111	(0)
ITAL	591	0.877	44.39	(3)	−1.733	−0.001	−0.568	(0)
JAMA	590	0.931	59.29	(3)	−5.669	−0.012	−1.722	(3)
JAPA	590	0.781	30.35	(3)	0.000	0.000	−0.001	(0)
JORD	363	0.745	21.27	(3)	2.755	0.014	0.639	(0)
KAZA	159	0.930	36.55	(3)	8.694	0.043	2.598	(3)
KENY	459	0.854	34.52	(3)	0.852	0.004	0.409	(0)
KORE	434	0.847	32.84	(3)	10.128	0.020	1.175	(0)
KYRG	134	0.904	24.66	(3)	4.772	0.029	1.043	(0)
LATV	171	0.697	13.20	(3)	−7.024	−0.035	−2.265	(3)

Table 4 (Continued)

Country	No. obs.	$AR(1)$ coefficient			Bilinear coefficient			
		$\hat{\alpha}$	$t(\alpha)$	signif	$\hat{\beta}$	$\tilde{\beta}$	$t(\beta)$	signif
(1)	(2)	(3)	(4)	(5)	(6)	(7)	(8)	(9)
LITH	167	0.837	19.73	(3)	−2.107	−0.009	−0.370	(0)
LUXE	591	0.751	23.90	(3)	46.486	0.056	4.563	(3)
MACE	159	0.799	16.66	(3)	−0.615	−0.005	−0.304	(0)
MADA	506	0.881	392.40	(3)	0.217	0.001	0.011	(0)
MALA	314	0.883	32.90	(3)	3.506	0.012	0.717	(0)
MALY	590	0.817	33.94	(3)	5.953	0.010	0.712	(0)
MALT	590	0.743	26.73	(3)	8.537	0.027	1.573	(0)
MAUT	246	0.749	17.64	(3)	1.725	0.011	0.507	(0)
MAUR	525	0.850	36.72	(3)	5.782	0.015	1.125	(0)
MEXI	591	0.947	71.98	(3)	3.795	0.005	1.051	(0)
MOLD	148	0.894	25.44	(3)	−0.209	−0.001	−0.125	(0)
MORO	590	0.761	28.39	(3)	−2.887	−0.008	−0.485	(0)
MOZA	153	0.887	23.51	(3)	0.306	0.002	0.167	(0)
NEPA	505	0.844	35.18	(3)	1.277	0.004	0.472	(0)
NETH	591	0.731	26.03	(3)	1.922	0.002	0.111	(0)
NICA	83	0.636	7.09	(3)	−39.927	−0.086	−0.765	(0)
NIGE	459	0.736	23.23	(3)	2.639	0.028	1.613	(0)
NIGR	554	0.829	34.70	(3)	3.412	0.015	1.735	(3)
NORW	591	0.819	34.10	(3)	−4.970	−0.007	−0.722	(0)
PAKI	591	0.768	29.11	(3)	−3.376	−0.006	−0.534	(0)
PANA	380	0.672	17.72	(3)	−5.090	−0.013	−0.844	(0)
PERU	591	0.997	43.77	(3)	27.078	0.065	10.088	(3)
PHIL	591	0.910	53.22	(3)	1.762	0.004	0.600	(0)
POLA	219	0.871	26.52	(3)	8.793	0.024	1.022	(0)
PORT	591	0.791	31.21	(3)	−8.363	−0.010	−0.649	(0)
ROMA	186	0.889	26.47	(3)	2.890	0.009	0.462	(0)
RUSS	170	0.983	41.13	(3)	16.815	0.108	10.427	(3)
SAMO	469	0.786	26.47	(3)	4.528	0.036	1.969	(3)
SAUD	313	0.770	21.44	(3)	5.499	0.017	0.990	(0)
SENE	459	0.802	28.49	(3)	4.460	0.031	2.239	(3)
SERB	146	0.917	27.27	(3)	4.713	0.034	2.371	(3)
SEYC	442	0.750	23.83	(3)	−0.877	−0.008	−0.625	(0)
SING	542	0.843	28.19	(3)	−1.377	−0.004	−0.122	(0)
SLOA	159	0.835	19.15	(3)	3.894	0.013	0.378	(0)
SLOE	172	0.765	25.86	(3)	−47.457	−0.119	−3.962	(3)
SOLO	328	0.767	21.59	(3)	−0.722	−0.003	−0.248	(0)
SOUT	591	0.913	53.44	(3)	−6.225	−0.009	−1.283	(0)

Table 4 (Continued)

Country	No. obs.	$AR(1)$ coefficient			Bilinear coefficient			
		$\hat{\alpha}$	$t(\alpha)$	signif	$\hat{\beta}$	$\tilde{\beta}$	$t(\beta)$	signif
(1)	(2)	(3)	(4)	(5)	(6)	(7)	(8)	(9)
SPAI	591	0.748	27.41	(3)	17.835	0.018	1.147	(0)
SRIL	591	0.770	29.33	(3)	0.233	0.001	0.315	(0)
STKI	323	0.721	18.25	(3)	−6.068	−0.022	−0.828	(0)
STLU	501	0.708	22.01	(3)	−1.825	−0.008	−0.218	(0)
SSAF	458	0.899	42.21	(3)	15.002	0.024	2.061	(3)
SURI	444	0.913	46.65	(3)	−0.070	0.000	−0.325	(0)
SWAZ	470	0.516	13.03	(3)	0.460	0.003	0.351	(0)
SWED	591	0.792	30.88	(3)	−21.449	−0.036	−3.327	(3)
SWIT	591	0.796	31.94	(3)	−3.960	−0.005	−0.111	(0)
THAI	495	0.874	39.60	(3)	10.180	0.020	1.287	(0)
TONG	194	0.667	11.36	(3)	−14.147	−0.058	−1.627	(0)
TRIN	590	0.801	32.40	(3)	1.753	0.003	0.112	(0)
TUNI	225	0.897	29.19	(3)	−7.236	−0.011	−0.408	(0)
TURK	447	0.919	47.03	(3)	−6.139	−0.013	−1.865	(3)
UGAN	169	0.872	21.54	(3)	7.979	0.052	1.361	(0)
UNIK	591	0.884	46.00	(3)	10.587	0.012	0.971	(0)
UNIS	591	0.794	30.66	(3)	−26.021	−0.026	−1.187	(0)
URUG	591	0.924	58.39	(3)	5.608	0.010	1.694	(3)
VENE	591	0.899	47.68	(3)	−4.098	−0.007	−0.791	(0)
VIET	131	0.892	23.03	(3)	2.750	0.007	0.229	(0)
ZAMB	250	0.880	28.15	(3)	12.067	0.032	1.245	(0)

Table 5 ML Kalman Filter forecast measures

Country	No. obs.	$\hat{\Phi}$	$\hat{\beta}^{0.90}_{\xi\mid\beta=0}$	$\hat{\delta}^{0.90}_{\hat{\beta}}(24)$	$\tau_{\max}$
(1)	(2)	(3)	(4)	(5)	(6)
ALBA	183	0.629	0.009	0.002	7
ARGE	591	0.904	−0.016	0.084	16
ARME	156	0.731	0.007	0.003	9
AUST	591	0.493	−0.057	0.042	4
BARB	483	0.566	−0.003	0.000	5
BELG	591	0.570	0.011	0.002	5
BENI	171	0.595	−0.003	0.000	6
BOLI	590	0.830	0.003	0.002	13
BOTS	370	0.784	−0.050	0.200	11
BRAZ	315	0.890	0.016	0.066	16

Table 5 (Continued)

Country (1)	No. obs. (2)	$\hat{\Phi}$ (3)	$\hat{\beta}^{0.90}_{\xi\|\beta=0}$ (4)	$\hat{\delta}^{0.90}_{\hat{\beta}}(24)$ (5)	$\tau_{\max}$ (6)
BULG	183	0.739	0.011	0.007	10
BURK	566	0.481	0.028	0.010	4
BURU	383	0.623	−0.015	0.006	6
CAMB	137	0.494	−0.009	0.001	4
CAME	455	0.730	0.045	0.106	9
CANA	590	0.623	−0.007	0.001	6
CAPE	171	0.501	0.050	0.034	4
CENT	299	0.610	0.000	0.000	6
CHAD	274	0.660	0.038	0.046	7
CHHK	304	0.535	−0.017	0.005	5
CHMC	218	0.608	0.000	0.000	6
COLO	591	0.781	0.000	0.000	11
CONG	496	0.644	0.014	0.006	7
COTE	550	0.633	0.005	0.001	7
CROA	243	0.898	0.014	0.065	16
CYPR	591	0.427	−0.009	0.001	3
CZEC	159	0.782	0.013	0.014	11
DENM	471	0.600	−0.020	0.009	6
DOMR	591	0.883	−0.002	0.001	16
EQUA	591	0.869	−0.008	0.013	15
EGYP	591	0.624	0.024	0.015	6
ELSA	591	0.697	0.015	0.009	8
ESTO	171	0.741	−0.070	0.277	10
ETHI	482	0.716	−0.011	0.006	9
FIJI	446	0.622	0.009	0.002	6
FINL	591	0.551	−0.007	0.001	5
FRAN	591	0.645	−0.014	0.006	7
GAMB	542	0.682	0.019	0.013	8
GEOR	147	0.524	0.043	0.028	5
GERM	183	0.409	−0.038	0.013	3
CHAN	517	0.801	0.002	0.000	12
GREE	591	0.549	−0.011	0.002	5
GREN	359	0.514	0.032	0.015	4
GUAT	591	0.825	−0.020	0.047	13
GUIN	241	0.586	−0.005	0.000	6
GUYA	143	0.436	0.008	0.001	4
HAIT	586	0.774	0.029	0.062	11
HOND	591	0.814	0.000	0.000	13

Table 5 (Continued)

Country	No. obs.	$\hat{\Phi}$	$\hat{\beta}^{0.90}_{\xi\mid\beta=0}$	$\hat{\delta}^{0.90}_{\hat{\beta}}(24)$	$\tau_{\max}$
(1)	(2)	(3)	(4)	(5)	(6)
HUNG	363	0.759	0.003	0.001	10
ICEL	279	0.801	−0.004	0.002	12
INDI	589	0.740	−0.010	0.006	10
INDI	459	0.879	0.003	0.002	15
IREL	111	0.750	−0.037	0.084	10
ISRA	591	0.835	0.001	0.000	14
ITAL	591	0.770	−0.001	0.000	11
JAMA	590	0.867	−0.015	0.047	15
JAPA	590	0.610	0.000	0.000	6
JORD	363	0.556	0.018	0.006	5
KAZA	159	0.867	0.055	0.583	15
KENY	459	0.729	0.005	0.001	9
KORE	434	0.717	0.026	0.032	9
KYRG	134	0.818	0.038	0.160	13
LATV	171	0.487	−0.045	0.026	4
LITH	167	0.701	−0.011	0.005	8
LUXE	591	0.567	0.072	0.097	5
MACE	159	0.638	−0.006	0.001	7
MADA	506	0.776	0.001	0.000	11
MALA	314	0.780	0.015	0.018	11
MALY	590	0.668	0.013	0.006	8
MALT	590	0.553	0.034	0.021	5
MAUT	246	0.561	0.015	0.004	5
MAUR	525	0.723	0.019	0.019	9
MEXI	591	0.897	0.006	0.012	16
MOLD	148	0.800	−0.001	0.000	12
MORO	590	0.579	−0.011	0.002	6
MOZA	153	0.787	0.003	0.001	12
NEPA	505	0.713	0.005	0.001	9
NETH	591	0.534	0.003	0.000	5
NICA	83	0.412	−0.111	0.110	3
NIGE	459	0.543	0.035	0.021	5
NIGR	554	0.687	0.019	0.014	8
NORW	591	0.670	−0.009	0.003	8
PAKI	591	0.590	−0.008	0.001	6
PANA	380	0.452	−0.016	0.003	4
PARA	591	0.706	0.000	0.000	9
PERU	591	0.998	0.083	57.523	19

Table 5 (Continued)

Country	No. obs.	$\hat{\Phi}$	$\hat{\beta}^{0.90}_{\xi\mid\beta=0}$	$\hat{\delta}^{0.90}_{\hat{\beta}}(24)$	τ_{max}
(1)	(2)	(3)	(4)	(5)	(6)
PHIL	591	0.829	0.005	0.003	13
POLA	219	0.760	0.030	0.061	10
PORT	591	0.625	−0.013	0.004	6
ROMA	186	0.790	0.012	0.012	12
RUSS	170	0.978	0.138	26.580	18
SAMO	469	0.620	0.047	0.055	6
SAUD	313	0.593	0.022	0.011	6
SENE	459	0.645	0.040	0.048	7
SERB	146	0.841	0.044	0.277	14
SEYC	442	0.563	−0.010	0.002	5
SING	542	0.711	−0.005	0.001	9
SLOA	159	0.698	0.017	0.012	8
SLOE	172	0.599	−0.153	0.489	6
SOLO	328	0.589	−0.004	0.000	6
SOUT	591	0.833	−0.012	0.019	13
SPAI	591	0.560	0.023	0.010	5
SRIL	591	0.593	0.001	0.000	6
STKI	323	0.520	−0.028	0.011	5
STLU	501	0.502	−0.010	0.001	4
SSAF	458	0.809	0.030	0.095	12
SURI	444	0.833	−0.001	0.000	13
SWAZ	470	0.267	0.004	0.000	2
SWED	591	0.629	−0.047	0.058	7
SWIT	591	0.633	−0.006	0.001	7
THAI	495	0.765	0.026	0.047	11
TONG	194	0.448	−0.074	0.059	4
TRIN	590	0.642	0.004	0.001	7
TUNI	225	0.805	−0.014	0.020	12
TURK	447	0.844	−0.017	0.044	14
UGAN	169	0.763	0.067	0.302	11
UNIK	591	0.782	0.015	0.018	11
UNIS	591	0.631	−0.033	0.029	7
URUG	591	0.854	0.013	0.029	14
VENE	591	0.808	−0.009	0.008	12
VIET	131	0.795	0.009	0.008	12
ZAMB	250	0.775	0.041	0.127	11

List of country names and codes

Country name	Code	Country name	Code
Albania	ALBA	Kyrgyz Republic	KYRG
Argentina	ARGE	Latvia	LATV
Armenia	ARME	Lithuania	LITH
Austria	AUST	Luxembourg	LUXE
Barbados	BARB	Macedonia, FYR	MACE
Belgium	BELG	Madagascar	MADA
Benin	BENI	Malawi	MALA
Bolivia	BOLI	Malaysia	MALY
Botswana	BOTS	Malta	MALT
Brazil	BRAZ	Mauritania	MAUT
Bulgaria	BULG	Mauritius	MAUR
Burkina Faso	BURK	Mexico	MEXI
Burundi	BURU	Moldova	MOLD
Cambodia	CAMB	Morocco	MORO
Cameroon	CAME	Mozambique	MOZA
Canada	CANA	Nepal	NEPA
Cape Verde	CAPE	Netherlands	NETH
Central African Rep.	CENT	Nicaragua	NICA
Chad	CHAD	Niger	NIGE
China, P.R.:Hong Kong	CHHK	Nigeria	NIGR
China, P.R.:Macao	CHMC	Norway	NORW
Colombia	COLO	Pakistan	PAKI
Congo, Dem. Rep. of	CONG	Panama	PANA
Côte d'Ivoire	COTE	Paraguay	PARA
Croatia	CROA	Peru	PERU
Cyprus	CYPR	Philippines	PHIL
Czech Republic	CZEC	Poland	POLA
Denmark	DENM	Portugal	PORT
Dominican Republic	DOMR	Romania	ROMA
Ecuador	EQUA	Russian Federation	RUSS
Egypt	EGYP	Samoa	SAMO
El Salvador	ELSA	Saudi Arabia	SAUD
Estonia	ESTO	Senegal	SENE
Ethiopia	ETHI	Serbia	SERB
Fiji	FIJI	Seychelles	SEYC
Finland	FINL	Singapore	SING
France	FRAN	Slovak Republic	SLOA
Gambia	GAMB	Slovenia	SLOE
Georgia	GEOR	Solomon Islands	SOLO
Germany	GERM	South Africa	SOUT
Ghana	CHAN	Spain	SPAI
Greece	GREE	Sri Lanka	SRIL
Grenada	GREN	St. Kitts and Nevis	STKI
Guatemala	GUAT	St. Lucia	STLU
Guinea-Bissau	GUIN	Sub-Saharan Africa	SSAF
Guyana	GUYA	Suriname	SURI
Haiti	HAIT	Swaziland	SWAZ
Honduras	HOND	Sweden	SWED
Hungary	HUNG	Switzerland	SWIT
Iceland	ICEL	Thailand	THAI
India	INDI	Tonga	TONG

Country name	Code	Country name	Code
Indonesia	INDI	Trinidad and Tobago	TRIN
Ireland	IREL	Tunisia	TUNI
Israel	ISRA	Turkey	TURK
Italy	ITAL	Uganda	UGAN
Jamaica	JAMA	United Kingdom	UNIK
Japan	JAPA	United States	UNIS
Jordan	JORD	Uruguay	URUG
Kazakhstan	KAZA	Venezuela	VENE
Kenya	KENY	Vietnam	VIET
Korea, Republic of	KORE	Zambia	ZAMB

References

Adam, K., Evans, G. W., & Honkapohja, S. (2006). Are hyperinflation paths learnable? *Journal of Economic Dynamics & Control, 30*, 2725–2748. doi:10.1016/j.jedc.2005.09.003.

Arce, O. J. (2009). Speculative hyperinflations and currency substitution. *Journal of Economic Dynamics & Control, 33*, 1808–1823. doi:10.1016/j.jedc.2009.03.007.

Bermingham, C. (2010). A critical assessment of existing estimates of US core inflation. *Journal of Macroeconomics, 32*, 993–1007. doi:10.1016/j.jmacro.2010.05.003.

Blankmeyer, E. (2012). The variance of a core-inflation index. *Atlantic Economic Journal, 40*, 211–212. doi:10.1007/s11293-012-9307-5.

Bodenstein, M., Erceg, C. J., & Guerrieri, L. (2008). Optimal monetary policy with distinct core and headline inflation rates. *Journal of Monetary Economics, 55*, S18–S33. doi:10.1016/j.jmoneco.2008.07.010 (Supplement).

Cecchetti, S. G. (1996). *Measuring short-run inflation for central bankers* (NBER Working Paper No. 5786). Cambridge Massachusetts. http://www.nber.org/papers/w5786.

Charemza, W., & Shah, I. H. (2013). Stability price index, core inflation and output volatility. *Applied Financial Economics Letters, 20*, 737–741. doi:10.1080/13504851.2012.739279.

Cristadoro, R. M., Forni, M., Reichlin, L., & Veronese, G. (2005). A core inflation index for the Euro area. *Journal of Money, Credit, and Banking, 37*, 539–560.

Eckstein, O. (1981). *Core inflation*. Englewood: Prentice-Hall.

Granger, C. W. J., & Anderson, A. P. (1978). *An introduction to bilinear models*. Gottingen: Vandenhoeck & Ruprecht.

Jha, S. K., Wand, P., & Yip, C. K. (2002). Dynamics in a transactions-based monetary growth model. *Journal of Economic Dynamics & Control, 26*, 611–635. doi:10.1016/S0165-1889(00)00072-5.

Kharin, Yu. S. (1996). *Robustness in statistical pattern recognition*. Dordrecht: Kluwer.

Quah, D., & Vahey, S. (1995). Measuring core inflation. *Econometrics Journal, 105*, 1130–1144.

Rich, R., & Steindel, C. (2007). A comparison of measures of core inflation. *Economic Policy Review, Dec*, 19–38. Fed Res Bnk NY

Roncalli, T. (1995). *Introduction à la programmation sous GAUSS, vol. 2, Applications à la finance at à l'econométrie*. Paris: RITME Informatique.

Schmitt-Grohé, S. (2005). Comment on "Limits to inflation targeting" by Christopher Sims. In B. S. Bernanke & M. Woodford (Eds.), *The inflation-targeting debate*. Chicago: The University of Chicago Press.

Silver, M. (2007). Core inflation: measurement and statistical issues in choosing among alternative measures. *IMF Staff Papers, 54*, 163–190. doi:10.1057/palgrave.imfsp.9450006.

Sims, C. (2005). Limits to inflation targeting. In B. S. Bernanke & M. Woodford (Eds.), *The inflation-targeting debate*. Chicago: The University of Chicago Press.

Siviero, S., & Veronese, G. (2011). A policy-sensible benchmark core inflation measure. *Oxford Economic Papers*, *63*, 648–672. doi:10.1093/oep/gpr016.

Vázquez, J. (1998). How high can inflation get during hyperinflation? A transactional cost demand for money approach. *European Journal of Political Economy*, *14*, 433–451. doi:10.1016/S0176-2680(98)00016-0.

Wynne, M. A. (2008). Core inflations: a review of some conceptual issues. *Review*, *May/June*, 205–228. http://research.stlouisfed.org/publications/review/08/05/part2/Wynne.pdf. Fed Res Bnk St Louis.

Currency Crises, Exchange Rate Regimes and Capital Account Liberalization: A Duration Analysis Approach

Mohammad Karimi and Marcel-Cristian Voia

Abstract This paper empirically analyzes the effects of exchange rate regimes and capital account liberalization policies on the occurrence of currency crises in 21 countries over the period of 1970–1998. We examine the changes of the likelihood of currency crises under the de jure, and the de facto exchange rate regimes. We also test whether the impact of the exchange rate regimes on currency stability would be different under free and restricted capital flows. Our findings show that the likelihood of currency crises changes significantly under the de facto regimes. However, the results are sensitive to the choice of the de facto exchange rate arrangements. Furthermore, in our sample, capital control policies appear to be helpful in preventing low duration currency crises. The results are robust to a wide variety of sample and models checks.

1 Introduction

The links between the incidence of currency crises and the choice of exchange rate regimes as well as the impact of capital market liberalization policies on the occurrence of currency crises have been subject of considerable debates in recent years. It is of great interest to assess how exchange rate arrangements and financial liberalization will affect episodes of crisis. Policy makers also seek to know what type of exchange rate regime is more sustainable and whether controlling capital flows in fact contributes to the stability of currencies.

Yet the literature is not clear on these issues and presents mixed views. Many economists argue that fixed exchange rates are a cause of currency crises while others find that the intermediate and/or flexible exchange regimes are more crisis prone.

M. Karimi
Department of Economics, Ottawa University, Ottawa, ON, Canada
e-mail: mkari096@uottawa.ca

M.-C. Voia (✉)
Department of Economics, Carleton University, 1125 Colonel By Drive, Loeb Building D885, Ottawa, ON K1S 5B6, Canada
e-mail: Marcel.Voia@carleton.ca

F. Schleer-van Gellecom (ed.), *Advances in Non-linear Economic Modeling*, Dynamic Modeling and Econometrics in Economics and Finance 17,
DOI 10.1007/978-3-642-42039-9_7, © Springer-Verlag Berlin Heidelberg 2014

The role of capital market liberalization is even more controversial. The common view in the literature blames high capital mobility as an underlying cause of currency crises, especially when combined with fixed exchange rates. However, several studies consider that capital mobility restrictions are responsible for crises—as a contributing factor behind the crises—and advocate financial liberalization. It is evident that, for the time being, there is no consensus on these topics and more research is required before the controversies can be settled.

The main purpose of this paper is to systematically examine what type of exchange rate regime is more susceptible to currency crises by investigating the data from twenty OECD countries and South Africa over the period of 1970–1998.[1] We adapt the empirical models of the determinants of currency crises, which were presented in Karimi and Voia (2011b) as benchmark models and examine how the likelihood of currency crises is influenced by the de jure and de facto exchange rate regimes. We also study the role of capital mobility and test for currency stability under free and restricted capital flows. We examine whether the hazard of speculative attack changes under the different combinations of exchange rate regimes and the presence or absence of capital controls.

We employ two prominent de facto exchange rate regime classifications in the literature, those of Reinhart and Rogoff (2004), and Levy-Yeyati and Sturzenegger (2005), to identify the actual exchange rate arrangements. Our index for the de jure exchange rate regimes is the IMF exchange rate classification. We also categorize capital mobility policies into restricted and open policies with the help of Chinn and Ito's (2005) index of financial openness.

We apply duration analysis to study the risk of a currency crisis occurrence under different exchange rate regimes and capital mobility policies. Duration models rigorously incorporate the time factor into the likelihood functions and allow to investigate how the amount of time that a currency has already spent in the tranquil state affects the stability of the currency. This feature helps us to capture the unobservable determinants of currency stability that are embodied in the baseline hazard functions. We apply semi-parametric hazard models to estimate the unrestricted baseline hazard of a currency exiting a tranquil state into a turbulent state. The hazard function is a nonlinear function of time and has some advantages over other conditional probability models as the logit and probit models that are widely used in the literature. These models do not require any distribution assumptions about the timing of failures and are capable to deal with both monotonic and non-monotonic duration dependence. Compared to other duration models, they are more realistic and can produce estimations that are more precise.

The nonlinear nature of duration specification is able to efficiently examine how the different exchange rate regimes or the presence and absence of capital controls can change the sensitivity of currency crises with respect to changes in a set of

[1]The analysis is restricted to OECD countries in order to minimize any specification bias due to unobserved heterogeneity. The time interval is also restricted to an observed period before the introduction of Euro in the European Union since 10 countries in our sample left their own national currencies and joined the Euro currency system from the beginning of 1999.

macroeconomic fundamentals and contagion channels. Furthermore, we use crisis episodes that are identified by extreme value theory to minimize the relevant concerns about the accuracy of crisis episodes dating. We apply several robustness checks, including running our models on two different crisis episode sets that are based on monthly and quarterly-type spells, to verify the reliability of our estimation results.

We find that there is a significant link between the choice of exchange rate regime and the incidence of currency crises in our sample. Nevertheless, the results are sensitive to the choice of the de facto exchange rate system. When we use Reinhart and Rogoff's (2004) de facto classification to categorize the exchange rate regimes, fixed exchange rate arrangements are the least susceptible to speculative attacks. However, when we rely on Levy-Yeyati and Sturzenegger's (2005) de facto classification, intermediate exchange rate regimes will experience the smallest number of currency crisis incidences. On the other hand, we find that the impact of capital account policies on the occurrence of currency crises, in our sample, demonstrates different results. While the baseline hazard of open-type capital accounts is lower than the baseline hazard of restricted-type capital accounts, when we enter our set of control variables to the models, the hazard of open-type capital accounts appears to be higher than the hazard of restricted-type capital accounts. This relation is more significant at the low duration crisis episodes.

The remainder of the paper is organized as follows. Section 2 looks at the exchange rate regimes classifications and briefly introduces the two de facto exchange rate regime classifications that we use. It also quickly reviews the empirical literature on the links between exchange rate regimes and the occurrence of currency crises. Section 3 reviews the empirical literature and presents the links between capital control policies and occurrences of currency crises. Section 4 describes the empirical methodology and data. Section 5 presents the main empirical results and robustness tests. Section 6 discusses the results and concludes.

2 Classification of Exchange Rate Regimes and Currency Crises

2.1 *Classifications*

Since the collapse of the Bretton Woods system, a large empirical literature has been developed to assess the performance of exchange rate regimes. The early literature—e.g. the influential work of Baxter and Stockman (1989)—compared the performance of key macroeconomic variables with fixed and flexible exchange rate arrangements. However, they found little significant difference across fixed and flexible regimes. There was a drawback in the way that they characterized the exchange rate regimes and this shortcoming affected negatively the early literature.

For many years, empirical studies relied on the International Monetary Fund's de jure classification of exchange rate regimes to measure the impact of exchange rate arrangements on economic performance.[2] This classification is a countries' self-declared index, which was published in the Fund's *Annual Report on Exchange Rate Arrangements and Exchange Restrictions*.[3] However, in a pioneering paper, Calvo and Reinhart (2002) noticed that in practice there is a substantial deviation between the officially reported and the actually prevailing exchange rate arrangements.[4] Therefore, the empirical results of those analyses based on the de jure classification could be misleading. This problem motivated researchers to devise alternative classifications to identify the de facto exchange rate regimes and categorize countries more accurately according to their actual practice rather than official statement.[5]

In this subsection, we briefly introduce two prominent alternative classifications in the literature: Reinhart and Rogoff (2004) and Levy-Yeyati and Sturzenegger (2005).[6] Reinhart and Rogoff (hereafter RR) rely on the IMF classification as their starting point and develop their own classification system based on a statistical analysis of the ex post behavior of exchange rates on the official, dual and/or parallel markets. For countries with only official rates they apply a broad variety of descriptive statistics (mostly exchange rate variability, variability with respect to the officially announced bands, and inflation) to verify whether the de jure classification is accurate. If not, they reclassify the exchange rate into the alternative categories. For countries with dual and/or parallel rates, they classify the exchange rate based on the market-determined rates, which they argue are important indicators of the underlying monetary policy.

RR classify the exchange rates regimes into fourteen fine categories. Nevertheless, these categories can be aggregated into three coarse branches: fixed, intermediate, and float. The fixed branch includes: (1) regimes with no separate legal tender, (2) regimes with a pre-announced peg or currency board arrangements, (3) regimes with a pre-announced horizontal band that is narrower than or equal to plus/minus two percent, and, (4) regimes with a de facto peg. The intermediate branch contains: (5) pre-announced crawling pegs, (6) regimes with a pre-announced crawling band that is narrower than or equal to plus/minus two percent, (7) de facto crawling

---

[2]Ideally, the exchange-rate system classification ought to be based on the degree to which a system in a particular category constrains domestic monetary policy independence (Tavlas et al. 2008).

[3]The de jure classification roughly distinguished between three broad categories: pegged, limited flexibility, and more flexible. These three coarse categories could be extended into fifteen fine subcategories that cover a continuum of exchange rates regimes from hard fixes to free floats.

[4]For example, several economies officially reported their currencies as pegs but often underwent frequent devaluations and, hence, in practice their regimes resembled a flexible more than a fixed. Alternatively, other countries officially committed to the flexible exchange rates, however, exhibited "fear of floating" and acted differently.

[5]To address this and a few other shortcomings, the IMF has adopted a modified classification system based on the Fund's members' de facto regimes since 1999. Bubula and Ötker-Robe (2002) provide more details.

[6]Tavlas et al. (2008) review the main methodologies that have been used to construct the de facto exchange rate regimes. They also survey the empirical literature that has been generated by the de facto classifications.

pegs, (8) regimes with a pre-announced crawling band that is wider than or equal to plus/minus two percent, (9) regimes with a de facto crawling band that is narrower than or equal to plus/minus two percent, (10) regimes with a de facto crawling band that is narrower than or equal to plus/minus five percent, (11) regimes with a moving band that is narrower than or equal to plus/minus two percent, and, (12) managed floating arrangements. Finally, the float branch includes: (13) freely floating exchange rates. The last category, (14) free falling regimes, can be reclassified into fixed, intermediate, or float on the basis of the provided chronologies.[7]

Levy-Yeyati and Sturzenegger (hereafter LYS) use cluster analysis and construct their alternative classification exclusively based on the official exchange rate and the evolution of foreign exchange reserves. They adopt the classic textbook definition of fixed and flexible exchange rates to classify the regimes. They categorize the exchange rate arrangements that are associated with low volatility in (1) nominal exchange rate level (σ_e) and, (2) changes in the nominal exchange rate ($\sigma_{\Delta e}$) but high volatility in international reserves (σ_R) as fixed exchange rate regimes, while arrangements with high volatility exchange rate levels and exchange rate movements but stable international reserves are defined as flexible exchange rate regimes.

LYS fine classification distinguishes five different regimes: (1) fixed regimes, (2) crawling pegs, (3) dirty floats, (4) floats, and, (5) inconclusive.[8] However, their coarse classification collapses into three categories: (1) fixed, (2) intermediate, and, (3) float. LYS purely rely on statistical methodology, hence, almost one third of the observations in their sample cannot be classified by their algorithm due to missing data or because the exchange rate was pegged to an undisclosed basket.

RR and LYS's de facto exchange rate regimes are very popular among alternative classifications and the series that they provide have been widely used in the empirical literature. The latest update of the RR dataset provides monthly de facto exchange rate regimes for 227 countries from January 1940 through December 2007, while the latest update of LYS dataset provides annual de facto exchange rate regimes for 183 countries from 1974 through 2004.

Both RR and LYS classifications have made a significant contribution to the de facto exchange regimes literature. Nevertheless, there are two concerns regarding the alternative classification. First, there is no empirical evidence on how to choose among the existing alternative systems. Second, there is no commonly accepted test—indeed few studies have been performed—to verify the reliability of these classifications and accordingly the studies that use them. In a recent paper, Eichengreen and Razo-Garcia (2011) investigate the disagreement between the de facto

[7]RR classify an exchange rate arrangement as a free falling regime if the 12-month inflation rate is equal to or exceeds 40 percent per annum. The regime is also considered to be free falling during the six months immediately following a currency crisis and there is a transition from a peg or a quasi-peg regime to a managed or independent float regime. See the Appendix in Reinhart and Rogoff (2004) for more details.

[8]Inconclusive regimes include those exchange rates that experience low volatility with respect to all three characteristics or for which there is no information about the classifying variables. Nearly two percent of the regimes were classified as inconclusive in the latest update of LYS.

exchange rate regimes.[9] They find that there is a good amount of agreement across the classifications; however, the disagreements are not negligible. Their results show that the disagreement is more pronounced in the case of emerging and developing countries.

2.2 *Exchange Rate Regimes and Currency Crises*

The wave of currency crisis incidences in the 1990's and early 2000's has stimulated the debates on the potential links between the choice of an exchange rate regime and the occurrence of crises. Fischer (2001) and Williamson (2002), among others, view fixed exchange rate regimes as crisis prone and argue that, in a world of integrated financial markets, rigid exchange rates are more susceptible to speculative attacks.

Yet during the major currency crisis events, intermediate exchange rate regimes (soft pegs and tightly managed floats) have been the main targets of speculative attacks. Therefore, some researchers suggest that such regimes are not viable and support for the "bipolar view" of exchange rate regimes. The proponents of the bipolar view claim that the intermediate regimes suffer from a lack of verification and transparency. Moreover, they argue that high capital mobility leaves little room for governments to follow inconsistent internal and external policies. Thus, in a world of free international capital mobility, countries will be forced to abandon the intermediate regimes and choose between the two extreme exchange rate arrangements: either hard pegs or freely floating regimes (see e.g., Eichengreen 1994; and Fischer 2001).

Nevertheless, many economists have challenged the bipolar view. Calvo and Reinhart (2002) demonstrated empirically that many intermediate regimes have not vanished and have maintained their existence. They point out that the bipolar systems do not necessarily enhance the credibility of monetary-exchange rate policies and can even destabilize the financial system. Williamson (2000, 2002) advocates intermediate regimes and proposes certain types of them (i.e. band, basket, and crawl) as the arrangements that can stabilize the real effective exchange rate and improve the sustainability of the exchange system. He argues these regimes can help prevent misalignments and provide greater flexibility to cope with shocks, whereas hard pegs and free floats can cause misalignments and damage the sustainability of the system.

Some researchers have empirically studied the links between the exchange rate regimes and the occurrence of currency crises. Ghosh et al. (2003) statistically examine the impact of exchange rate regimes on currency crises for the IMF country members from 1972 to 1999. Using the IMF's du jure exchange rate regimes and their own constructed de facto classification, they find that crises are more likely under floating regimes.

[9]They use data from three popular classification schemes: RR, LYS, and Bubula and Ötker-Robe (2002), which has been extended by Habermeier et al. (2009).

Bubula and Ötker-Robe (2003) investigate the links between the exchange rate regime and the incidence of currency crises among IMF country members from 1990 to 2001. Their logit model estimation results, obtained on the basis of the de facto exchange rate regimes of Bubula and Ötker-Robe (2002), provide some support for the bipolar view. During their sample period, the likelihood of crises for the intermediate regimes was significantly higher than that of hard pegs and floating regimes.

Rogoff et al. (2004) and Husain et al. (2005), using the de facto classification of Reinhart and Rogoff (2004), estimate the probability of currency crises for IMF country members. According to their results, over the 1970 to 2000 period, currency crises tended to occur more frequently in the intermediate regimes. Applying an alternative measure of currency crises, they find floating regimes have a significantly lower risk of entering into a crisis compared to pegs and intermediate regimes.

Haile and Pozo (2006) apply probit models to test whether the exchange regime in place has an impact on the vulnerability of countries to currency crises. Their sample includes 18 developed countries from 1974 to 1998. When they use Levy-Yeyati and Sturzenegger's (2005) de facto exchange rate regimes, their results show that the de facto exchange arrangements play no role in determining crisis periods. However, when they use the IMF de jure classification, they find that the probability of currency crises is higher for the declared pegged regimes than for intermediate or floating regimes.

Esaka (2010a, 2010b) examines how the de facto exchange rate regimes affect the occurrence of currency crises in 84 countries from 1980 to 2001. His probit model estimation results, obtained by employing the de facto classification of Reinhart and Rogoff (2004), demonstrate no significant increase in the likelihood of currency crises for the intermediate regimes compared with the hard pegs and free floating regimes (Esaka 2010a). He finds pegged regimes significantly decrease the likelihood of currency crises compared with floating regimes (Esaka 2010b). He also found that hard pegs with liberalized capital account significantly decrease the probability of currency crises compared to the floating and intermediate regimes with capital control.

3 Capital Markets Liberalization and Currency Stability

The link between capital markets liberalization and macroeconomic instability is one of the key topics in international economics. Many economists and policymakers believe that large and volatile capital flows make the international financial system unstable and cause currency crises. In their view, the liberalization of international capital flows, especially when combined with fixed exchange rates, will lead to financial disruptions (see e.g., Radelet and Sachs 2000; Stiglitz 2002).

On the other hand, capital mobility restrictions may also undermine the stability of the financial system and contribute to the occurrence of crises. Imposing capital controls will induce investment irreversibility, result in a net capital outflow, and worsen financial instability (Dooley and Isard 1980). Moreover, restricted capital accounts can create distortions, signify inconsistent policies, and exhibit the potential vulnerabilities of the financial system, which may induce capital flight and trigger currency crises (Bartolini and Drazen 1997).

In addition to the lack of consensus on the links between capital market liberalization and the occurrence of currency crises, the potential interdependence of capital account policies with the choice of exchange rate regime makes the issue at stake even more complicated. It is widely recognized that under high capital mobility, monetary policies cannot easily focus on both maintaining fixed exchange rates and accommodating with real shocks effectively. This is usually referred to as the "impossible trinity".[10] It points to the argument that policymakers in open economies may concentrate only on two of three conflicting objectives: capital mobility, monetary independence, and the stable fixed exchange rate.[11] This argument implies there could be interdependence between the choices of exchange rate regimes and capital account policies.

As a direct implication of the impossible trinity, one can expect, due to the current trend of financial liberalization, monetary policies will increasingly become inconsistent with the sustainability of fixed exchange rates and make this type of exchange rate arrangement more crisis prone (this conclusion is incompatible with the bipolar view). Furthermore, wide financial and trade integration, rapid financial innovations, and deep financial developments have gradually reduced the effectiveness of capital controls and consequently the monetary policy-exchange rate stability dilemma is now evident even in the countries that are willing to impose capital controls.

Several studies empirically investigate the impact of capital control policies on insulating countries from the macroeconomic instability and currency crises. Edwards (1989) investigates the role of capital controls in 39 devaluation episodes for 24 developing countries from 1961 to 1982. His findings show that these countries typically employed intensified capital control programs in the year before the devaluation to slow down the unavoidable balance of payment crises. Demirgüç-Kunt and Detragiache (1997) estimate the probability of a systemic crisis for both industrial and developing countries over the period of 1980–1994. Their results indicate that capital account liberalization can contribute to the macroeconomic instability and the occurrence of banking crises.

On the other hand, Glick and Hutchison (2005) study the link between capital controls and currency stability for 69 emerging and developing countries from 1975

[10] However, Lavoie (2001) counters the impossible trinity claim and argues that even under capital mobility can maintain their monetary policy autonomy. Partially based on his argument, Frenkel and Rapetti (2007) analyze the macroeconomic evolution of Argentina during the (2001) crisis and question the validity of impossible trinity.

[11] Obstfeld and Taylor (2005) elaborate the role of impossible trinity on the evolution of the international financial system.

to 1997. Their probit estimation results show that restrictions on capital flows are unable to efficiently protect countries from currency crises. Their findings provide no evidence that countries with high capital mobility are more prone to speculative attacks. Glick et al. (2006) address concerns about self-selection bias and attempt to revise their earlier work accordingly.[12] The outcome of their analysis suggests that even after controlling for the sample selection bias, countries with liberalized capital accounts experience a lower likelihood of speculative attacks. Glick and Hutchison (2010) present a new version of their earlier study. They expand the time coverage from 1975 to 2004 and apply duration-adjusted measures of capital control intensity to allow for changes in control programs over time. Their results re-emphasize their previous findings and assert that countries with less restrictive capital controls and more liberalized financial markets appear to be less vulnerable to speculative pressures.

A possible cause of these mixed empirical results could be attributed to the complexity of properly measuring the degree of openness or restrictions on cross-border financial transactions. The underlying source of data for the conventional measures of quantifying financial openness is based upon the IMF's de jure classifications, which are published in the *Annual Report on Exchange Arrangements and Exchange Restrictions* (*AREAER*). However, this information is overly aggregated to fully capture the dynamics of actual capital controls. Moreover, it is almost impossible to distinguish between de jure and de facto controls on capital account transactions. Consequently, the indices that are constructed to quantify the capital account restrictions, especially those that are dichotomous, fail to account for the intensity of capital controls. It is well known that measuring the extent of openness on capital account transactions is very complicated.

Nonetheless, many studies rely on the IMF's *AREAER* attempt to quantify the degree of financial openness and measure the impact or determinants of capital controls. Chinn and Ito (2005) present an index for measuring the degree of capital account openness. The Chinn-Ito index is based on a five-year moving average of the de jure binary dummy variables that codify the tabulation of restriction on cross-border transactions. This index attempts to measure the intensity of capital controls. The latest update of this index covers 182 countries for the period of 1970–2009. The index is constructed in such a way that the series has a mean of zero and country values range from -1.844 to 2.478, where higher values indicate a greater intensity of restrictions on capital account transactions. Chinn and Ito (2008) provide details on how their index is constructed and compare it with other existing measures in the literature.

[12] Self-selection bias points to the non-random choice of capital control programs. Countries that are facing considerable amount of pressure in their exchange markets are more likely to impose capital control programs and accordingly a positive correlation between capital controls and speculative attacks will be observed.

4 Data and Methodology

4.1 Data

This paper analyzes the incidence of currency crises in 21 countries with the help of an unbalanced panel of quarterly data over the period of 1970 through 1998. The dataset does not cover a longer period, since 10 countries in our sample left their own national currencies and joined the Euro currency system from the beginning of 1999. The countries in our sample include: Australia, Austria, Belgium, Canada, Denmark, Finland, France, Greece, Iceland, Ireland, Italy, Japan, the Netherlands, New Zealand, Norway, Portugal, South Africa, Spain, Sweden, Switzerland, and the UK. These countries share common similarities and provide higher frequency data for our empirical models.

Episodes of currency crises come from Karimi and Voia (2011a). These episodes correspond to the extreme values of exchange market pressure indices. The indices are constructed on the basis of monthly and quarterly data. Accordingly, two different types of crisis episodes are obtained: monthly-type and quarterly-type.[13]

We borrow the empirical models of the determinants of currency crises from Karimi and Voia (2011b) as the benchmark and examine how the likelihood of currency crises change under the de jure and the de facto exchange rate regimes. We also study the role of capital mobility and test for currency stability under free and restricted capital flows. In particular, we investigate the impact of different combinations of exchange rate arrangements and capital controls on the hazard of speculative attacks. The unobservable determinants of currency stability are captured with the help of baseline hazards. They will be identified non-parametrically by applying a semi-parametric mixed proportional hazard model. The following subsection presents a brief discussion of the estimation methodology.

Our index for the de jure exchange rate regimes is the same as the IMF's classification. For the choice of a de facto regimes index, we have some options available. However, there is no systemic methodology to choose and/or evaluate the existing alternative systems. Moreover, as Eichengreen and Razo-Garcia (2011) point out, during periods of currency volatility, different de facto classifications tend to produce different results. Hence, they advise investigators to be particularly careful when attempting to link the de facto regimes to financial crises. Thus, considering the time coverage of the de facto regimes, we employ both RR and LYS classifications in an attempt to capture the probable discrepancies. We adopt the coarse classification of RR and LYS and divide the exchange rate arrangements into three categories: (1) fixed, (2) intermediate, and; (3) floating regimes.[14] On this ground, we

[13]Since little monthly data is available to run our empirical models, the monthly incidences of crisis are expanded to contain the relevant quarters. Hence, the monthly-type crisis episodes suggest that at least one month within that quarter is recognized as the incidence of a crisis.

[14]RR and LYS datasets are respectively available at: http://www.carmenreinhart.com/research/publications-by-topic/exchange-rates-and-dollarization/, and: http://www.utdt.edu/ver_contenido.php?id_contenido=4643&id_item_menu=8006.

construct the categorical variables of exchange rate regimes: $Fix_{i,t}$, $Intermediate_{i,t}$, and $Float_{i,t}$.[15] Each country (i) at the time t is assigned to one of these categories based on RR or LYS classifications.

We utilize the Chin-Ito index as our measure of capital account restrictions. This index, to some extent, can capture the intensity of capital mobility restrictions and enjoys a wide coverage across countries and time. On the basis of this index, we construct a dummy variable for capital controls ($CapControls_{i,t}$). A capital account is classified as open—$CapControls_{i,t}$ takes the value of one—if the value of the Chinn-Ito index is more than the average of similar countries during that period of time.[16] Otherwise, it is classified as restricted and $CapControls_{i,t}$ takes the value of zero.[17]

To examine the impact of different exchange rate regimes under the presence or absence of capital controls, we combine the exchange rate classifications with the capital account policies and categorize our sample into six different regimes (three different exchange rate classifications with two capital account choices). Consequently, we construct two series of six categorical variables (one for RR-based and the other for LYS-based classifications), which are introduced in Section 5.

Finally, we should address the potential problem of reverse causation. This paper deals with the impact of exchange rate regimes and capital account policies on the occurrence of currency crises, not the other way around. To mitigate the potential problem of reverse causality (the impact of crises on exchange rate and capital regimes), we use lagged variables. Hence, the exchange rate regimes and capital account openness variables enter into the models with at least a one-period lags. This remedy to the potential problem of reverse causality is also useful to treat the potential interdependence between the choice of exchange rate regimes and capital account liberalization policies. In order to deal with this concern, we recognize and control for the duration of the policy mix composed of the exchange rate regimes and capital control programs. It is in line with the recent studies in the literature.

4.2 Methodology

We adopt duration models to tackle our research question. In particular, we model the hazard function by applying a semi-parametric specification. The hazard function is a nonlinear function of time and has some advantages over the logit and probit models that are widely used in the literature. First, our models can estimate

[15] Since we have two indexes for exchange rate regimes, RR and LYS, we construct two series of categorical variables.

[16] All countries in our sample are categorized as advanced economies except for South Africa and some years in case of Greece and Portugal, which are categorized as emerging economies. The average value of the Chinn-Ito index for industrialized countries equals 0.257, 0.804, and 2.152 over the periods of 1970–1979, 1980–1989, and 1990–1999, respectively.

[17] The Chinn-Ito index dataset is available at http://web.pdx.edu/~ito/Chinn-Ito_website.htm.

the impact of time-varying covariates on currency stability. They are also able to evaluate whether the duration of time spent in tranquil periods has any significant influence on the probability of exit into turbulent episodes. Second, these models can accommodate the censored observations. Third, while probit and logit models require strong assumptions about the distribution of the time to failure and implicitly imply the monotonic hazard function, semi-parametric hazard models are able to capture the real relationship between the probability of an exit and the duration of tranquil states, which in many real situations is not monotonic.

In what follows, we briefly introduce the basic setting of duration analysis and present the Cox mixed proportional hazard model. A detailed and comprehensive statistical discussion of duration models can be found in Kalbfleisch and Prentice (2002) and Klein and Moeschberger (2010). Also, Kiefer (1988) and Lancaster (1990) provide econometrics applications and the related technicalities.

Let T be a nonnegative random variable denoting the time to a failure event—e.g. a currency exits a tranquil state and entering into a crisis state. The cumulative probability distribution is $F(t) = \Pr(T \leq t)$, and the survivor function is given by $S(t) = \Pr(T > t) = 1 - F(t)$, where t is time, and $\Pr(T > t)$ is the probability that the timing of the failure event, T, is greater than t. The survivor function indicates the probability that a currency still remains in a tranquil state beyond time t. One can alternatively describe the time to exit using a hazard function (or the instantaneous probability) of exits. The hazard is a measure of the probability that a currency will exit the tranquil state in time t, given that it has survived up to time t. The hazard function can be defined as:

$$h(t) = \lim_{\Delta t \to 0} \frac{\Pr(t + \Delta t > T > T + t \mid T > t)}{\Delta t} = \frac{f(t)}{S(t)} = \frac{-d \ln S(t)}{dt}, \quad (1)$$

where, $f(t)$ denotes the probability density function associated with $F(t)$.

Equation (1) specifies that there is a one-to-one mapping between the probability density function, the cumulative distribution function, the survivor function, and the hazard function. Given one of these functions that describe the probability distribution of failure times, the others are completely determined. However, in the literature, it is more common to think in terms of the hazard rather than the traditional density and cumulative distribution functions. Hazard functions can be specified by parametric, semi-parametric, and non-parametric models. In this paper, we employ a semi-parametric hazard specification for its comparative advantages over other hazard specifications in tracking the baseline hazard (see Huynh et al. 2012). A semi-parametric hazard approach assumes time's distribution is non-parametric, but the effect of covariates is still parameterized.

A very well known way to represent the hazard function is to write it as:

$$h_j(t) = h_0(t)\varphi\big(x_j(t), \beta\big). \quad (2)$$

This method is called proportional hazards because subject j faces the hazard that is multiplicatively proportional to the baseline hazard. The popular Cox (1972) article uses this technique and assumes the covariates multiplicatively shift the baseline hazard function.[18] The Cox model leaves the baseline hazard, $h_0(t)$, unspecified and assumes all subjects at risk face the same baseline hazard, which is a restricted assumption. This innovation lets the Cox models enjoy the important advantage of not requiring any assumptions on the distribution of the time of failures (or the shape of the hazard over time) and helps these semi-parametric models to be robust to misspecification of the baseline hazard. In fact, the baseline hazard, $h_0(t)$, will be canceled out in building the likelihood function. This model presents the ratio of hazard rates for subject j to subject k as:

$$\frac{h_j(t)}{h_k(t)} = \frac{h_0(t)\varphi(x_j(t), \beta)}{h_0(t)\varphi(x_k(t), \beta)} = \frac{\varphi(x_j(t), \beta)}{\varphi(x_k(t), \beta)}. \tag{3}$$

Therefore, one can write the conditional probability of ith observation that fails at the time t_i, given all of the n observations have exited by time t_n, as:

$$\frac{h_i(t)}{\sum_{i=1}^{n} h_i(t)}. \tag{4}$$

Thus, the likelihood function will be:

$$L\{\beta \mid (t_1, x(t_1)), \ldots, (t_n, x(t_n))\} = \prod_{j=1}^{n} \left(\frac{\varphi(x_j(t), \beta)}{\sum_{i=j}^{n} \varphi(x_i(t), \beta)} \right) \tag{5}$$

and the estimation of coefficients, β_x, can be obtained conditional on the failure times.

The resulted estimated hazard is robust to misspecifications because of the flexibility of the non-parametric baseline hazard. However, to check if the parameters of the model are consistently estimated we use different testing procedures. First, a key requirement for using a Cox proportional hazard model is to have proportional hazards. The test used for proportionality of hazards is based on Schoenfeld residuals. The test suggests that if the covariates are generating proportional hazards we will observe zero slopes when we do a generalized regression of the scaled Schoenfeld residuals on functions of time. Second, we use the cumulative Cox-Snell residuals (which are residuals that are summed over the observations within a subject) to assess the fit of the model. As the Cox-Snell residuals are distributed as unit exponential, the estimated cumulative hazard of the Cox-Snell residuals should look as a 45-degree line for a perfect fit. Due to space constraints, the tables for the Schoenfeld residuals and the graphs of Cox-Snell residuals are not reported in the paper but available upon request.

---

[18] The most common specification of $\varphi(\cdot)$ is in exponential form. Hence, the hazard can be represented as: $h_j(t) = h_0(t)\exp(x_j(t), \beta)$, which is convenient to deal with non-negative values of $\varphi(\cdot)$ and has computational feasibility.

5 Empirical Results

In this section, first, we empirically investigate the links between the risk of a currency crisis and the choice of exchange rate regimes. Then, we evaluate the impact of capital mobility on the stability of exchange rates. Finally, we examine how the likelihood of currency crises changes under different combinations of exchange rate regimes and capital controls.

5.1 Exchange Rate Regimes and Currency Crises

As a first step, we find how the incidences of different exchange rate regimes are distributed across our sample. As Table 1 presents, the IMF de jure system classifies major portions of the sample as the intermediate regimes compared to the fixed and floating arrangements. The same pattern is even more pronounced under the RR de facto classification (it should not be surprising knowing that RR relies on the IMF classification). However, LYS de facto system assigns more quarters to the corner regimes—fixed or floats—than the intermediate regimes.

In the next step, we figure out how the monthly and quarterly-type of currency crisis episodes are jointly scattered with the exchange rate arrangements and calculate the unconditional probability of a currency crisis under different exchange rate regimes. From the reported results in Table 2, it is evident that when the regimes are categorized based upon the de jure classification the differences between the calculated probabilities for currency crisis incidences under different exchange rate regimes are negligible. Yet the probabilities that are calculated under the de facto classifications show significant results, but differ according to the chosen classification. When regimes are categorized by the LYS classification, the intermediate exchange rate arrangements are the least susceptible regime to the speculative attacks. However, when regimes are categorized by the RR classification, the fixed arrangements are the most sustainable exchange rates. To verify that the results are statistically significant and not random or due to differences in sample sizes, we run Chi-square independence test (not reported) and log-rank test. Both tests produce

Table 1 Incidence of Exchange Rate Regimes under different classifications

	de jure (IMF)		de facto (LYS[a])		de facto (RR)	
	Quarters	Share (%)	Quarters	Share (%)	Quarters	Share (%)
Fix	696	28.57	796	45.64	615	25.25
Intermediate	1040	42.69	344	19.72	1654	67.90
Float	700	28.74	604	34.63	167	6.86
Total	2436	100.00	1744	100.00	2436	100.00

[a]LYS classification starts from 1974 and contains several unclassified observations

Table 2 Unconditional probability of crisis under different Exchange Rate Regime classifications

	Monthly-type			Quarterly-type		
	IMF	LYS	RR	IMF	LYS	RR
Fix $(t-1)$	9.91	9.57	5.63	7.18	5.87	4.03
Intermediate $(t-1)$	10.14	7.87	12.11	6.24	4.37	8.3
Float $(t-1)$	10.39	12.69	6.10	7.79	10.02	4.27
Log-rank test[a]	1.36	5.81	18.63	1.39	11.70	11.25
P-value	0.51	0.06	0.00	0.50	0.00	0.00

Probabilities are calculated by dividing the number of crises under a particular regime to the total number of regime-quarters. All numbers are in percent, except for the Long-rank test results
[a]The null hypothesis of log-rank test is whether the hazard functions are equal across different groups

similar results and confirm our findings. The same structure is observed for both the monthly and quarterly-type spells.

In addition, to obtain a visual understanding of the dynamics of the hazards under different exchange rate regimes, we present the smoothed estimations of the non-parametric hazards in Figure 1. In the diagrams of this figure, the vertical axis measures the probability that a currency exits a tranquil state and enters into a crisis state, while the horizontal axis represents the successive number of quarters spent in tranquility. The presented diagrams reconfirm the pattern that we observed in Table 2. When regimes are categorized based on the RR classification, the hazard of crises is at the highest level for the intermediate regimes. However, when regimes are classified under the LYS alternative system, the outcomes inverted and the intermediate regimes enjoy the lowest probability of attack (especially for the quarterly-type spells). Similar to the Table 2 results, the hazards that are built upon the IMF classification do not show a clear pattern. It should also be useful to mention that the observed non-monotonic nature of the hazards in Figure 1 validates our choice for the semi-parametric Cox proportional hazard models. Next, the exchange rate regime categorical variables are introduced into the models. We run four different models for each monthly and quarterly-type spells. Variables in models 1 and 2 are contemporaneous while in models 3 and 4 they are lagged by one quarter. In models 1 and 3 the variables of each country are measured in real levels, while all time-varying variables in models 2 and 3 are measured relative to the reference countries—Germany or the U.S. The results related to the RR and LYS classifications are presented in Tables 3 through 6.[19] The results related to the IMF classification are not statistically significant whether monthly or quarterly-type models are being used, and hence are not reported here.

[19]From what we perceived in Figure 1, the Fix exchange rate regimes are chosen as the base in our RR-based categorical variables while the Intermediate exchange rate regimes are assigned as the base for LYS-based categorical variables.

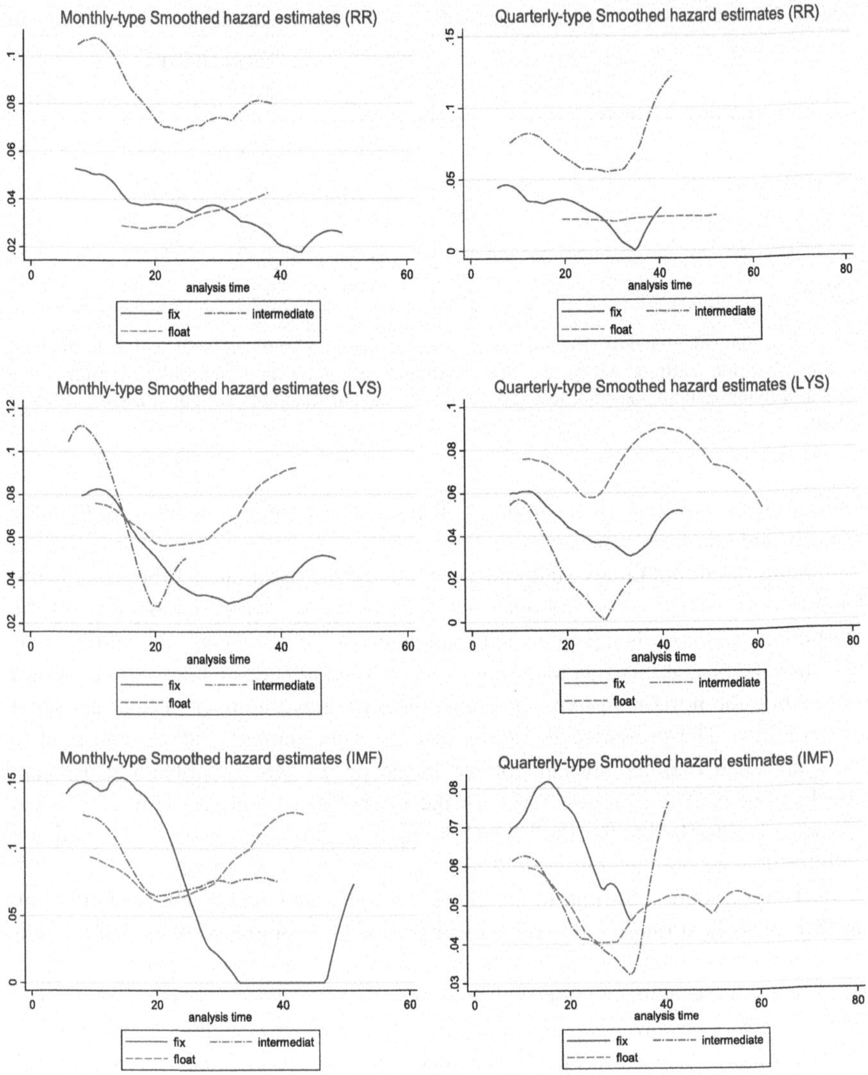

Fig. 1 Monthly and quarterly-type smoothed hazards under different exchange rate regimes

Tables 3 and 4 present the estimation results for monthly-type models of RR and LYS classifications.[20] An examination of the reported results in Table 3 reveals that the hazard of fixed exchange rate regimes is significantly lower than the hazard of intermediate exchange rate regimes in all of the four RR-based monthly-type models. The hazard of fixed regimes is also lower than the hazard of float regimes;

[20]The models are interacted with different linear and non-linear time functions. The presented estimation results are the outcome of the interaction with a logarithmic form of time.

Table 3 Cox proportional hazard estimation (monthly-type spells) under RR de facto classifications

Variable	Contemporaneous		Lagged	
	Model (I)	Model (II)	Model (III)	Model (IV)
Fix is the base				
Intermediate	1.37* (1.9)	1.5* (1.79)	1.11* (1.65)	1.15* (1.71)
Float	1.69* (1.91)	2.15** (2.27)	0.68 (078)	1.22 (1.28)
Unemployment volatility	0.03 (1.38)	0.04** (2.22)	0.05** (2.03)	0.07** (2.45)
Previous crises				0.4* (1.68)
Size of economy	0.83** (2.39)		0.76** (2.1)	1.04** (2.52)
Whole period GDP growth		0.02** (2.1)	0.02 (1.38)	0.01 (1.25)
GDP growth rate	0.00 (−0.05)	−0.01 (−0.17)	−0.06 (−0.79)	−0.07 (−0.87)
Inflation	0.26*** (3.21)	0.23** (2.3)	0.04 (0.45)	0.2** (2.24)
Unemployment rate	0.00 (0.11)	−0.02 (−0.98)	−0.01 (−0.4)	−0.03 (−1.15)
Share price index growth	−0.03*** (−3.34)	−0.01 (−0.64)	−0.01 (−0.62)	0.01 (0.7)
Real effective exchange rate	0.00 (−0.17)	0.01 (0.31)	0.01* (1.71)	0.01** (2.53)
Money growth	−0.03** (−2.44)	−0.05*** (−3.4)	0.01 (0.48)	−0.01 (−0.61)
Real domestic credit growth	0.04*** (3.61)	0.05*** (3.6)	0.02 (1.23)	0.02* (1.68)
Trade openness	0.29 (0.69)	0.04** (2.59)	0.13 (0.28)	0.03 (1.58)
Current account/GDP	−0.02 (−0.46)	0.00 (−1.63)	−0.09* (−1.96)	0.00 (−0.61)
Capital account/GDP	0.28 (0.84)	0.00 (−0.42)	−0.74 (−0.85)	0.00 (−0.51)
Financial account/GDP	0.07** (2.3)	0.00** (−2.08)	−0.11* (−1.92)	0.00*** (−2.88)
Budget deficit/GDP	0.02** (2.54)	0.00 (0.16)	0.02 (1.38)	0.00 (−0.13)
Trade linkages	0.11** (2.43)	0.14** (2.47)	0.12** (2.28)	0.14* (1.88)

Table 3 (Continued)

Variable	Contemporaneous		Lagged	
	Model (I)	Model (II)	Model (III)	Model (IV)
Financial linkages	−0.02	−0.02	0.00	0.00
	(−0.93)	(−0.88)	(−0.1)	(−0.13)
Macroeconomic similarities	0.05	0.06	0.01	−0.03
	(1.03)	(1.00)	(0.11)	(−0.38)
Log likelihood	−184.78	−136.55	−185.68	−132.83

The values in parentheses below estimates are the corresponding z-statistics
***, (**), [*] imply estimates are significant at 1, (5), and [10] percent

Table 4 Cox proportional hazard estimation (monthly-type spells) under LYS de facto classifications

Variable	Contemporaneous		Lagged	
	Model (I)	Model (II)	Model (III)	Model (IV)
Intermediate is the base				
Fix	1.45**	0.89	1.15	1.82**
	(2.19)	(1.16)	(1.64)	(2.3)
Float	1.03*	0.83	1.3**	1.39*
	(1.66)	(1.1)	(1.97)	(1.72)
Unemployment volatility	0.07**	0.03	0.06**	0.09**
	(2.51)	(1.44)	(2.27)	(2.52)
Previous crises				−0.04
				(−0.21)
Size of economy	1.44**		0.92*	1.34**
	(2.47)		(1.71)	(2.4)
Whole period GDP growth		0.01	−0.00	−0.01
		(0.48)	(−0.06)	(−0.35)
GDP growth rate	−0.05	−0.15	0.04	−0.03
	(−0.44)	(−1.41)	(0.32)	(−0.31)
Inflation	0.48***	0.13	0.3**	0.44***
	(4.06)	(0.87)	(2.28)	(2.78)
Unemployment rate	0.03	0.01	0.03	0.03
	(1.17)	(0.36)	(1.05)	(0.92)
Share price index growth	−0.03**	0.00	−0.02	0.02
	(−2.65)	(0.2)	(−1.16)	(1.08)
Real effective exchange rate	0.01	0.00	0.02**	0.03***
	(1.27)	(−0.01)	(2.16)	(3.33)
Money growth	−0.03**	−0.04**	−0.04	−0.03
	(2.39)	(−2.46)	(−0.94)	(−1.17)

Table 4 (Continued)

Variable	Contemporaneous		Lagged	
	Model (I)	Model (II)	Model (III)	Model (IV)
Real domestic credit growth	0.04***	0.05***	0.02	0.03*
	(3.41)	(3.04)	(1.01)	(2.14)
Trade openness	0.3	0.07***	0.36	0.04
	(0.58)	(3.3)	(0.61)	(1.09)
Current account/GDP	0.01	0.00	−0.1*	0.00
	(0.13)	(−1.41)	(−1.66)	(−0.73)
Capital account/GDP	0.12	0.00	−1.82	0.00
	(0.38)	(0.26)	(−1.22)	(−0.86)
Financial account/GDP	0.06*	0.00*	−0.12**	0.00**
	(1.7)	(−1.79)	(−2.04)	(−2.46)
Budget deficit/GDP	0.02**	0.00	0.01	0.00
	(2.37)	(0.05)	(0.89)	(−1.45)
Trade linkages	0.12**	0.14**	0.18**	0.21***
	(2.32)	(2.07)	(2.18)	(3.4)
Financial linkages	−0.01	0.01	0.01	0.01
	(−0.25)	(0.48)	(0.59)	(0.5)
Macroeconomic similarities	0.05	0.08	−0.06	−0.08
	(1.16)	(1.18)	(−0.76)	(−1.6)
Log likelihood	−118.28	−91.9	−124.01	−95.29

The values in parentheses below estimates are the corresponding z-statistics
***, (**), [*] imply estimates are significant at 1, (5), and [10] percent

however, in two of these models the difference is statistically significant. On the other hand, the results of Table 4 show that the hazard of intermediate exchange rate regimes is significantly lower than the hazard of fixed exchange rate regimes in two of the four LYS-based monthly-type models. Furthermore, in three of these models, the hazard of intermediate regimes is significantly lower than the hazard of float regimes. In both Tables 3 and 4, some control variables (trade linkages, inflation, unemployment volatility, and the financial account ratio to GDP) are repeatedly significant in all models and have the expected sign.

Tables 5 and 6 provide the estimation results for quarterly-type models of RR and LYS classifications. The results that they present are similar to those reported in Tables 3 and 4. Table 5 shows that when the episodes of currency crisis are identified with lower frequency data, the hazard of fixed exchange rate regimes is significantly lower than the hazard of intermediate exchange rate regimes in three of the four RR-based models. However, in all LYS-based quarterly-type models, the hazard of intermediate regimes is significantly lower than that of fixed and floating exchange regimes. Among the control variables, inflation is statistically significant in most of the models.

Table 5 Cox proportional hazard estimation (quarterly-type spells) under RR de facto classifications

Variable	Contemporaneous		Lagged	
	Model (I)	Model (II)	Model (III)	Model (IV)
Fix is the base				
Intermediate	1.2* (1.7)	1.95* (1.95)	1.45* (1.75)	0.43 (0.47)
Float	1.27 (1.52)	1.99* (1.87)	1.51 (1.5)	0.17 (0.16)
Unemployment volatility	0.01 (0.26)	0.00 (−0.18)	0.00 (0.04)	0.02 (0.64)
Previous crises				−0.4 (−1.21)
Size of economy	0.39 (1.1)		0.44 (1.11)	0.36 (0.83)
Whole period GDP growth		0.00 (0.33)	0.00 (−0.02)	0.01 (0.49)
GDP growth rate	−0.04 (−0.41)	0.02 (0.2)	0.07 (0.81)	0.06 (0.6)
Inflation	0.2** (2.04)	0.16* (1.85)	0.1 (1.15)	0.21* (1.66)
Unemployment rate	0.01 (0.39)	0.1 (0.49)	0.03 (1.23)	0.04 (1.45)
Share price index growth	−0.01 (−1.62)	0.00 (−0.34)	−0.02* (−1.68)	0.00 (−0.51)
Real effective exchange rate	0.00 (−0.41)	−0.01 (−0.24)	0.01 (1.44)	0.01* (1.74)
Money growth	−0.02 (−1.43)	−0.04*** (−2.61)	−0.01 (−0.45)	0.00 (−0.02)
Real domestic credit growth	0.03* (1.92)	0.01 (0.02)	0.04 (1.54)	0.04 (1.35)
Trade openness	0.01 (0.31)	0.02 (0.83)	0.18 (0.4)	0.04* (1.9)
Current account/GDP	0.02 (0.72)	0.00 (0.43)	0.02 (0.36)	0.00 (0.05)
Capital account/GDP	0.19 (0.39)	0.00 (0.8)	−2.28** (−2.19)	0.00 (0.8)
Financial account/GDP	−0.01 (−0.39)	0.00 (0.24)	−0.02 (−0.32)	−0.02 (−0.71)
Budget deficit/GDP	0.01 (1.19)	0.00 (0.73)	0.00 (0.45)	0.00 (0.49)
Trade linkages	0.1 (0.84)	0.15 (1.55)	0.13 (1.25)	0.22 (1.35)

Table 5 (Continued)

Variable	Contemporaneous		Lagged	
	Model (I)	Model (II)	Model (III)	Model (IV)
Financial linkages	0.00 (−1.41)	0.00 (−0.95)	0.00 (−1.46)	0.00 (−0.84)
Macroeconomic similarities	0.05 (0.43)	−0.15 (−0.16)	0.05 (0.43)	−0.09 (−0.56)
Log likelihood	−124.79	−106.2	−114.47	-95.74

The values in parentheses below estimates are the corresponding z-statistics
***, (**), [*] imply estimates are significant at 1, (5), and [10] percent

Table 6 Cox proportional hazard estimation (quarterly-type spells) under LYS de facto classifications

Variable	Contemporaneous		Lagged	
	Model (I)	Model (II)	Model (III)	Model (IV)
Intermediate is the base				
Fix	3.14*** (3.17)	2.57*** (3.21)	2.89** (2.09)	3.53*** (3.56)
Float	2.8*** (2.94)	2.8*** (3.82)	3.33** (2.48)	2.74*** (3.34)
Unemployment volatility	0.04 (1.46)	0.01 (0.41)	0.03 (0.9)	0.02 (0.83)
Previous crises				−0.76** (−2.18)
Size of economy	0.74* (1.68)		0.76 (1.15)	0.68 (1.5)
Whole period GDP growth		−0.01 (−0.63)	−0.04 (−1.61)	−0.02 (−0.96)
GDP growth rate	0.04 (0.36)	0.05 (0.48)	0.08 (0.65)	0.17* (1.95)
Inflation	0.49*** (4.05)	0.3*** (2.8)	0.42** (2.7)	0.27** (2.06)
Unemployment rate	0.04 (1.44)	0.03 (1.07)	0.12*** (3.067)	0.6** (2.27)
Share price index growth	−0.01 (−1.26)	0.01 (0.44)	−0.02** (−2.09)	−0.02 (−1.39)
Real effective exchange rate	0.02 (1.47)	0.01 (0.98)	0.02** (2.27)	0.02** (2.23)
Money growth	−0.05* (−1.79)	−0.05*** (−2.9)	−0.01 (−0.25)	−0.02 (−0.9)

Table 6 (Continued)

Variable	Contemporaneous		Lagged	
	Model (I)	Model (II)	Model (III)	Model (IV)
Real domestic credit growth	−0.01 (−0.19)	0.00 (−0.03)	0.06** (2.04)	0.05* (1.85)
Trade openness	0.03* (1.79)	0.02 (0.81)	1.11 (1.15)	0.03 (1.14)
Current account/GDP	−0.01 (−0.15)	0.00 (1.02)	−0.02 (−0.24)	0.00 (−0.36)
Capital account/GDP	0.16 (0.51)	0.00 (1.02)	−3.06** (−2.27)	0.00 (0.55)
Financial account/GDP	−0.01 (−0.56)	0.00 (0.38)	−0.03 (−0.46)	0.00 (0.95)
Budget deficit/GDP	0.01 (0.81)	0.00 (0.86)	0.00 (0.81)	0.00 (−0.98)
Trade linkages	0.19 (1.02)	0.35*** (2.9)	0.19 (1.33)	0.09 (0.84)
Financial linkages	0.00 (−0.6)	0.00 (0.01)	0.00 (−1.07)	0.00 (−0.76)
Macroeconomic similarities	−0.02 (−0.13)	−0.12 (−1.08)	0.08 (0.62)	0.11 (1.02)
Log likelihood	−76.1	−75.9	−67.35	−66.49

The values in parentheses below estimates are the corresponding z-statistics
***, (**), [*] imply estimates are significant at 1, (5), and [10] percent

It is clear that in our sample there is a statistically significant link between the choice of an exchange rate regime and the occurrence of currency crises. Nevertheless, the results are sensitive to the choice of the de facto exchange rate system. Fixed exchange rate regimes are the least susceptible exchange arrangement to speculative attacks, if the exchange rate regimes are determined by the RR classification, while the intermediate exchange rate regimes will experience the least number of currency crisis incidences, if the exchange rate regimes are determined with the help of the LYS classification. The Akaike Information Criterion (AIC) indicates a better fit of the data for all LYS-based models compared to RR-based models. However, determining the outcome of which the de facto system is more appropriate, definitely, requires a methodology that is more comprehensive and, ideally, looks to determine how close these systems are to the "true" regimes.

We run several robustness tests to verify whether our adopted methodology is appropriate and the obtained results are consistent. In the first step, we run four different models for each of monthly and quarterly-type spells. The observed consistency of the results is a sign of the stability of the models and the reliability of the results. Then, we run the Schoenfeld and Cox-Snell residuals tests to check whether the hazards are truly proportional and if the estimated model has a good fit, hence, if applying Cox models is appropriate. The test results (not reported) show that almost on all monthly and quarterly-type models (both RR-based and LYS-based) all covariates are proportional and, thus, confirm that it is appropriate to use a proportional hazard model. Also, the model specification test suggests a good fit.

We also checked the sensitivity of our results with respect to the tied spells and ran our models with two alternative methods: the Efron and marginal calculations. The obtained results (not reported) from both methods are similar and do not indicate any significant issue related to the tied spells. Finally, we examined our results for the existence of unobservable heterogeneity. The test results did not show any unobservable heterogeneity between the countries in our sample. This is due to the homogeneity of the sample.

5.2 Capital Mobility and Currency Crises

We start our investigation by examining the types of capital accounts, which are categorized with the help of the Chinn-Ito index, and figuring how the restricted and open-type of capital accounts have been distributed across our sample. As Table 7 presents, open and restricted-type of capital accounts have almost an equal share in our sample. However, the unconditional probability of currency crisis episodes with different types of capital accounts shows that more incidences of speculative attack have taken place during the periods of time that are categorized as restricted-type of capital accounts. We also run log rank test and Chi-square independence test (not reported) and verify that observed differences between the calculated probabilities of currency crises for different types of capital accounts are statistically significant. Table 7 reports the results.

Figure 2 visualizes the hazards of currency crises for different types of capital accounts. The presented diagrams confirm the observed pattern in Table 7 for both

Table 7 Distribution of Capital Account Type and incidences of currency crisis

	Chinn-Ito index		Monthly-type spells		Quarterly-type spells	
	Quarters	Share (%)	Probability	Stcox	Probability	Stcox
Restricted	1116	48.69	0.12		0.08	
Open	1176	51.31	0.08	−0.32**	0.06	−0.28*
				(−2.34)		(−1.73)
Log-rank test[a]			6.42		3	
p-value			0.01		0.08	

[a]The null hypothesis of log-rank test is whether the hazard functions are equal across different groups
For the Cox proportional hazard estimations, the restricted-type is the base
The values in parentheses below estimates are the corresponding z-statistics
***, (**), [*] imply estimates are significant at 1, (5), and [10] percent

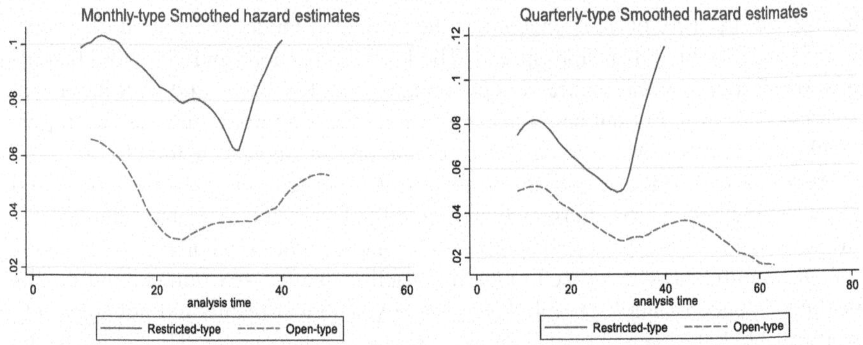

Fig. 2 Monthly and quarterly-type smoothed hazards under different exchange rate regimes

monthly and quarterly-type spells. We also run the Cox proportional hazard model (the restricted-type model been chosen as the base) without any control variables. The results (reported in Table 7) are in line with our previous findings and indicate that the baseline hazards of open-type capital accounts are lower than the baseline hazards of restricted-type. However, when we apply the Cox proportional hazard models with our set of control variables, strikingly, the obtained results will be different from what we found in Table 7 and Figure 2.

According to the results presented in Tables 8 and 9, the hazards of open-type capital accounts are higher than the hazards of restricted-type capital accounts, although this relation is statistically significant only in the monthly-type spells (models 1, 2, and 3). Similar to our previous estimation results, some control variables such as inflation, trade linkages, unemployment volatility, and financial account ratio to GDP, are repeatedly significant and have the expected sign.

The impact of capital account policies on the occurrence of currency crises demonstrates different results in our sample. While baseline hazard of open-type capital accounts are lower than the baseline hazard of restricted-type capital accounts, when we enter the set of control variables to our models, the hazard of open-type capital accounts appear to be higher than the hazard of restricted-type capital accounts. This relation is often statistically significant when the episodes of currency crises are identified with higher frequency—monthly—data. It can be interpreted as a sign that capital control policies could help in preventing low duration—milder—crises. The obtained results are robust to a variety of samples and models. We ran different models and received consistent results for both monthly and quarterly-type models. The results of the Schoenfeld residuals test show that a few covariates in some models do not individually pass the proportionality test; however, all models jointly pass the proportionality test. The results on Cox-Snell residuals show a good fit. We also did sensitivity checks for the tied spells and found no significant differences between the results of the Efron and the marginal calculations. Finally, we test our results for the existence of unobservable heterogeneity. Again, the test results do not show any unobservable heterogeneity at the country level in our sample.

Table 8 Cox proportional hazard estimation (monthly-type spells) under capital mobility

Variable	Contemporaneous		Lagged	
	Model (I)	Model (II)	Model (III)	Model (IV)
Restricted-type is the base				
Open-type	0.87*	0.89*	0.77**	0.46
	(1.67)	(1.82)	(1.9)	(1.02)
Unemployment volatility	0.04**	0.04**	0.05***	0.07***
	(1.65)	(2.33)	(2.79)	(2.88)
Previous crises				0.19
				(1.08)
Size of economy	0.72**		0.59*	0.9**
	(1.97)		(1.77)	(2.29)
Whole period GDP growth	0.02	0.03***	0.02*	0.02
	(1.15)	(2.66)	(1.76)	(1.47)
GDP growth rate	0.03	−0.04	−0.04	−0.09
	(0.4)	(−0.53)	(−0.62)	(−1.11)
Inflation	0.33***	0.4***	0.09	0.24***
	(3.9)	(3.27)	(1.25)	(2.64)
Unemployment rate	0.00	−0.01	0.01	−0.01
	(−0.38)	(−0.36)	(0.23)	(−0.44)
Share price index growth	−0.03***	−0.01	−0.01	0.001
	(−3.41)	(−0.61)	(−0.9)	(0.51)
Real effective exchange rate	0.01	0.04	0.02**	0.02***
	(0.94)	(0.94)	(2.59)	(3.38)
Money growth	−0.01	−0.05***	0.01	−0.01
	(−0.95)	(−2.89)	(0.56)	(−0.6)
Real domestic credit growth	0.03**	0.05***	0.02	0.03*
	(2.09)	(3.33)	(1.32)	(1.98)
Trade openness	−0.12	0.04**	−0.04	0.03*
	(−0.27)	(2.5)	(−0.1)	(1.65)
Current account/GDP	−0.03	0.00*	−0.1***	0.00
	(−0.62)	(−1.94)	(−3.28)	(−0.45)
Capital account/GDP	0.14	0.00	−0.72	0.00
	(0.43)	(−0.14)	(−1.37)	(−0.48)
Financial account/GDP	0.09*	0.00**	−0.12*	0.00**
	(1.79)	(−2.24)	(−1.81)	(2.46)
Budget deficit/GDP	0.01	0.00	0.01	0.00
	(1.62)	(0.2)	(0.66)	(0.24)
Trade linkages	0.12**	0.16***	0.14**	0.14*
	(2.3)	(3.05)	(2.15)	(1.84)

Table 8 (Continued)

Variable	Contemporaneous		Lagged	
	Model (I)	Model (II)	Model (III)	Model (IV)
Financial linkages	0.01 (−0.32)	0.00 (−0.05)	0.01 (0.19)	0.01 (0.25)
Macroeconomic similarities	0.03 (0.61)	0.04 (0.76)	0.02 (0.45)	−0.03 (−0.45)
Log likelihood	−152.25	−136.67	−152.02	−131.22

The values in parentheses below estimates are the corresponding z-statistics
***, (**), [*] imply estimates are significant at 1, (5), and [10] percent

Table 9 Cox proportional hazard estimation (quarterly-type spells) under capital mobility

Variable	Contemporaneous		Lagged	
	Model (I)	Model (II)	Model (III)	Model (IV)
Restricted-type is the base				
Open-type	0.21 (0.41)	0.61 (1.23)	0.44 (0.77)	0.72 (1.47)
Unemployment volatility	0.05* (1.92)	0.03 (1.08)	0.5* (1.8)	0.05 (1.46)
Previous crises	−0.26 (−1.12)		−0.5* (−1.9)	−0.45 (−1.28)
Size of economy	0.51 (1.38)	0.23 (0.63)	0.44 (1.15)	0.32 (0.9)
Whole period GDP growth		0.01 (0.6)		0.01 (0.61)
GDP growth rate	0.02 (0.17)	0.04 (0.57)	0.08 (0.95)	0.05 (0.61)
Inflation	0.23*** (2.85)	0.25** (2.32)	0.2** (2.1)	0.31** (2.44)
Unemployment rate	0.02 (1.09)	0.02 (0.76)	0.6** (2.29)	0.04** (1.98)
Share price index growth	−0.01 (−1.19)	0.00 (−0.34)	−0.01 (−1.31)	−0.01 (−0.58)
Real effective exchange rate	0.01 (0.81)	0.00 (0.15)	0.02** (2.32)	0.01*** (3.63)
Money growth	−0.01 (−1.02)	−0.02* (−1.76)	−0.02 (−0.88)	−0.03 (−1.56)
Real domestic credit growth	0.03* (1.71)	−0.01 (0.34)	0.06*** (2.81)	0.07*** (3.84)

Table 9 (Continued)

Variable	Contemporaneous		Lagged	
	Model (I)	Model (II)	Model (III)	Model (IV)
Trade openness	0.24 (0.66)	0.02 (1.25)	0.3 (0.06)	0.04* (1.93)
Current account/GDP	0.00 (0.1)	0.00 (0.07)	0.02 (0.28)	0.00 (0.18)
Capital account/GDP	0.08 (0.21)	0.00 (1.2)	−2.21** (−2.05)	0.00 (0.97)
Financial account/GDP	−0.01 (−0.27)	0.00 (0.45)	0.00 (−0.05)	0.00 (0.27)
Budget deficit/GDP	0.01 (0.96)	0.00 (0.4)	0.01 (0.74)	0.00 (1.02)
Trade linkages	0.13 (1.11)	0.19** (2.24)	0.17* (1.73)	0.19* (1.71)
Financial linkages	0.00 (−0.78)	0.00 (−0.17)	0.00 (−1.59)	0.00 (−1.37)
Macroeconomic similarities	0.03 (0.25)	−0.06 (−0.55)	0.05 (0.47)	−0.07 (−0.61)
Log likelihood	−132.42	−112.35	−117.76	−98.29

The values in parentheses below estimates are the corresponding z-statistics
***, (**), [*] imply estimates are significant at 1, (5), and [10] percent

6 Conclusion

In this paper, we investigated whether there is a link between the choice of exchange rate regimes and the occurrence of currency crises. Our adopted methodology is duration analysis and the incidences of currency crisis come from 21 countries over the period 1970–1998. With the help of mixed proportional hazard models, we tested how the likelihood of currency crises changes under the de jure and de facto exchange rate classifications. We also examined the role of capital mobility on the sustainability of the currencies.

Our data indicate that there exists a meaningful link between the choice of exchange rate regime and the occurrence of currency crises. Nevertheless, the results are sensitive to the choice of the de facto exchange rate classification. While RR-based models show that fixed exchange rate arrangements are the least susceptible to speculative attacks, LYS-based models introduce the intermediate exchange rate regimes less crises prone than other arrangements. As far as a crucial improvement has not been made to assess how appropriate the current de facto exchange rate arrangements are, the sensitivity to the choice of classification does remain in evaluating the role of the exchange rate arrangements in occurrence of currency crises. In the meantime, researchers may rely on the characteristics of individual countries and scrutinize the monetary system of the countries under surveillance to determine more precisely the classification of their exchange rate regimes.

The data also show that the impact of capital account policies on the occurrence of currency crises takes different directions. While the baseline hazard of open-type capital accounts is lower than the baseline hazard of restricted-type capital accounts, when we enter our set of control variables into the models, the hazard of open-type capital accounts appears to be higher than the hazard of restricted-type capital accounts. This relation is more significant for low duration crisis episodes and can be interpreted as a sign that capital control policies could help preventing milder currency crises.

Acknowledgements We thank Carmen Reinhart, Kenneth Rogoff, Ethan Ilzetzki, Eduardo Levy-Yeyati, Federico Sturzenegger, Menzie Chinn, and Hiro Ito for making their data public. We also tank Marc Lavoie, Mario Seccareccia, Ehsan Choudhri, and Raul Razo-Garcia for their useful comments. All remaining errors are ours.

References

Bartolini, L., & Drazen, A. (1997). Capital account liberalization as a signal. *The American Economic Review*, *87*(1), 138–154.

Baxter, M., & Stockman, A. (1989). Business cycles and the exchange-rate regime: some international evidence. *Journal of Monetary Economics*, *23*(3), 377–400.

Bubula, A., & Ötker-Robe, I. (2002). *The evolution of exchange rate regimes since 1990: evidence from De Facto policies* (IMF Working Paper 02/155). Washington, DC: The International Monetary Fund.

Bubula, A., & Ötker-Robe, I. (2003). *Are pegged and intermediate exchange rate regimes crisis prone?* (IMF Working Paper 03/229). Washington, DC: The International Monetary Fund.

Calvo, G., & Reinhart, C. (2002). Fear of floating. *The Quarterly Journal of Economics*, *117*(2), 379–408.

Chinn, M. D., & Ito, H. (2005). *What matters for financial development? Capital controls, institutions, and interactions* (NBER Working Paper 11370).

Chinn, M. D., & Ito, H. (2008). A new measure of financial openness. *Journal of Comparative Policy Analysis*, *10*(3), 309–322.

Cox, D. R. (1972). Regression models and life tables (with discussion). *Journal of the Royal Statistical Society, Series B*, *34*, 187–220.

Demirgüç-Kunt, A., & Detragiache, E. (1997). *The determinants of banking crises: evidence from industrial and developing countries* (Policy Research Working Paper 1828). Washington, DC: The World Bank.

Dooley, M., & Isard, P. (1980). Capital controls, political risk, and deviations from interest rate parity. *Journal of Political Economy*, *88*(2), 370–384.

Edwards, S. (1989). *Real exchange rates, devaluation and adjustment: exchange rate policy in developing economies*. Cambridge: MIT Press.

Eichengreen, B. (1994). *International monetary arrangements for the 21st century*. Washington, DC: The Brookings Institution.

Eichengreen, B., & Razo-Garcia, R. (2011). *How reliable are de facto exchange rate regime classifications?* (NBER Working Paper 17318).

Esaka, T. (2010a). Exchange rate regimes, capital controls, and currency crises: does the bipolar view hold? *Journal of International Financial Markets, Institutions & Money*, *20*(1), 91–108.

Esaka, T. (2010b). De facto exchange rate regimes and currency crises: are pegged regimes with capital account liberalization really more prone to speculative attacks? *Journal of Banking & Finance*, *34*(11), 1109–1128.

Frenkel, R., & Rapetti, M. (2007). *Argentina's monetary and exchange rate policies after the convertibility regime collapse* (Mimeo). Center for Economic and Policy Research.

Fischer, S. (2001). Exchange rate regimes: is the bipolar view correct? *The Journal of Economic Perspectives*, *15*(2), 3–24.

Ghosh, A., Gulde, A. M., & Wolf, H. (2003). *Exchange rate regimes: choices and consequences*. Cambridge: MIT Press.

Glick, R., & Hutchison, M. (2005). Capital controls and exchange rate instability in developing economies. *Journal of International Money and Finance*, *24*(3), 387–412.

Glick, R., & Hutchison, M. (2010). *The illusive quest: do international capital controls contribute to currency stability?* (Working Paper 2010-15). Federal Reserve Bank of San Francisco.

Glick, R., Guo, X., & Hutchison, M. (2006). Currency crises, capital-account liberalization, and selection bias. *Review of Economics and Statistics*, *88*(4), 698–714.

Habermeier, K., Kokenyne, A., Veyrune, R., & Anderson, H. (2009). *Revised system for the classification of exchange rate arrangements* (IMF Working Paper No. 09/211).

Haile, F. D., & Pozo, S. (2006). Exchange rate regime and currency crises: an evaluation using extreme value theory. *Review of International Economics*, *14*(4), 554–570.

Husain, A. M., Mody, A., & Rogoff, K. S. (2005). Exchange rate regime durability and performance in developing versus advanced economies. *Journal of Monetary Economics*, *52*(1), 35–64.

Huynh, K. P., Petrunia, R. J., & Voia, M. C. (2012). Initial financial conditions, unobserved heterogeneity and the survival of nascent Canadian manufacturing firms. *Managerial and Decision Economics*, *33*, 109–125.

Kalbfleisch, J. D., & Prentice, R. L. (2002) *Wiley series in probability and statistics. The statistical analysis of failure time data* (2nd ed.). New York: Wiley

Karimi, M., & Voia, M. C. (2011a). *Identifying extreme values of exchange market pressure* (Carleton Economic Papers with number 11-10). Carleton University.

Karimi, M., & Voia, M. C. (2011b). *Empirics of currency crises: a duration analysis approach* (Carleton Economic Papers with number 11-11). Carleton University.

Kiefer, D. (1988). Economic duration data and hazard functions. *Journal of Economic Literature*, *26*(2), 646–679.

Klein, J. P., & Moeschberger, M. L. (2010). *Survival analysis: techniques for censored and truncated data. Statistics for biology and health* (2nd ed.). Berlin: Springer.

Lancaster, T. (1990). *The econometric analysis of transition data*. Cambridge: Cambridge University Press.

Lavoie, M. (2001). The reflux mechanism in the open economy. In L. P. Rochon & M. Vernengo (Eds.), *Credit, interest rates and the open economy: essays on horizontalism*. Cheltenham: Edward Elgar.

Levy-Yeyati, E., & Sturzenegger, F. (2005). Classifying exchange rate regimes: deeds vs. words. *European Economic Review*, *49*(6), 1603–1635.

Obstfeld, M., & Taylor, A. M. (2005). *Global capital markets: integration, crisis, and growth*. Cambridge: Cambridge University Press.

Radelet, S., & Sachs, J. (2000). The onset of the East Asian financial crisis. In P. Krugman (Ed.), *Currency crises* (pp. 105–162). Chicago: University of Chicago Press.

Reinhart, C., & Rogoff, K. S. (2004). The modern history of exchange rate arrangements: a reinterpretation. *The Quarterly Journal of Economics*, *119*(1), 1–48.

Rogoff, K. S., Husain, A. M., Mody, A., Brooks, R., & Oomes, N. (2004). *Evolution and performance of exchange rate regimes* (IMF Occasional Paper 229). Washington, DC: The International Monetary Fund.

Stiglitz, J. E. (2002). Capital market liberalization and exchange rate regimes: risk without reward. *The Annals of the American Academy of Political and Social Science*, *579*, 219–248.
Tavlas, G., Dellas, H., & Stockman, A. C. (2008). The classification and performance of alternate exchange rate systems. *European Economic Review*, *52*(6), 941–963.
Williamson, J. (2000). *Exchange rate regimes for emerging markets: reviving the intermediate options*. Washington: Peterson Institute Press, Peterson Institute for International Economics.
Williamson, J. (2002). The evolution of thought on intermediate exchange rate regimes. *Annals of the American Academy of Political and Social Science*, *579*, 73–86.

The manufacturer's authorised representative in the EU is Springer Nature Customer Service Centre GmbH, Europaplatz 3, 69115 Heidelberg, Germany. If you have any concerns regarding our products, please contact ProductSafety@springernature.com

Printed and bound by CPI Group (UK) Ltd, Croydon, CR0 4YY
25/09/2026
02222869-0010